Sustainability Sciences in Asia and Africa

Sustainable Agriculture and Food Security

Series Editor

Rajeev K. Varshney, State Agricultural Biotechnology Centre, Centre of Crop and Food Innovation, Food Futures Institute, Murdoch University, Perth, Australia

This book series support the global efforts towards sustainability by providing timely coverage of the progress, opportunities, and challenges of sustainable food production and consumption in Asia and Africa. The series narrates the success stories and research endeavors from the regions of Africa and Asia on issues relating to SDG 2: Zero hunger. It fosters the research in transdisciplinary academic fields spanning across sustainable agriculture systems and practices, post- harvest and food supply chains. It will also focus on breeding programs for resilient crops, efficiency in crop cycle, various factors of food security, as well as improving nutrition and curbing hunger and malnutrition. The focus of the series is to provide a comprehensive publication platform and act as a knowledge engine in the growth of sustainability sciences with a special focus on developing nations. The series publishes mainly edited volumes but some authored volumes. These volumes have chapters from eminent personalities in their area of research from different parts of the world.

Manish K. Pandey • Alison Bentley •
Haile Desmae • Manish Roorkiwal •
Rajeev K. Varshney
Editors

Frontier Technologies for Crop Improvement

Springer

Editors
Manish K. Pandey
Center of Excellence in Genomics and Systems Biology (CEGSB)
International Crops Research Institute for the Semi-Arid Tropics (ICRISAT)
Hyderabad, India

Alison Bentley
International Maize and Wheat Improvement Center (CIMMYT)
Mexico City, Mexico

Haile Desmae
International Maize and Wheat Improvement Center (CIMMYT)
Thies, Senegal

Manish Roorkiwal
Khalifa Center for Genetic Engineering and Biotechnology
United Arab Emirates University
Al-Ain, United Arab Emirates

Rajeev K. Varshney
Murdoch's Centre for Crop and Food Innovation, State Agricultural Biotechnology Centre, Food Futures Institute
Murdoch University
Murdoch, WA, Australia

ISSN 2730-6771 ISSN 2730-678X (electronic)
Sustainability Sciences in Asia and Africa
ISSN 2730-6798 ISSN 2730-6801 (electronic)
Sustainable Agriculture and Food Security
ISBN 978-981-99-4672-3 ISBN 978-981-99-4673-0 (eBook)
https://doi.org/10.1007/978-981-99-4673-0

This Springer imprint is published by the registered company Springer Nature Singapore Pte Ltd.
The registered company address is: 152 Beach Road, #21-01/04 Gateway East, Singapore 189721, Singapore

Preface to the Series

The global food production systems require expanding their capacity to produce almost twice the current levels to safeguard the food security of the burgeoning population worldwide. More than 800 million suffering from undernourishment worldwide poses a great risk to the attainment of sustainable development goal (SDG) 2 of the UN that targets "End hunger, achieve food security and improved nutrition and promote sustainable agriculture" within the next 7 years. The challenge is further exacerbated by the rising weather extremities and unpredictability in rainfall patterns and pest-pathogen dynamics associated with global climate change that has a profound negative impact on agricultural productivity and farm incomes worldwide. Also, the future targets of food production should be secured in resource-constrained agricultural settings and with least environment footprint, thus calling for sustainable innovations in agri-farming systems and enhanced participation of women in agriculture. The challenge to reduce hunger is alarming in the case of developing nations particularly Asia and Africa that house the largest proportion of people suffering from malnutrition and other nutrition-related issues. Furthermore, the agri-food systems in Asia and Africa are severely constrained by the subsistence nature of farming, declining land and other agricultural resources, increasing environmental pollution, soil and biodiversity degradation, and climate change. Therefore, this book series, "Sustainable Agriculture and Food Security" has been planned to support the global efforts towards sustainability by providing timely coverage of the progress, opportunities, and challenges of sustainable food production and consumption in Asia and Africa. The series narrates the success stories and research endeavors from the regions of Africa and Asia on issues relating to SDG 2: Zero Hunger. It fosters research in transdisciplinary academic fields spanning sustainable agriculture systems and practices, post-harvest, and food supply chains. The focus of the series is to provide a comprehensive publication platform and act as a knowledge engine in the growth of sustainability sciences with a special focus on developing nations.

Throughout history, agriculture has served as the cornerstone of human civilization, supplying not only sustenance but also other resources required for our survival. As the human population burgeoned, so did the demand for food, compelling the development of more efficient agricultural practices. The mid-twentieth century bore witness to the Green Revolution, a transformative period in the agricultural industry marked by substantial growth in crop production. This period saw the introduction of high-yielding dwarf crop varieties and the widespread application of synthetic fertilizers and pesticides, resulting in the doubling of agricultural production across many parts of the globe. Despite these achievements, the Green Revolution was not without its drawbacks, as it raised concerns about the environment and the long-term sustainability of these agricultural practices. In recent years, a growing emphasis has been placed on sustainable agriculture capable of ensuring food security without compromising the environment. This shift has spurred the development and application of frontier technologies such as genomics-assisted breeding, genetic engineering, gene editing, all aimed at enhancing crop improvement and promoting sustainable agriculture.

Genomics-assisted breeding (GAB) has emerged as a powerful tool, harnessing the power of genomic information to pinpoint genes associated with desirable crop traits. Through the rapid advancements in sequencing technologies and bioinformatics tools, scientists can now expeditiously sequence crop genomes and uncover the genes responsible for critical traits like yield, disease resistance, and nutritional quality. This invaluable knowledge paves the way for the creation of new crop varieties with improved traits through GAB methods namely marker-assisted selection, forward breeding, genomic selection, etc. GAB has not only revolutionized the development of superior varieties in major crops but also in orphan crops, which, despite being crucial in developing countries, have received relatively little research focus. In Asia and Africa, plant breeders have utilized/are utilizing GAB approaches to develop high-yielding, resilient crops capable of thriving in harsh environments while catering to the unique needs of these regions. Plant breeding community is now moving towards applying sequencing-based breeding approach.

On the other hand, genetic engineering encompasses the direct manipulation of an organism's DNA to introduce or modify specific traits. This powerful approach (GM, genetic modification) has led to the development of numerous successful crop varieties, such as pest-resistant Bt cotton, herbicide-tolerant soybean and maize, and vitamin A-fortified Golden Rice. Today, genetically engineered crops are cultivated on 190.4 million hectares across 29 countries, including 24 developing and 5 industrial nations. These crops contribute substantially to food security, sustainability, climate change mitigation, and the betterment of up to 17 million biotech farmers and their families worldwide. Very recently, genome/gene-editing (GE) approach is becoming popular to generate and use genetic variation via precise edits within the genomes of important food crop species. By embracing more flexible policies in GM- and GE-crops, many governments and regulatory bodies are fostering the growth of these technologies, ensuring they can be utilized to address pressing challenges like climate change and food insecurity.

In view of the above, this book, *Frontier Technologies for Crop Improvement* edited by Manish Pandey, Alison Bentley, Haile Desmae, Manish Roorkiwal, and me, offers a comprehensive exploration of the innovative technologies/approaches, such as gene bank genomics, forward breeding, multiomics, sequence-based breeding, genomic selection, gene editing, and bioinformatics that have transformed the landscape of crop improvement. With contributions from renowned researchers from various countries, the book encompasses an array of chapters that delve into the intricacies of these cutting-edge technologies, their applications, and their challenges. I firmly believe the book will provide readers with a thorough understanding of the technologies that are shaping the future of agriculture, highlighting their potential to address the growing demand for sustainable and environmentally friendly practices while ensuring food security for the global population.

I wish to extend my sincere thanks and gratitude to the Springer staff, particularly Aakanksha Tyagi, Senior Editor (Books), Life Sciences, and Naren Aggarwal, Editorial Director, Medicine, Biomedical and Life Sciences Books Asia, for their constant support for the accomplishment of this compendium. The cooperation received from my senior colleagues such as David Morrisson, Peter Davies, Daniel Murphy, and my laboratory colleagues—Vanika Garg, Abhishek Bohra, and Anu Chitikineni from Murdoch University (Australia) is also gratefully acknowledged. I would like to thank my family members—Monika Varshney, Prakhar Varshney, and Preksha Varshney for their love and support in discharging my duties as Series Editor.

Rajeev K. Varshney

International Chair, Agriculture and
Food Security;
Director, WA State Agricultural
Biotechnology Centre;
Director, Centre for Crop and Food
Innovation,
Food Futures Institute,
Murdoch University,
Murdoch, Western Australia, Australia

Preface

Technologies and innovations are key drivers for progress made by humans worldwide in all spheres of life. The descriptive technologies bring a drastic change in the way of functioning, thus leading to a multi-fold increase in efficiency and outcome. These technologies have significantly impacted many sectors and elevated them to the forefront of business. Agriculture is one industry that has benefited a lot from innovations and technologies. Investments in agriculture have helped in developing a few frontier technologies, but better adoption has typically only been observed in the private sector. Agriculture is core to the industry as well as human civilization and sustainability of the society. Therefore, more investment is required for enhancing the quality and quantity of the produce for ensuring food and nutritional security. Emphasis should be more on developing and deploying newer frontier technologies in order to bring a major shift in agriculture in the coming decade. Some of these areas include the digitalization of agriculture process and food chain, mechanization and automation in agriculture, artificial intelligence in agriculture, shortening varietal development duration using speed breeding and genomics, genomic selection breeding for achieving higher genetic gain, sequence-based/haplotype-based breeding for faster value addition during varietal development, modern practices for reducing post-harvest losses and storage, quality seed availability to farmers and industry, and food processing and food safety.

This book compiles information on the most recent technologies for crop improvement, along with their current status and future strategy in place. The technologies relating to genetic gain, nutrition, and food safety received the most attention. The book contains a total of 10 chapters contributed by experts from various countries such as Austria, Australia, Egypt, China, India, Kenya, Lebanon, Mexico, Morocco, Pakistan, Peru, Senegal, the United Arab Emirates, and the United States (see List of Contributors). Each chapter covered a particular enabling technology and forward-looking strategy. The book will be a valuable resource for updating the scientific community, academicians, policymakers, and other stakeholders of global agriculture about the rapid advancements in the various areas of agricultural biotechnology and genomics-assisted breeding. The global research and academics community engaged in the field will be the primary beneficiary while research scholars/students and industry stakeholders from technology development will be the secondary audience.

The editors are grateful to all the authors for their quality contributions. The editors are also grateful to the series (Naren Aggarwal, Aakanksha Tyagi and Sanchi Bhimrajka) editor Prof. Rajeev K. Varshney who is also part of the editorial team of this book. Editors also would like to thank the team of Springer Nature for their patience and follow-ups namely (*name to be included such as Aakanksha, Mahalakshmi & others*). We are also thankful to our family members, friends, collaborators, and research group members for their patience and support. We hope that the information provided in this book will be useful to researchers, academicians, students, policymakers including funders.

Hyderabad, India — Manish K. Pandey
Mexico City, Mexico — Alison Bentley
Thies, Senegal — Haile Desmae
Al-Ain, United Arab Emirates — Manish Roorkiwal
Murdoch, Australia — Rajeev K. Varshney

Contents

Editors and Contributors

About the Editors

Manish K. Pandey completed his PhD in Plant Genetics from Osmania University while working in ICAR-Indian Institute of Rice Research (IIRR), Hyderabad, India. He also had post-doctoral research experience at the University of Georgia, USA, for a period of 2 years. He is currently leading the Groundnut and Pigeonpea Genomics, Prebreeding and Bioinformatics research at the International Crops Research Institute for the Semi-Arid Tropics (ICRISAT), Hyderabad, India. He is also Adjunct Associate Professor in the University of Southern Queensland (USQ), Australia. His key contributions include reference genome sequence for diploid progenitor and both the subspecies of cultivated tetraploid groundnut, gene expression atlas, low to high density SNP genotyping assays, diagnostic markers for key traits, quality control panel, marker-assisted pyramiding, and genomic selection in groundnut. Through genomics-based breeding, his team developed three high oleic groundnut varieties which are released for cultivation in six states of India. He has published more than 150 scientific articles in various journals of international repute including *Nature Genetics*, *PNAS-USA*, *Journal of Advanced Research*, *Genome Biology*, and *Molecular Plant*. His efforts have been recognized at international level as he has been invited to deliver presentations in several international conferences, review the proposals for international funding agencies and he is collaborating with a large number of scientists at international level. Recognizing his contribution, he has now been inducted in many scientific academies of India.

Alison Bentley completed her undergraduate and post-graduate degrees at the University of Sydney, Australia. She worked at NIAB, Cambridge for 12 years, working on translational wheat research aiming to move fundamental scientific breakthroughs into tangible impacts for the agri-food sector. She is currently Director of CIMMYT's Global Wheat Program. Her work combines genetics and genomics to develop and deliver new tools and technology to improve plant breeding. Her core research addresses two major questions facing crop production: how can we adapt crops to fluctuating and changing climates to ensure food security, and how can we produce crops with reliable yield and product quality while limiting their environmental and economic cost. Addressing these two applied research questions requires a breadth and scale of scientific research and reflecting this, her current work spans the targeting of specific genes regulating plant processes to the development of agronomically informed breeding strategies.

Haile Desmae obtained his PhD from the University of Queensland (UQ), Brisbane, Australia, in 2007. He worked as a post-doctoral and research fellow at UQ for close to 6 years before joining ICRISAT in 2013. Currently, he is the Groundnut Breeder at CIMMYT based in Senegal. He develops breeding populations for specific traits of importance, plans and implements field trials to identify high-yielding groundnut varieties adapted to WCA. He closely works with national breeding programs in the region for enhanced technology generation and dissemination. He has contributed for the release of several improved varieties. He also works with farmers and private entrepreneurs to enhance quality seed production and marketing of improved groundnut varieties. He has authored and coauthored more than 50 articles.

Manish Roorkiwal is Genomic Breeding Lead at Khalifa Center for Genetic Engineering and Biotechnology (KCGEB), United Arab Emirates University, UAE and leading efforts to optimize and establish a crop breeding strategy for the UAE. With a basic background in molecular genetics and applied genomics, Manish has over 15 years of research experience. At the core of his work is the improvement of crop productivity of legumes in marginal environments using modern genetics and breeding approaches, including genomic selection and GWAS. Manish has strong interest in the area of modern breeding approaches such as genomic selection and next-generation sequencing-based re-sequencing and low-cost genotyping for enhancing the use of markers in routine breeding. He is known for leading the efforts in developing cost-effective genotyping platform enabling use of markers in routine breeding program. Manish has also contributed to developing and releasing several molecular breeding products for drought tolerance and disease-resistant chickpea for commercial cultivation in India.

Rajeev K. Varshney is an agricultural research scientist specializing in genomics and molecular breeding with 20+ years of service in international agriculture while working in India, Germany, Australia, Mexico, and several countries in Africa. At present, he is serving at Murdoch University (Australia) as a Director, Centre for Crop and Food Innovation; Director, State Agricultural Biotechnology Centre; and International Chair in Agriculture and Food Security with the Food Futures Institute. He is an Honorary or Adjunct Professor with more than 10 universities/organizations in China, Australia, Africa, and India. Varshney is a globally recognized leader for his work on genome sequencing, cataloguing and utilizing genetic diversity, genomics-assisted breeding, seed systems, and capacity building in developing countries. He has made centrally important contributions towards improving food and nutrition security in India and several countries in Africa and Asia by assembling genomes, developing genomic resources and integrating genomic technologies in crop improvement programs in many tropical crops, and delivering several superior crop varieties to some of the world's poorest farmers. His research group at present at Murdoch University is working on improving wheat, legume, and horticultural crops for a range of agronomic, and abiotic stress tolerance traits by developing and deploying novel genomics and prebreeding approaches such as pangenomics, haplotype cataloguing, and functional genomics approaches.

List of Contributors

Naglaa Abdallah Faculty of Agriculture, Cairo University, Giza, Egypt

Sumita Acharjee Department of Agricultural Biotechnology, Assam Agricultural University, Jorhat, Assam, India

Muzamil Baig International Crops Research Institute for the Semi-Arid Tropics (ICRISAT), Hyderabad, India

Swapnil S. Baraskar International Crops Research Institute for the Semi-Arid Tropics (ICRISAT), Hyderabad, India

Rutwik Barmukh International Crops Research Institute for the Semi-Arid Tropics (ICRISAT), Hyderabad, India

Michael Baum International Center for Agricultural Research in the Dry Areas (ICARDA), Rabat, Morocco

Alison Bently International Maize and Wheat Improvement Center (CIMMYT), Mexico City, Mexico

Mamta Bhattacharjee Department of Agricultural Biotechnology, Assam Agricultural University, Jorhat, India

Karma L. Bhutia Dr. Rajendra Prasad Central Agricultural University (RPCAU), Pusa, India

Rajaguru Bohar CGIAR Excellence in Breeding (EiB), International Maize and Wheat Improvement Center (CIMMYT), c/o ICRISAT, Hyderabad, India

Deekshitha Bomireddy International Crops Research Institute for the Semi-Arid Tropics (ICRISAT), Hyderabad, India

Sonal Channale University of Southern Queensland (USQ), Queensland, Australia

Palak Chaturvedi Molecular Systems Biology Lab (MOSYS), Department of Functional and Evolutionary Ecology, University of Vienna, Vienna, Austria

Sunil Chaudhari World Vegetable Center, South Asia, c/o ICRISAT, Hyderabad, India

Jyotsna Dayma Department of Agricultural Biotechnology, Assam Agricultural University, Jorhat, India

Haile Desmae International Maize and Wheat Improvement Center (CIMMYT), Thies, Senegal

Susanne Dreisigacker International Maize and Wheat Improvement Center (CIMMYT), Carretera, México-Veracruz, Mexico

Nourhan Fouad International Center for Agricultural Research in the Dry Areas (ICARDA), Cairo, Egypt

Nourhan Fouad International Center for Agricultural Research in the Dry Areas ICARDA, Cairo, Egypt

Ashish Gautam Central University of Karnataka, Kalaburagi, India

Arindam Ghatak Molecular Systems Biology Lab (MOSYS), Department of Functional and Evolutionary Ecology, University of Vienna, Vienna, Austria

Sonali Habade IRRI, South Asia Regional Center (ISARC), Varanasi, India

Aladdin Hamwieh International Center for Agricultural Research in the Dry Areas (ICARDA), Cairo, Egypt

Zhonghu He Department of Plant Sciences, Quaid-i-Azam University, Islamabad, Pakistan

Tawffiq Instanbulli International Center for Agricultural Research in the Dry Areas (ICARDA), Terbol, Lebanon

Kommineni Jagadeesh International Crops Research Institute for the Semi-Arid Tropics (ICRISAT), Hyderabad, India

Harsha Vardhan Rayudu Jamedar International Crops Research Institute for the Semi-Arid Tropics (ICRISAT), Hyderabad, India

Moctar Kante International Potato Center (CIP), Lima, Peru

Rakesh Kumar Central University of Karnataka, Kalaburagi, India

Shiv Kumar International Center for Agricultural Research in the Dry Areas (ICARDA), Rabat, Morocco

Tapan Kumar International Center for Agricultural Research in the dry Areas (ICARDA), Amlaha, India

Anil Kumar International Crops Research Institute for the Semi-Arid Tropics (ICRISAT), Hyderabad, India

Hannele Lindqvist-Kreuze International Potato Center (CIP), Lima, Peru

Fouad Maalouf International Center for Agricultural Research in the Dry Areas (ICARDA), Terbol, Lebanon

Mallana Gowdra Mallikarjuna Division of Genetics, ICAR-Indian Agricultural Research Institute, New Delhi, India

Swapnil Meshram Department of Agricultural Biotechnology, Assam Agricultural University, Jorhat, India

Richard Mott University College London, London, UK

Yogesh Dashrath Naik Dr. Rajendra Prasad Central Agricultural University (RPCAU), Pusa, India

Vinay Nandigam IRRI, South Asia Hub, ICRISAT Campus, Patancheru, India

Spurthi N. Nayak University of Agricultural Sciences, Dharwad, India

Vidya Niranjan RV College of Engineering, Bengaluru, India

Manish K. Pandey Center of Excellence in Genomics and Systems Biology (CEGSB), International Crops Research Institute for the Semi-Arid Tropics (ICRISAT), Hyderabad, India

Neha Pandey Department of Agricultural Biotechnology, Assam Agricultural University, Jorhat, India

Ovais Peerzada International Crops Research Institute for the Semi-Arid Tropics (ICRISAT), Hyderabad, India

Iro Pierides Molecular Systems Biology Lab (MOSYS), Department of Functional and Evolutionary Ecology, University of Vienna, Vienna, Austria

Somashekhar Punnuri College of Agriculture, Georgia, USA

Khaled Radwan Agricultural Genetic Engineer Research Institute (AGERI), Agricultural Research Center (ARC), Giza, Egypt

Prasanna Rajesh Agricultural College, Acharya NG Ranga Agricultural University, Bapatla, India

Awais Rasheed Chinese Academy of Agricultural Sciences (CAAS), and CIMMYT-China office, Beijing, China

Manish Roorkiwal Khalifa Center for Genetic Engineering and Biotechnology, United Arab Emirates University, Al-Ain, United Arab Emirates

Jana Schwarzerova Molecular Systems Biology Lab (MOSYS), Department of Functional and Evolutionary Ecology, University of Vienna, Vienna, Austria

Ramachandran Senthil International Crops Research Institute for the Semi-Arid Tropics (ICRISAT), Hyderabad, India

Trushar M. Shah International Institute of Tropical Agriculture (IITA), Nairobi, Kenya

Priya Shah International Crops Research Institute for the Semi-Arid Tropics (ICRISAT), Hyderabad, India

Vinay Sharma International Crops Research Institute for the Semi-Arid Tropics (ICRISAT), Hyderabad, India

Uma Maheshwar Singh IRRI, South Asia Regional Center (ISARC), Varanasi, India

Vikas Kumar Singh IRRI, South Asia Hub, ICRISAT Campus, Hyderabad, India

Kuldeep Singh International Crops Research Institute for the Semi-Arid Tropics (ICRISAT), Hyderabad, India

Pallavi Sinha IRRI, South Asia Hub, ICRISAT Campus, Hyderabad, India

Krishna Tesman Sundaram IRRI, South Asia Hub, ICRISAT Campus, Hyderabad, India

Sawsan Tawkaz International Center for Agricultural Research in the Dry Areas (ICARDA), Terbol, Lebanon

Mahendar Thudi Dr. Rajendra Prasad Central Agricultural University (RPCAU), Pusa, India

Rajeev K. Varshney Murdoch's Centre for Crop and Food Innovation, State Agricultural Biotechnology Centre, Food Futures Institute, Murdoch University, Murdoch, WA, Australia

Raguru Pandu Vasanthi S. V. Agricultural College, Acharya N.G Ranga Agricultural University (ANGRAU), Tirupati, India

H. V. Veerendrakumar International Crops Research Institute for the Semi-Arid Tropics (ICRISAT), Hyderabad, India

Mani Vetriventhan International Crops Research Institute for the Semi-Arid Tropics (ICRISAT), Hyderabad, India

Xingjun Wang Shandong Academy of Agricultural Sciences (SAAS), Shandong, China

Wolfram Weckwerth Vienna Metabolomics Center (VIME), University of Vienna, Vienna, Austria

Xianchun Xia Chinese Academy of Agricultural Sciences (CAAS), and CIMMYT-China office, Beijing, China

Shuang Zhang Molecular Systems Biology Lab (MOSYS), Department of Functional and Evolutionary Ecology, University of Vienna, Vienna, Austria

Chuanzhi Zhao Shandong Academy of Agricultural Sciences (SAAS), Jinan, Shandong, China

Chapter 1
Introduction: Frontier Technologies for Crop Improvement

Manish K. Pandey, Alison Bentley, Haile Desmae, Manish Roorkiwal, and Rajeev K. Varshney

Abstract The last two decades have witnessed the rapid development and application of several frontier technologies for crop improvement, which have brought speed, precision and cost-effectiveness in making selection decisions for improved breeding lines with better genetics. A few such technologies to be mentioned are accurate and efficient germplasm characterization of diverse genebank accessions, high-throughput sequencing and genotyping, rapid generation advancement, modern sequencing-based trait mapping and gene discovery followed by identification of superior haplotypes, genomic selection, gene editing, forward breeding and multi-omics approaches including better bioinformatics tools/software. While there is still scope for improving phenotyping protocols for various traits, especially the complex ones, the above-mentioned frontier technologies provide huge opportunities in improving the precision and speed in developing new cultivars with future traits to ensure the sustainability of different crop plants. The integration and use of these technologies on a large scale using a common platform to provide flawless support to

---

M. K. Pandey (✉)
Center of Excellence in Genomics and Systems Biology (CEGSB), International Crops Research Institute for the Semi-Arid Tropics (ICRISAT), Hyderabad, India
e-mail: Manish.Pandey@icrisat.org

A. Bentley
International Maize and Wheat Improvement Center (CIMMYT), Mexico City, Mexico
e-mail: a.bentley@cgiar.org

H. Desmae
International Maize and Wheat Improvement Center (CIMMYT), Thies, Senegal
e-mail: h.s.desmae@cgiar.org

M. Roorkiwal
Khalifa Center for Genetic Engineering and Biotechnology, United Arab Emirates University, Al-Ain, United Arab Emirates
e-mail: mroorkiwal@uaeu.ac.ae

R. K. Varshney
Murdoch's Centre for Crop and Food Innovation, State Agricultural Biotechnology Centre, Food Futures Institute, Murdoch University, Murdoch, WA, Australia
e-mail: rajeev.varshney@murdoch.edu.au

M. K. Pandey et al. (eds.), *Frontier Technologies for Crop Improvement*, Sustainable Agriculture and Food Security, https://doi.org/10.1007/978-981-99-4673-0_1

crop improvement programs is still a challenge for many crop species, but will be accomplished sooner or later.

Keywords Frontier technologies · Speed breeding · Genetic gain · Productivity · Genomics-assisted breeding

1.1 Introduction

The importance of agriculture to global economic growth and the livelihood of households makes it one of the most important sectors in the world. It serves as the primary source of human food, animal feed, and raw materials for industry. Agricultural products are traded worldwide and generate substantial income. In sub-Saharan Africa, for example, agriculture employs more than 50% of the labour force and contributes about 15% of the GDP (OECD/FAO 2016). Governments and donors have, therefore, prioritized expenditures in agricultural research for development in order to increase the sector's productivity. A significant area of impact has been crop improvement. Breeders have historically relied heavily on the conventional breeding (crop improvement) approach to develop improved varieties with high yields. The approach is based on the art and science of creating variations to select improved varieties based on phenotypic diversity. The efforts have led to the release and adoption of several improved varieties for various crops, which have considerably improved the standard of living for smallholder households in developing countries and the profitability of farming/agribusinesses in developed countries. Albeit considerable disparity between countries, as an illustration, we know how improved varieties, in combination with fertilizer and pesticides, almost doubled the productivity of crops like wheat, rice, and others during the green revolution, which helped many farmers overcome poverty (Pingali 2012). In general, about 50% of the increase in crop productivity over the past century can be attributed to plant breeding, with the other 50% coming from improved crop management, such as fertilizers, irrigation, and weeding (Evenson and Gollin 2003).

Improved varieties remain the cheapest and the main inputs to improve agricultural productivity to feed and meet the nutritional needs of the growing population, which is expected to hit about 10 billion by 2050 (UN 2017). Hence, the development of improved varieties for food security and sustainable agriculture is considered a major element in tackling poverty and rural displacement. Crop improvement has a high return on investment value (Renkow and Byerlee 2010). The development of varieties with improved productivity and yield stability contributes to food security; drought-resistant varieties that lower production costs and increase the viability of marginal agribusinesses enhance economic benefits; varieties with high-nutrient contents improve nutrition; and varieties that are less reliant on pesticides or more efficient in their use of water and nutrients benefit the environment.

Conventional breeding, although still commonly used today, may not meet the ever-increasing food demand in the face of climate change and variability. This is due to practical limitations including the slow process of taking more than 10 years to develop and commercialize a variety and heavy reliance on natural variability which may not be available for some important traits. In recent years, massive efforts were invested in developing technologies and innovations for improving the accuracy and efficiency of the breeding process. Consequently, the last two decades have witnessed the rapid development and use of several frontier technologies for crop improvement, which have brought speed, precision and cost-effectiveness in making selection decisions for improved breeding lines with better genetics. Hence, aided by such progress, the conventional breeding approach has evolved to integrated (modern) breeding, which has provided ample opportunities for developing high-yielding improved varieties that meet farmer preferences and consumer/market needs.

This book, divided into ten chapters, presents a review and vision on use of the frontier technologies covering important crop improvement areas including exploiting genebanks, modern bioinformatics tools, contemporary genotyping platforms, rapid generation advancement (speed breeding) approach, multi-omics, sequence-based breeding, forward breeding, genomic selection and gene editing. Each chapter, authorship led by subject matter experts, discusses the technologies and their practical applications in plant breeding and presents future strategies. However, the authors have emphasized that the combined use of these frontier technologies has enormous potential to positively impact crop plant genetic improvement to develop improved varieties in a faster and more precise way. A common platform coupled with open breeding informatics involving different stakeholders and active support from donors is expected to realize the impacts on the improvement of various crop species. A brief overview of chapters 2–10 is provided below.

1.2 Linking of Genebank to Breeding and Food Security

This chapter authored by Singh et al. discusses the role of genebank in conserving and contributing to making available trait diversity for breeding future-ready crops. It provides detailed information on efforts towards germplasm collection, conservation, and characterization as well as making these available to breeders for use in breeding new varieties. This chapter also discusses the impact of genebank's diverse germplasm in achieving higher yield and genetic gain, adaptation to climate challenges and stresses, increasing nutritional compounds and consumer/industry-preferred traits in several crops. A high-throughput large-scale phenotypic assessment for key traits as well as the use of multi-omic tools, including high-throughput genotyping, are suggested as new frontier technologies to improve our understanding of genetic diversity in germplasm collection in genebank and their use in diversifying the primary and more specifically cultivated gene pool. The chapter also discusses access to germplasm and the impact of genebank by contributing to the sustainability of world agriculture.

1.3 Bioinformatics for Plant Genetics and Breeding Research

The volume and frequency of data generation have increased multiple folds which need to be well documented and analysed using efficient software to get an insight to a specific biological question. This chapter authored by Naik et al. discusses the use of bioinformatics and computational tools to make good sense of the tremendous wealth of data generated each day as part of plant genetics, genomics, and breeding research. The field of bioinformatics and computational biology have witnessed huge advancements in developing and deploying the most advanced software or tools in the last two decades as the field needed to match the speedy developments in sequencing and genotyping data generation platforms. The different bioinformatics and computational tools are providing great support to molecular genetics and genomics research, right from developing reference genomes to molecular breeding product in farmers' fields.

1.4 Evolution in the Genotyping Platforms for Plant Breeding

Genotyping of diverse germplasm and breeding materials is required for performing genetic diversity analysis, genome-wide association study, hybridity confirmation, genetic mapping, foreground selection using Kompetitive Allele Specific PCR (KASP) markers for selected traits, background selection for genome recovery, genomic selection and genetic purity. There is no 'one-size-fits-all' solution, and this chapter led by Rasheed et al. provides detailed information on the efforts around developing cost-effective, high-throughput and breeding-oriented crop genotyping platforms.

1.5 Rapid Generation Advancement for Accelerated Plant Improvement

Faster varietal replacement using new varieties with improved genetics is key to achieving higher genetic gains in farmers' fields (Varshney et al. 2019; Pandey et al. 2020). This milestone can only be achieved if the breeding cycle of the new varieties can be shortened using modern technologies such as rapid generation advancement (RGA) to support speed breeding. This chapter led by Hamwieh et al. explores different such technologies which can help in breeding new varieties faster. Some of these technologies include shuttle breeding, double haploidy, speed breeding, genome editing, marker-assisted selection and genomic selection.

1.6 Multi-Omics for Crop Improvement

The last couple of decades have witnessed sharply increased adoption of different new technologies in the area of different omics sciences such as genomics, transcriptomics, proteomics, metabolomics and phenomics. Most of such studies are done using a single approach or sometimes using two approaches, however, it's very rare to see research wherein all the omics platforms are used for trait understanding and gene discovery. The results are going to be more reliable upon integrating these omics platforms together, which will also ensure faster deployment in crop breeding. This chapter led by Chaturvedi et al. provides the current progress, opportunities and challenges in this direction. These authors also termed this as PANOMICS approach which is very much essential in moving forward while deploying more data-driven and science-based crop breeding to develop future-ready crops.

1.7 Sequence-Based Breeding for Crop Improvement

Sequencing technologies have enabled high-quality genome sequence assemblies, pangenomes, high-density genetic maps, various marker genotyping platforms, and trait discovery in most of the crop plants. This chapter led by Sinha et al. discusses the significant improvement in breeding programs utilizing modern genomic tools and technologies. The scientific community is continuously witnessing the success stories of new improved varieties being successfully bred utilizing genomics-assisted backcrossing and, more recently, genomic selection. It is necessary to continuously work on population improvement after every breeding cycle in order to obtain long-term genetic gains and hasten the genetic gains in crops. In this context, the chapter proposes a sequencing-based breeding approach involving parental selection, enhancement of genetic diversity of breeding materials, forward breeding for early generation selection, and genomic selection using sequencing/genotyping technologies. This chapter also emphasizes the integration of other modern technologies such as speed breeding technology which allows for four to six generations per year and has great potential to further accelerate the pace of delivering genetic gain in farmers' fields.

1.8 Forward Breeding for Efficient Selection

Crop improvement coupled with the modern plant breeding approaches such as genomic-assisted breeding is a proven solution to meet the food security of a rapidly growing global population. This chapter led by Bohar et al. emphasizes combining the power of genomic selection and foreground selection for key traits into the breeding pipeline by employing low-cost genotyping solutions. Currently, multiple SNPs marker-based genotyping platforms are available for deployment in an array of applications in crop improvement. A shared genotyping platform coupled with open breeding informatics involving different stakeholders and active support from donors will make genotyping cost-effective with a quicker turnaround. This will address several constraints faced by public breeding programs to employ forward breeding. Currently available forward-breeding genomic resources in the low-mid density genotyping platform space are also covered in detail in this chapter, with special emphasis on shared services.

1.9 Genomic Selection in Crop Improvement

Genomic Selection (GS) is the most promising approach for improving multiple traits including those with complex genetic control in addition to accommodating the benefits of marker-assisted selection. This breeding approach is still going through the maturity phase and may become one of the futuristic breeding methods not just for breeding varieties but also for population improvement. This chapter led by Veerendrakumar et al. provides information on such progress made during the last decade in multiple crops. This chapter also briefly discusses the methodology, current progress, its advantages, different genomic prediction models as well as factors affecting prediction accuracy. Suggestions are also made to integrate other frontier technologies together such as genomic selection with rapid generation advancements and single-seed chipping-based genotyping for achieving greater benefits.

1.10 Genetic Engineering: A Powerful Tool for Crop Improvement

The applications of genetic engineering including genome editing are important as they can complement modern breeding activities to mitigate the effects of changing environments and boost crop production. This chapter led by Bhattacharjee et al. emphasizes on the importance of genetically modified (GM) technology which delivered multiple plant attributes such as herbicide resistance, tolerance against pests and pathogens, and nutritional enhancement. The chapter also briefly discusses technologies, namely, Zinc Finger Nucleases (ZFNs), Transcription Activator-Like Effector Nucleases (TALENs), and Clustered Regularly Interspaced Short Palindromic Repeats (CRISPR)/Cas. The chapter also provides information on success stories in many plant species already commercialized, namely, soybean, papaya, maize, cotton, common bean, sweet potato, cowpea, etc. Furthermore, the technology holds immense promise to achieve the UN's sustainable development goals

(SDGs) to fight hunger, attain food security, enhance nutrition, and promote sustainable agriculture.

1.11 Summary and Outlook

In summary, this book presents articles in leading frontier technologies for crop improvement. As mentioned above, some of above technologies are already under use for crop improvement in some crop species. It is anticipated that with costs going down on several of these technologies, their adoption will be enhanced. Furthermore, some new areas such as single cell genomics, systems biology and synthetic biology are emerging for crop improvement. However, optimisation of these approaches is still in infancy for majority of crops. We envisage the successful utilization of some of these approaches for crop improvement in coming years.

References

Evenson RE, Gollin D (2003) Assessing the impact of the green revolution, 1960-2000. Science 300:758–762

OECD/FAO (2016) Agriculture in sub-Saharan Africa: prospects and challenges for the next decade. In: OECD-FAO Agricultural Outlook. OECD Publishing, Paris. 2016–2025. https://doi.org/10.1787/agr_outlook-2016-5-en

Pandey MK, Pandey AK, Kumar R, Nwosu V, Guo B, Wright G, Bhat RS, Chen X, Bera SK, Yuan M, Jiang H, Faye I, Radhakrishnan T, Wang X, Liang X, Liao B, Zhang X, Varshney RK, Zhuang W (2020) Translational genomics for achieving higher genetic gains in groundnut. Theor Appl Genet 133:1679–1702

Pingali PL (2012) Green revolution: impacts, limits, and the path ahead. PNAS 109(31): 12302–12308. https://doi.org/10.1073/pnas.0912953109

Renkow M, Byerlee D (2010) The impacts of CGIAR research: a review of recent evidence. Food Policy 35:391–402

UN (2017) The World Population Prospects: The 2017 Revision, UN Department of Economic and Social Affairs. https://www.un.org/en/desa/world-population-projected-reach-98-billion-2050-and-112-billion-2100

Varshney RK, Pandey MK, Bohra A, Singh VK, Thudi M, Saxena RK (2019) Toward sequence-based breeding in legumes in the post-genome sequencing era. Theor Appl Genet 132(3): 797–816

Chapter 2
Linking of Genebank to Breeding and Food Security

Kuldeep Singh, Ramachandran Senthil, Ovais Peerzada, Anil Kumar, Swapnil S. Baraskar, Kommineni Jagadeesh, Muzamil Baig, and Mani Vetriventhan

Abstract Genebanks have the responsibility of collecting, maintaining, characterizing, evaluating, documenting and distributing plant genetic resources for research, education and breeding purposes globally. About 7.4 million germplasm accessions are conserved ex situ in the genebanks globally. Efficient use of germplasm in crop improvement is depending on the availability of accession-level information on the traits of interest. For the majority of accessions, only basic passport and characterization data are available, while data on unique traits is generally lacking that limits their utilization in crop improvement. Development of germplasm diversity and trait-specific subsets enhanced availability of accessions-level information. Researchers can search in the global plant genetic resources database called Genesys PGR which contains passport data, characterization and evaluation data sets and trait-specific subsets developed on various crops (https://www.genesys-pgr.org/). The impact of germplasm for contributing to increased yield, adaptation, nutrition and improved health and sustainable agriculture has been demonstrated in many crops. There are many instances where a single plant genetic resource has proved to have large commercial value by conferring a specific trait. With the availability of new technologies such as high-throughput large-scale phenotypic assessment for key traits and use of multi-omic tools could accelerate rapid identification of traits and genes for breeding improved cultivars. This chapter details about ex situ germplasm conservation, discovering climate resilient germplasm following different approaches such as diversity and trait-specific subsets, focused germplasm

K. Singh (✉) · R. Senthil · O. Peerzada · A. Kumar · M. Baig · M. Vetriventhan
Genebank, International Crops Research Institute for the Semi-Arid Tropics (ICRISAT), Patancheru, Hyderabad, Telangana, India
e-mail: Kuldeep.singh@icrisat.org

S. S. Baraskar · K. Jagadeesh
Genebank, International Crops Research Institute for the Semi-Arid Tropics (ICRISAT), Patancheru, Hyderabad, Telangana, India

Professor Jayashankar Telangana State Agricultural University (PJTSAU), Hyderabad, Telangana, India

M. K. Pandey et al. (eds.), *Frontier Technologies for Crop Improvement*, Sustainable Agriculture and Food Security, https://doi.org/10.1007/978-981-99-4673-0_2

identification strategy, molecular characterization of germplasm and trait discovery, access to germplasm and the impact of genebank contributing to the global agriculture sustainability.

Keywords Plant Genetic Resources · Germplasm · Ex situ genebank · Conservation · Breeding · Genomics · Food Security

2.1 Introduction

Plant genetic resources (PGR) are the key to crop improvement and have an important role to play to address food security and nutrition. The unprecedented rate of biodiversity loss is one of the major challenges that erode the resilience of the agricultural system and threatens food and nutrition security. Although the number of plant species used for food by pre-agricultural humans is estimated to be around 7000, only a small fraction (~250) of them has been fully domesticated (von Wettberg et al. 2020). The trends in diversity across crops and regions with a 50-year perspective indicate the increased dominance of mega crop varieties in agricultural landscapes and displacement of traditional landraces. This trend occurs at a faster rate in Asia than in Africa (Gatto et al. 2021). The main focus of Sustainable Development Goal 2.5 is to maintain the genetic diversity of crops and their wild relatives in the genebanks, ensuring access to that diversity following international laws. To minimize the biodiversity loss due to the replacement of landrace by improved varieties (Gepts 2006; Van de Wouw et al. 2009; Khoury et al. 2014), national and international genebanks have been established to conserve and distribute germplasm globally for sustainable agriculture. Globally, about 7.4 million accessions are conserved ex situ in the genebanks. Germplasm resources with information on key traits aid in the selection and their use in crop improvement. The trait of importance to most users includes productivity, stress tolerance, and quality traits. However, for the majority of germplasm accessions, only basic passport and characterization data are available, while data on unique traits is generally lacking. This chapter details about ex situ germplasm conservation, discovering climate-resilient germplasm following different approaches such as diversity and trait-specific subsets, focused germplasm identification strategy, molecular characterization of germplasm and trait discovery, access to germplasm and the impact of genebank contributing to the global agriculture sustainability.

2.2 Ex Situ PGR Conservation

Plant genetic resources for food and agriculture (PGRFA) comprise the diversity of genetic materials, including landraces, breeding or modern cultivars, genetic stocks, crop wild and weedy relatives, that can be used now or in the future for food and agriculture. PGR includes any materials of plant origin including reproductive and vegetative propagating materials, which contain the functional units of heredity. The biological diversity of the PGR is mainly conserved within or away from their natural habitats called in situ and ex situ conservation, respectively. Global threats to PGRFA in situ and on farm have increased in the last few decades because of many reasons including global climate change and the increased impact of human activities. Therefore, ex situ conservation in the genebank is the most common approach either semi-controlled (field genebank) or under controlled conditions (seeds, tissues, seedlings, pollens, and DNA). Seed genebanks are the easiest way to store germplasm at low temperature, while field genebank is for the conservation of genetic resources under normal growing conditions. To safeguard against the loss of plant biological diversity, intensive collection of different crop species was undertaken by the global community (Upadhyaya et al. 2010). As a result, over 7.4 million PGR conserved ex situ in over 1750 genebanks globally (https://www.fao.org/wiews) and the International Agricultural Research Centres (IARC) conserve about 10% of the total accessions accounts (Table 2.1). As of December 31, 2021, the 11 IARC Centers conserve over 722,000 accessions of crop, forage, and tree germplasm and make them available under the standard material transfer agreement (SMTA). These IARCs account for about 94% of the germplasm distributed within the guidelines of the International Treaty on Plant Genetic Resources for Food and Agriculture (Plant Treaty). During the 15 years (January 2007 to December 2021), the IARC's genebanks and breeding programs distributed over six million samples under 61,000 SMTAs.

The conservation of plant genetic resources (PGR) has gained significant importance. This is demonstrated by the impressive number of nations that have ratified the Convention on Biological Diversity, endorsed the International Undertaking on Plant Genetic Resources, or both. Despite this very encouraging development, many genebanks face financial and operational difficulties. According to the FAO report 'State of the World's Plant Genetic Resources for Food and Agriculture' (FAO 1998), many genebanks may not at present be capable of performing their basic conservation role. In the case of seed genebanks, where the technology of storing germplasm samples is relatively easy to apply under most operational circumstances, the problems relate more to resource constraints that impact the performance of essential operations. This is critical in the case of the core activities of maintaining the viability and genetic integrity of the stored accessions, as well as sufficient stocks, to meet user demands. Consequently, the importance of efficient and cost-effective genebank management has increased over the years and has become a decisive element in the long-term ex situ conservation of PGR.

2.2.1 Seed Genebank

Seed genebanks conserve crop diversity mostly in the form of seeds. Every genebank in the world follows some basic core activities/operations such as germplasm collection and acquisition, conservation, distribution, characterization,

Table 2.1 Plant genetic resources for food and agriculture conserved and made available by the International Agricultural Research Centres (IARC) genebanks

CGIAR Centre	Crop	Number of accessions	CGIAR Centre	Crop	Number of accessions available with SMTA
AfricaRice	Rice	19,696	ICRAF	Multipurpose trees	6744
Bioversity	Banana	1682		Fruit trees	8246
CIAT	Beans	37,934	ICRISAT[a]	Chickpea	20,838
	Forages	22,662		Groundnut	15,360
	Cassava	5965		Pigeon pea	13,559
CIMMYT	Maize	28,694		Pearl millet	24,663
	Wheat	135,021		Small millets	11,797
CIP	Andean roots and tubers	1173		Sorghum	42,880
	Potato	7367	IITA	Cowpea	17,051
	Sweet potato	6143		Cassava	3184
ICARDA	Lentils	14,295		Maize	1561
	Grass pea	4301		Miscellaneous legumes	6747
	Forages	25,358		Banana	392
	Faba bean	9594		Yam	5929
	Chickpea	15,230	ILRI	Forages and fodder	3918
	Barley	31,843	IRRI	Rice	127,413
	Pea	4593			
	Wheat	41,967			

Source: Global Crop Diversity Trust/CGIAR Online Reporting Tool, covering the period up to December 31, 2021

[a]ICRISAT genebank data as on Jan 2023

regeneration, viability testing and monitoring, safety duplication and documentation. New accessions are collected or assembled to enrich the diversity of the genebank collections considering the geographical and taxonomical gaps in the collection. A comprehensive technical guide on collecting plant genetic resources providing many practical and managerial suggestions has been published (Guarino 1995). It is important that collected or harvested germplasm material is processed as soon as possible to avoid loss in viability or decrease in longevity. Seed moisture content (SMC) is one of the most important factors determining longevity of the stored seeds. Before the seeds are stored, they should be properly dried and the seed moisture content should be accurately determined. A small change in SMC can greatly affect the storage life of the seeds (Roberts 1973). Different SMC determination methods and equipment are available, the principles and methodology of which are presented by Ellis et al. (1985), and the procedures by Hanson (1985). The recommended levels of SMC are between 3% and 7% for long-term conservation

(FAO–IPGRI. 1994). The genebank curator has to accurately assess the initial viability of each accession to be stored and monitor the viability of an accession during its storage life to reduce or avoid the loss of genetic diversity within and between the accessions. Details on genebank standards for viability monitoring were proposed by a panel of experts and were subsequently endorsed by the FAO Commission on Plant Genetic Resources (FAO 2014). Knowing the precise storage behaviour of the species is essential for a seed sample before it can be stored, in order to ensure its optimum storage conditions are used, that is, the optimum moisture content and storage temperature. A protocol to determine seed storage behaviour has been published by IPGRI (Hong and Ellis 1996), in conjunction with a seed storage behaviour compendium (Hong et al. 1996), which contains storage behaviour information on more than 7000 species. When the optimum seed moisture content is accurately determined and the seeds have been packaged, they should be stored at the best available temperature. The genebank standards recommend a preferred temperature of −18 °C or below for long-term storage also called as base collections and 5–10 °C for medium-term storage also known as active collections. This two-tiered storage concept of the base collection for long-term storage and active collection for accessions for distribution or research is largely based on experiences with the storage of orthodox seeds. Regeneration of accessions is one of the most crucial processes involved in genebank management, since during regeneration accessions are particularly vulnerable to loss or change of genetic diversity. It is also a costly process in which practical compromises are frequently made, the consequences of which might only be observed much later. It was for these reasons that IPGRI published a scientific background paper for the regeneration and multiplication of germplasm resources in seed genebanks (Breese 1989). Regeneration in genebanks is carried out after the accessions show viability that is below threshold level or the quantity of accession reaches a critical level after which it cannot be distributed. Most genebanks have computerized documentation systems which greatly facilitate the storage and maintenance of data, as well as its retrieval. A helpful overview of the various aspects of genebank documentation can be found in the guidebook for genetic resources documentation (Painting et al. 1995). Most of the routine genebank operations described above generate information which is key to the efficient functioning of the genebank operations.

2.2.2 *Field Genebanks and in Vitro Conservation*

Germplasm from clonal crops which are either vegetatively propagated and/or do not produce seeds, or for species with short-lived recalcitrant seeds are usually conserved in the field genebanks and/or as in vitro conservation. The field genebank has limitations regarding efficiency, cost, security and long-term maintenance. In vitro conservation involves maintenance of explants in a protected environment, aseptic plants and supports safe and easy international exchange of plant materials and lower conservation cost. Techniques for collecting species which produce recalcitrant seeds have been developed which enable the collector to grow the material in vitro, under aseptic conditions. This approach will allow germplasm collections to be made in remote areas (e.g. for highly recalcitrant cacao seeds), or when the transport of the collected fruits would become prohibitively expensive (e.g. coconut collecting in the South Pacific) where the target species would not have seeds or other storage organs to be collected. A good overview of such techniques has been presented by Withers (1995).

2.2.3 Cryopreservation

The cryopreservation technique ensures long-term and safe storage of those species which are difficult to conserve as seed. This can be achieved by storing the samples at ultra-low temperatures, either above −150 °C or at −196 °C liquid nitrogen. For several species (e.g. potato, apple, banana and cassava), procedures have been developed which allow this technique to be applied routinely for conservation (Engelmann and Takagi 2000). Cryopreservation is also being used for the long-term storage of orthodox seed having short longevity. Cryopreservation mostly involves the two-step cooling process which is based on the induction of explant 'vitrification' during a very fast decrease in temperature. Vitrification of cells and tissues is the physical process, which avoids intracellular ice crystallization, during ultra-freezing, by the transition of the aqueous solution of the cytosol into an amorphous, glassy state. As a result of this process, plant tissues are protected from damage and remain viable during their long-term storage at −196 °C. For different plant species, a number of vitrification-based techniques have been developed such as vitrification, encapsulation-dehydration, encapsulation-vitrification, desiccation (Reed 2008) and, more recently, droplet vitrification and D/V cryoplate (Yamamoto et al. 2011; Niino et al. 2013) but the techniques are continuously modified and improved to produce higher plant recovery rates, to expand the number of the cryopreserved species and, above all, working on the species, which are still hard to process with the cryopreservation. Cryopreservation techniques are now used for plant germplasm storage in many institutes around the world (Niino 2006; Malik et al. 2012).

2.2.4 DNA Banking

DNA banking is an efficient and long-term method to conserve the genetic information. DNA banks are now considered as a means of complimentary conservation. DNA storage is particularly useful for those species that cannot be conserved in traditional seed or field genebanks nor conserved in situ due to high risk in that area. DNA storage has so far been undertaken with objectives other than conservation in mind, usually to allow genetic material to be made readily available for molecular applications, for distribution or training. The DNA Data Bank of Japan (DDBJ, http://www.ddbj.nig.ac.jp) (Mashima et al. 2017) is a public database of nucleotide sequences established at the National Institute of Genetics (NIG, https://www.nig.ac.jp/nig). Since 1987, the DDBJ has been collecting annotated nucleotide sequences as its traditional database service. The data at DDBJ primarily accumulated via submissions of sequence data by the researchers. This endeavour has been conducted in collaboration with GenBank (Benson et al. 2017) at the National Center for Biotechnology Information (NCBI) and with the European Nucleotide Archive (ENA) (Toribio et al. 2017) at the European Bioinformatics Institute (EBI). The collaborative framework is called the International Nucleotide Sequence Database Collaboration (INSDC) (Cochrane et al. 2016) and the product database from this framework is called the International Nucleotide Sequence Database (INSD). In 2020, the DDBJ accepted 6836 submissions of annotated nucleotide sequences, and 59.3% were submitted by Japanese research groups. The DDBJ has periodically released all public DDBJ/ENA/GenBank nucleotide sequence data in the flat-file format.

Plant DNA Bank in Korea (PDBK) is responsible for collection of the Korean vascular plants and useful plants mainly in East Asia and establishing the genomic DNA database from those plants with its voucher information. The PDBK is one of the largest plant genomic DNA bank in the world and has various plant genomic DNAs including about 2950 domestic species and many foreign species in its collection that mostly belong to Korean endemic, rare, and endangered plant species. The PDBK is having approximately 22,000 accessions of the purified and concentrated genomic DNA from other countries in East Asia. All the DNA materials are well characterized and handled according to the standard procedures and are being dispatched under the material transfer agreement for research purpose. PDBK dispenses approximately 500 accessions of genomic DNAs per year to researchers globally with very high purity of each DNAs. The quality is monitored randomly by the qualitative and quantitative tests (https://pdbk.korea.ac.kr/about.asp).

2.3 Safety Duplication

Safety duplication involves duplication of a genetically identical sub-sample of the accession to mitigate the risk of its partial or total loss caused by natural or man-made catastrophes. The safety duplicates are genetically identical to the base collection and are referred to as the secondary most original sample (Engels and Visser 2003). Safety duplicates include both the duplication of material and its related information, and are deposited in a base collection at a different location, usually in another country. The location is chosen to minimize possible risks and provides the best possible storage facilities. Safety duplication is generally under a 'black-box' approach. This means that the repository genebank has no entitlement to the use and distribution of the germplasm. It is the depositor's responsibility to ensure that the deposited material is of high quality, to monitor seed viability over time and to use their own base collection to regenerate the collections when they begin to lose viability. The germplasm is not touched without permission from the depositor and is only returned on request when the original collection is lost or destroyed.

The Svalbard Global Seed Vault (SGSV) in Norway is an example of a secure facility for safety duplication of crop genetic resources. SGSV is the world's largest safety stock for seeds from the earth's diversity of cultivated crops. Located far beyond the Arctic Circle and 130 m deep inside a frozen mountain, permafrost provides an environmentally friendly solution to long-term secure conservation of crop diversity as a safety duplicate that is only accessed in case of disaster or loss of the samples from the main safety backup. The vault can hold 4.5 million seed samples of crop diversity. The seeds are stored at −18 °C which is required for optimal storage of the seeds and the seeds are stored and sealed in custom-made three-ply foil packages. The packages are sealed inside boxes and stored on shelves inside the vault. The low temperature and moisture levels inside the SGSV ensure low metabolic activity, keeping the seeds viable for long periods of time (https://www.croptrust.org/our-work/svalbard-global-seed-vault).NordGen. Together with the Norwegian Ministry of Agriculture and Food, the organization Global Crop Diversity Trust (GCDT) is responsible for the operation of the SGSV. It offers free storage of seed specimens conserved by international, national, regional genebanks as well as institutions and organizations. Ownership of the seeds never changes. They are stored under so-called black box conditions, which means, among other things, that only the institution that puts in the seeds can take them out. The SGSV currently conserves more than 1.1 million seed samples of 5934 species that have been deposited by 89 national and international genebanks worldwide (https://seedvault.nordgen.org/).

2.4 Germplasm Exchange

The introduction of germplasm for conservation and use is an important function for most genebanks. At the same time, many genebanks also play an important role in distributing germplasm samples to potential users, thus linking conservation directly with use. As germplasm is never free of pests and diseases, great care has to be given to quarantine aspects to avoid the transfer of harmful pathogens together with the germplasm. When exchanging germplasm accessions, the curator has to adhere to existing plant quarantine regulations for both legal and biological reasons. Furthermore, the curator can actively contribute to the safe exchange of germplasm samples by following the technical guidelines which are jointly being produced by FAO and IPGRI. Since the early 1990s, the availability of germplasm has become more restricted. Several countries have introduced access legislation, as part of the implementation of the Convention on Biological Diversity (CBD), and many have implemented the Prior Informed Consent provision of the CBD. The latter requires a mutual agreement on the conditions under which the germplasm material is allowed to be taken out of the country. Both these measures have led to the development and use of material transfer agreements and germplasm acquisition agreements which spell out the conditions under which germplasm can be used and acquired.

2.5 Discovering Climate-Resilient Germplasm

Characterization of germplasm following the crop-specific descriptors provide first-hand information for selection of desirable germplasm based on the traits of interest for which the data is available. The global plant genetic database called Genesys is an online platform on PGR conserved in genebanks globally, contains passport data, characterization and evaluation data sets and trait-specific subsets developed on various crops (https://www.genesys-pgr.org/).

2.5.1 Germplasm Diversity and Trait-Specific Subsets

Ex situ germplasm collections have grown enormously in size and number over the years as a result of global efforts to conserve plant genetic resources for food and agriculture (Odong et al. 2013). The larger size of the germplasm collection and limited information on traits of importance have been highlighted as significant issues hindering their effective utilization in crop improvement programs (Gollin et al. 2000; Koo and Wright 2000; FAO 2010). To overcome this situation, a small set of accessions be selected from the collection containing as much genetic diversity as possible and these types of selections would offer a good starting point when targeting new traits of interest. Considering this, Frankel (1984) proposed a 'core collection' which would 'represent with a minimum of repetitiveness, the genetic diversity of a crop species and its relatives.' A core collection consists of a limited set of accessions (about 10%) derived from an existing germplasm collection, chosen to represent the genetic spectrum in the whole collection. The available data on the geographic origin, specific plant characteristics, trait data, and molecular data are utilized to develop core subsets. There are many methods and free software packages available such as PowerCore (Kim et al. 2007), CoreHunter (De Beukelaer et al. 2018), ccChooser (Studnicki et al. 2012), and MSTRAT (Gouesnard et al. 2001) and GenoCore (Jeong et al. 2017) are few examples that could help to construct core subsets using molecular marker data, genetic distances, phenotypic traits, geographic origin, or integration of these various data types. The accessions remaining after selecting core accessions are considered as the *reserve collection* (Brown 1989). Due to reduced size, the core collections can be evaluated extensively and more economically for important traits. Following this approach, core collections have been constituted in several crops species, including rice (Yan et al. 2007), groundnut (Upadhyaya et al. 2003), pearl millet (Upadhyaya et al. 2009a), sorghum (Grenier et al. 2001), and other crops. In many cases, the germplasm collections conserved by most of the genebanks are very large in size. For example, the size of the ICRISAT sorghum core collection is 2242 accessions that was developed from 22,473 accessions (Grenier et al. 2001), which is still large in size and limits their utilization. To overcome this, Upadhyaya and Ortiz (2001) developed the concept of mini-core collection (10% of core or 1% of the entire collection). Following this approach, mini-core collection has been developed in many crops including rice (Agrama et al. 2009), sorghum (Upadhyaya et al. 2009b), chickpea (Upadhyaya and Ortiz 2001), and other crops (Table 2.2). The important point is that core collection should be dynamic, not static; thus a periodic review and modification of the core collection is necessary considering the increase in size and information of collection, to add new diversity.

Once the core and/or mini core are available, researchers would have a manageable number of accessions to evaluate extensively and identify new variability and traits combinations. For example, the evaluation of 242 accessions of sorghum mini core resulted in the identification of promising germplasm sources resistance to biotic stress (70 accessions), abiotic stress (12 accessions), and other traits such as bioenergy (13 accessions) and nutritional traits (27 accessions) (Upadhyaya et al. 2019). Similarly, in the groundnut mini-core collection (184 accessions), 28 accessions were identified as resistant to abiotic stress and 30 to biotic stress (Upadhyaya et al. 2014a); and in the chickpea mini-core collection, 40 accessions were reported as resistant to abiotic stress and 31 to biotic stress (Upadhyaya et al. 2013a). When we require additional or new source variability for a given trait, researchers can refer back to the clusters from which the core collection accession came to select similar accessions from the entire collection. This approach will increase the probability of identifying specific traits from a large ex situ collection.

2.5.2 Focused Identification of Germplasm Strategy (FIGS)

FIGS is a tool that supports researchers to identify promising germplasm traits-specific sources from large ex situ collections more accurately and efficiently. The FIGS approach matches plant traits with geographic and agro-climatic information of the places where germplasm accessions were collected as the environment strongly influences natural selection, thus it increases the chances of finding the adaptive trait of interest. The main aim of this method is to develop trait-specific subsets rather than capturing all the genetic variation present in the genetic resources. Thus, it is one of the efficient strategies to explore and sort out the plant genetic resources for climate change adaptive traits. FIGS can be developed either by following filtering and modelling strategies. FIGS following filtering requires a deep understanding of the ecology and the optimal conditions of the expression of the traits under study and how these conditions affect the crop. Filters can be applied in the search process to narrow down from a large collection to a small subset considering geographic locations of a given stress occurrence, climatic conditions favouring stress occurrence, and long-term-climatic and/or soil characteristics of the

Table 2.2 Core and mini-core subset developed in different field crops globally

Crop	Core/ mini core	Number of accessions used	No. of traits used	Number of accessions in core/mini core	Reference
Rice	Core	4310	50 phenotypic traits and 36 SSRs	932	(Zhang et al. 2011)
Sorghum	Core	22,474	21	2247	(Grenier et al. 2001)
Sorghum	Core	33,100	7	3475	(Prasada Rao and Ramanatha Rao 1995)
Groundnut	Core	14,310	14	1704	(Upadhyaya et al. 2003)
Groundnut	Core	7432		831	(Holbrook et al. 1993)
Groundnut (Valencia)	Core	630	26	77	(Dwivedi et al. 2008)
Groundnut	Asian Core		15	504	(Upadhyaya et al. 2005)
Soyabean	Core	15,558	18	1600	(Oliveira et al. 2010)
A worldwide bread wheat	Core	3942	38 SSRs	372	(Balfourier et al. 2007)
Pearl millet	Core	16,063	11	1600	(Bhattacharjee et al. 2007)
Pearl millet (augmented)	Core	20,844	22	2094	(Upadhyaya et al. 2009a)
World sesame	Core	1724	17	172	(Mahajan et al. 2007)
West African yam *Dioscorea* spp.	Core	1724	18	172	(Mahalakshmi et al. 2007)
USDA rice	Core	18,412	14	1790	(Yan et al. 2007)
Korean sesame	Core	2246	12	475	(Kang et al. 2006)
Pigeon pea	Core	12,153	14	1290	(Reddy et al. 2005)
Iberia peninsula common beans	Core	388	34	52	(Rodiño et al. 2003)
Safflower	Core	5522	12	570	(Dwivedi et al. 2005)
China sesame	Core	4251	14	453	(Xiurong et al. 2000)
Indian mung bean	Core	1532	38	152	(Bisht et al. 1998)

(continued)

Table 2.2 (continued)

Crop	Core/ mini core	Number of accessions used	No. of traits used	Number of accessions in core/mini core	Reference
Perennial *Medicago*	Core	1100	50	200	(Basigalup et al. 1995)
Annual *Medicago*	Core	1240	16	211	(Diwan et al. 1995)
Saccharum spontaneum	Core	342	11	75	(Tai and Miller 2001)
Chickpea	Core	3350		505	(Hannan et al. 1994)
Chickpea	Core	16,991	13	1956	(Upadhyaya et al. 2001)
Finger millet	Core	5940	14	622	(Upadhyaya et al. 2006b)
Foxtail millet	Core	1474	23	155	(Upadhyaya et al. 2009c)
Proso millet	Core	833	20	106	(Upadhyaya et al. 2011a)
Barnyard millet	Core	736	21	89	(Upadhyaya et al. 2014b)
Kodo millet	Core	656	20	75	(Upadhyaya et al. 2014b)
Little millet	Core	460	20	56	(Upadhyaya et al. 2014b)
Rice	Mini core	1794	26 phenotypic traits and 70 molecular markers	217	(Agrama et al. 2009)
Sorghum	Mini core	2247	21	242	(Upadhyaya et al. 2009b)
Japanese rice landraces	Mini core	236	32 SSRs	50	(Ebana et al. 2008)
Pearl millet	Mini core	2094	12	238	(Upadhyaya et al. 2011b)
Chickpea	Mini core	1956	16	211	(Upadhyaya and Ortiz 2001)
Pigeon pea	Mini core	1290	16	146	(Upadhyaya et al. 2006a)
Groundnut	Mini core	1704	34	184	(Upadhyaya et al. 2002)
Groundnut	Mini core	831	16	112	(Holbrook and Dong 2005)
Finger millet	Mini core	5940	18	80	(Upadhyaya et al. 2010)
Foxtail millet	Mini core	1474	21	35	(Upadhyaya et al. 2011c)

Table 2.3 A few examples of promising germplasm sources identified following FIGS approach for biotic and abiotic stress tolerance

S. No	Crop	Trait	Reference
1.	Wheat	Powdery mildew (*Blumeria graminis* (DC) Speer f.sp. *tritici*)	(Bhullar et al. 2009)
2.	Wheat	Powdery mildew (*Blumeria graminis* (DC) Speer f.sp. *tritici*)	(Vikas et al. 2020)
3.	Wheat	Sunn pest (*Eurygaster intergriceps* put.)	(Bouhssini et al. 2009)
4.	Wheat	Russian wheat aphid (*Diuraphis noxia* Kurdj.)	(Bouhssini et al. 2011)
5.	Wheat	Stem rust (*Puccinia graminis* Pers.)	(Endresen et al. 2012)
6.	Wheat	Stripe (yellow) rust (*Puccinia striiformis*)	(Bari et al. 2014)
7.	Barley	Net blotch (*Pyrenophora teres* Drechs.)	(Endresen et al. 2011)
8.	Faba bean	Drought tolerance	(Khazaei et al. 2013)

collection site, etc. When evaluation data is available for adaptive traits, FIGS can explore the mathematical relationship between the adaptive trait of interest and the long-term climatic and/or soil characteristics of collection sites to choose a small set from a large collection. Further, the small FGIS set can be evaluated to identify promising germplasm sources for use in crop improvement. A few examples of promising germplasm sources identified following FIGS approach for biotic and abiotic stress tolerance are presented in Table 2.3.

2.5.3 Molecular Characterization and Trait Discovery

Advances in genome sequencing technologies have made a significant contribution to the next-generation genebanking for the efficient conservation and enhanced use of germplasm in crop improvement. Genomics and gene editing technological interventions could enable a new era of de novo domestication through the introduction of domestication genes into non-domesticated plants (Van Tassel et al. 2020). Large-scale high-density genotyping helps in understanding the genetic diversity and population structure of the germplasm collection and linking DNA sequence variants to the phenotypes of interest (Varshney et al. 2021). There are several large-scale genotyping efforts in different crops. For example, in chickpea, 3366 accessions including 3171 cultivated and 195 wild species accessions were sequenced at an average coverage of around 12×, and constructed a pan-genome to describe the genomic diversity of chickpea (Varshney et al. 2021). This study identified superior haplotypes for improvement-related traits in landraces that can be introgressed into elite breeding lines through haplotype-based breeding, and also found targets for purging deleterious alleles through genomics-assisted breeding and/or gene editing (Varshney et al. 2021). In wheat, Sansaloni et al. (2020) sequenced about 80,000 wheat accessions using DArTseq technology and identified over 300,000 high-quality SNPs and SilicoDArT markers, provides great opportunity for developing wheat varieties utilizing allelic diversity missing in the current breeding program. In rice, resequencing of a core collection of 3000 accessions originating from 89 countries resulted in the identification of about 29 million single nucleotide polymorphisms (SNPs), 2.4 million small indels, and over 90,000 structural variations that contributed to within and between-population variation (3000 Rice Genome Project 2014; Wang et al. 2018). The phylogenetic analysis based on SNP data confirmed the presence of five varietal groups in *O. sative* gene pool, namely, indica, aus/boro, basmati/sadri, tropical japonica and temperate japonica, and also suggest several subpopulations that correlate with genographic locations (3000 Rice Genome Project 2014; Wang et al. 2018). In addition, using pan-genome analysis, over 10,000 novel full-length protein-coding genes and also presence-absence variations were reported (Wang et al. 2018). From the USDA soybean collection, 14,430 soybean accessions were selected from the whole set of about 22,000 were genotyped using the Illumina Infinium SoySNP50K BeadChip (Bandillo et al. 2015). The results indicated that the accessions originating from Japan were relatively homogenous and distinct from the Korean accessions, while both Japanese and Korean accessions diverged from the Chinese accessions. The GWAS performed using 12,000–13,000 accessions identified SNPs signals for seed protein and oil (Bandillo et al. 2015), and also for ten key phenotypic descriptive traits (Bandillo et al. 2017). Such large-scale genotyping of genebank collections support in gene discovery, genomic prediction, genome-wide association mapping, marker development, and other applications.

2.5.4 Contribution of Plant Genetic Resources for Global Food Security and Nutrition, and Environmental and Economic Benefits

Breeding of high-yielding, resistance/tolerance to biotic and abiotic stresses, and climate-resilient crops is important for meeting the food demand of the increasing population globally. Plant genetic resources contribute significantly for addressing the food security, malnutrition and environmental sustainability. Impact of germplasm for contributing to increased yield, adaptation, nutrition and improved health and sustainable agriculture have been demonstrated in many crops. There are many instances where a single plant genetic resources has proved to have large commercial value by conferring a specific trait. Well-known examples include *Rht1* and *Rht2* dwarfing genes in wheat, the dwarfing genes of the green revolution, originated in Japan, by crossing a semi-dwarf wheat variety called Daruma with American high-yielding variety to produce Norin 10, which was further used to develop number of semi-dwarf cultivars. The dwarfing alleles are named *Rht1* (*Rht-B1b*) and *Rht2* (*Rht-D1b*) (Gaur et al. 2020). In rice, the semi-dwarfing gene, *sd1* first identified in the Chinese variety 'Dee-geo-woo-gen' was utilized to develop the semi-dwarf cultivars such as Taichung Native 1 (TN1) and IR8, and later it formed the basis for the development of new high-yielding, semi-dwarf cultivars (Spielmeyer et al. 2002). The semi-dwarfing gene in rice (sd1) is a recessive allele that confers lodging resistance through shortened culm and highly responsive to nitrogenous fertilizers. Groundnut is an important oil seed crop, originated in southern Bolivia to northern Argentina region of South America. Recent study revealed that the contribution of a wild species accession, *Arachis cardenasii* GKP 10017 originating from Bolivia for the development of groundnut cultivars resistant to foliar fungal disease. The ICRISAT genebank assembled the GKP 10017 accession from USDA-ARS, registered as ICG 8216. From ICRISAT it reached globally and contributed as a source for developing groundnut cultivars resistance to late leaf spot and rust in Africa, Asia, Oceania, and the Americas, and provided widespread improved food security and environmental and economic benefits (Bertioli et al. 2021). Table 2.4 shows a few examples of ICRISAT-supplied germplasm that impacted global crop productivity.

Globally, the burden of malnutrition in all its forms remains a major challenge to the humanity. Thus, there is an urgent need to transform food systems to sustainably deliver better quality diets for improved nutrition and health. Breeding staple crops by mainstreaming nutrition as a key component could deliver biofortified crop cultivars for different nutrients. Globally, HarvestPlus program focuses on biofortification of major staples (rice, wheat, maize, beans, cassava, sweet potato, and pearl millet) through conventional plant breeding methods to increase the micronutrients content of staple food crops, works with several CGIAR research centres and national agriculture research systems in collaboration. Between 2004 and 2022, 262 biofortified cultivars in 12 crops have been released in 30 countries. For example, in pearl millet, utilizing the intra-population variability within ICTP 8203, the high Fe and Zn biofortified varieties of pearl millet 'Dhanshakti' and 'Chakti' were released in India and Africa (Rai et al. 2014; Govindaraj et al. 2019). The ICTP 8203 is a large-seeded and high-yielding open-pollinated variety derived from *iniadi* landrace from northern Togo, bred at ICRISAT, Patancheru. Currently, India is growing >70,000 ha of biofortified pearl millet, and many more cultivars are under various stages of testing for a possible release. There are several such examples on the impact of germplasm globally for addressing food security and nutrition.

Landraces, crop wild relatives, and specifically adapted ecotypes are generally heterogeneous, adapted to specific local environments, and often low/or no market preference, they can be endowed with rich sources of genes for crop improvement. Advances in plant genomics is opening a new era in germplasm research such as deployment of desirable alleles originating from the germplasm (landrace) in the crop improvement programs. For instance, genes can now be edited in situ such that alleles conferring desirable traits or phenotype can be reintroduced into elite cultivars without disturbing the genetic background that confers valuable traits including

Table 2.4 Germplasm lines that impacted ICRISAT mandate crops productivity globally

Crop	Accessions	Origin	Contribution	Selection from germplasm directly released as variety
Pigeon pea	ICP 7035	India	Source for resistance to sterility mosaic disease (SMD) and a large seed size	*Kamica* in Fiji, *Guimu 4* in China, *JK Sweety* in India and ICP 7035 in both Nepal and Philippines
Pigeon pea	ICP 8863	India	Source for resistance to *fusarium* wilt	*Maruthi* in India
Chickpea	ICC 4958	India	Donor for drought tolerance, used as parents in chickpea improvement for drought tolerance	Transferred through MAS in several varieties recently
Groundnut	ICG 12991	India	Source of resistance to rosette virus disease	Baka in Malawi, as Serenut 4 T in Uganda, Nematil in Mozambique and Msandile in Zambia
Sorghum	IS 2205	India	Source for resistance to shoot fly and stem borer resistance	Used as national check in India for shoot fly and stem borer resistance
Sorghum	IS 33844	India	It is an excellent maldandi-type with large and lustrous grains and high yield (predominant post-rainy sorghum landrace in Maharashtra and Karnataka states of India). This was selected from a germplasm collection from Maharashtra by ICRISAT genebank staff in 1989.	Parbhani Moti in India
Pearl millet	IP 17862	Togo	An *Iniadi* pearl millet landrace was the important source material for the development of improved cultivars	ICTP 8203, MP 124, PCB 138 in India; *Okashana 1* and *Okashana 2* in Namibia and Nyankhombo in Malawi.
Barnyard millet	IEc 542	Japan	High grain and fodder yielding, most popular in Uttarakhand, India	PRJ 1 in India

yield, quality and stress tolerance traits. A few examples of genes that contributed to enhancing productivity, quality and stress tolerance in crop cultivars are listed in Table 2.5.

Table 2.5 Examples of genes that contributed to enhancing productivity, quality and stress tolerance in crop cultivars

S. No	Crop	Genes	Traits	Reference
1	Rice	*Sd1*	Semi-dwarf	(Spielmeyer et al. 2002)
2	Wheat	*Rht-B1b (Syn. Rht1)*	Semi-dwarf	(Peng et al. 1999)
3	Wheat	*Rht-D1b (Syn. Rht2)*	Semi-dwarf	(Peng et al. 1999)
4	Sorghum	*Sh1*	Seed shattering	(Lin et al. 2012)
5	Sorghum	*SbWRKY*	Seed shattering	(Tang et al. 2013)
6	Sorghum	*Wx*	Endosperm texture	(McIntyre et al. 2008, Sattler et al. 2009)
7	Sorghum	*Ma1/SbPRR37*	Maturity	(Murphy et al. 2011)
8	Sorghum	*Ma3*	Maturity	(Childs et al. 1997)
9	Sorghum	*Ma6*	Maturity	(Murphy et al. 2014)
10	Sorghum	*SbSUC9*	Maturity	(Upadhyaya et al. 2013b)
11	Sorghum	*LD*	Maturity	(Upadhyaya et al. 2013b)
12	Sorghum	*SbMED12*	Maturity	(Upadhyaya et al. 2013b)
13	Sorghum	*Dw1/Sbht9.1*	Plant height	(Hilley et al. 2016)
14	Sorghum	*Dw2*	Plant height	(Hilley et al. 2017)
15	Sorghum	*Dw3*	Plant height	(Multani et al. 2003)
16	Sorghum	*bmr2*	Brown midrib	(Saballos et al. 2012)
17	Sorghum	*bmr6*	Brown midrib	(Saballos et al. 2009)
18	Sorghum	*bmr12*	Brown midrib	(Sattler et al. 2012)
19	Sorghum	*Glossy 15*	Shoot fly resistance	(Satish et al. 2009); (Aruna et al. 2011)
20	Sorghum	*Rf1/SbPPR13*	Fertility	(Klein et al. 2005)
21	Sorghum	*Rf2*	Fertility	(Madugula et al. 2018)
22	Sorghum	*Rf6*	Fertility	(Praveen et al. 2015)
23	Sorghum	YELLOW SEED1 (Y1)	Grain mould resistance	(Nida et al. 2019)
24	Sorghum	YELLOW SEED3 (Y3)	Grain mould resistance	(Nida et al. 2019)
25	Barley	*Short clum1 (hcm1)*	Short clum	(Lundqvist et al. 1997)
26	Rye	*Dw1 (Ddwl)*	Dwarf	(Tenhola-Roininen and Tanhuanpää 2010)

2.6 Access to Genebank Collection

The legal landscape for biodiversity and genetic resources has changed dramatically over the last 40 years, and continues to evolve. Arguably, the biggest changes that took place were the shift from the common heritage concept in the International Undertaking (1983) to national sovereignty in the Convention on Biological Diversity (1993) and then countries choosing to exercise that national sovereignty to create an international multilateral system for PGRFA under the Plant Treaty (2004). Many peoples' perceptions of genetic resources and their value have been influenced by advances in science and technology and their potential for commercial exploitation. One policy response was to strengthen intellectual property laws to protect commercial investments. This development catalysed a call for national control over genetic resources that are relied upon as 'inputs' into research and development chains with commercial potential. These policy responses are, in turn, influencing the ability of research and development organizations (both public and private) to access and use genetic resources conserved in genebanks worldwide and also to exploit new technologies, and share benefits created through their work. Genebanks worldwide are mostly facilitating access to their collections through the Multilateral System (MLS) of the International Treaty on Plant Genetic Resources for Food and Agriculture (ITPGFRA).

2.6.1 The International Treaty on Plant Genetic Resources for Food and Agriculture (ITPGRFA)

To address PGRFA in the post-CBD era, the FAO drafted and adopted the International Treaty on Plant Genetic Resources for Food and Agriculture (ITPGRFA, www.fao.org/plant-treaty), which came into force on 29 June 2004 (FAO 2002). The objectives of the ITPGRFA are very similar to those of the CBD but focus on the conservation and sustainable use of PGRFA and the sharing of the benefits arising from their use (FAO 2002). PGRFA are defined as: "any genetic material of plant origin of actual or potential value for food and agriculture" (FAO 2002). The ITPGRFA confirms the sovereign rights of countries over their genetic resources but aims to facilitate the exchange of PGRFA by the establishment of a Multilateral System of Access and Benefit-Sharing (MLS) in which PGRFA are exchanged under a Standard Material Transfer Agreement (SMTA), instead of under the prior informed consent and mutually agreed terms prescribed by the CBD.

The MLS is a global pool of PGRFA, meant to facilitate access to these PGRFA as well as to achieve fair and equitable sharing of the benefits arising from their utilization. PGRFA may be added to this pool by countries and the institutions under their control, by natural and legal persons in the contracting parties and by international institutes (Manzella 2013). The MLS does not extend to all PGRFA but covers a set of 35 food crops and 29 forages, which are listed in Annex I of the ITPGRFA. The selection of this set of crops and forages was based on criteria of food security and interdependence and was a negotiated compromise between countries favouring the inclusion of all PGRFA and countries favouring the inclusion of only a limited number of crops (Visser 2013). According to Article 11 of the ITPGRFA, the MLS is to include all PGRFA of the food crops and forages listed in Annex I that are "under the management and control of the Contracting Parties and in the public domain" (FAO 2002). PGRFA that belong to the food crops and forages listed in Annex I but do not fulfil the other conditions are not automatically included in the MLS but can be included on a voluntary basis by natural and legal persons holding these PGRFA. Access to materials in the MLS under the SMTA is granted only for their use in research, breeding and training for food and agriculture; other uses are explicitly excluded (FAO 2002). With regard to benefit sharing, the Contracting Parties to the ITPGRFA recognize that facilitated access itself is an important benefit, but also underline the importance of other forms of benefit sharing, such as the exchange of information, technology transfer, capacity building, and the sharing of commercial benefits. If material received under an SMTA is used to create PGRFA that are not freely available for research and breeding by others, the recipients must pay 0.77% of the sales of those PGRFA (or 0.5% of all sales of PGRFA belonging to the same crop) to an international benefit-sharing fund (www.fao.org/plant-treaty/areas-of-work/benefit-sharing-fund), which is used to support conservation and sustainable utilization of PGRFA. The Contracting Parties to the ITPGRFA undertake to include in the MLS those PGR of the crops and forages in Annex I that are in the public domain and under their management. However, even if material is not part of the MLS, providers of PGR can distribute their material under the SMTA.

2.6.2 *Article 15 of ITPGRFA*

Article 15 deals with ex situ collections of PGRFA held by the CGIAR genebanks and other international institutions. The Treaty called on the CGIAR Centres to sign agreements with the Governing Body to bring their collections under the Treaty. PGRFA listed in Annex I that are held by the CGIAR Centres are to be made available as part of the MLS. In 2006, all Centres of the CGIAR System holding collections of Plant Genetic Resources for food and agriculture (PGRFA) signed Agreements with the Governing Body of the International Treaty on Plant Genetic Resources for Food and Agriculture (the Treaty) placing them in-trust collections of PGRFA within the purview of the Treaty. In accordance with these Agreements, all shipments of PGRFA of crops listed in Annex 1 to the Treaty (shipments of PGRFA under the Multilateral System) were subjected to the terms and conditions of the Standard Material Transfer Agreement (SMTA) adopted by the Governing Body of the International Treaty on Plant Genetic Resources for Food and Agriculture in June 2006. The CGIAR centres make more than 750,000 accessions available under the MLS (FAO 2019).

2.7 Summary

The key to sustainable agriculture is genetic material that is better adapted to withstand biotic and abiotic stresses. In order to address present and forthcoming threats to food and nutritional security, it is imperative to preserve the genetic diversity that is especially crucial. One of the primary concerns, however, is the enormous rate of biodiversity loss, which threatens food and nutrition security, weakens the agricultural system's resilience, and jeopardises crop improvement. Hence, national and international genebanks that hold more than 7.5 million accessions of crops have been established in order to reduce the biodiversity loss caused by the replacement of landraces by improved cultivars. Efficient use of germplasm in crop improvement is depending on the availability of accession-level information on the traits of interest. Thus, core collection, mini-core collection, and FIGS approaches have been created to successfully find novel variations and trait recombinants. Molecular characterization, which includes activities like high-density genotyping, phylogenetic analysis, pan-genome analysis, etc., can be further combined with the selection of germplasm and the construction of subsets and mining novel alleles for use in traits improvement. Impact of germplasm for contributing to increased yield, adaptation, nutrition and improved health and sustainable agriculture have been demonstrated in many crops. There are many instances where a single plant genetic resource has proved to have large commercial value by conferring a specific trait. With the availability of new technologies such as high-throughput large-scale phenotypic assessment for key traits and use of omic tools could accelerate rapid identification of traits and genes for breeding improved cultivars.

References

Agrama HA, Yan WG, Lee F, Fjellstrom R, Chen MH, Jia M, McClung A (2009) Genetic assessment of a mini-core subset developed from the USDA rice genebank. Crop Sci 49: 1336–1346

Aruna C, Bhagwat VR, Madhusudhana R, Sharma V, Hussain T, Ghorade RB, Khandalkar HG, Audilakshmi S, Seetharama N (2011) Identification and validation of genomic regions that affect shoot fly resistance in sorghum [Sorghum bicolor (L.) Moench]. Theor Appl Genet 122: 1617–1630

Balfourier F, Roussel V, Strelchenko P (2007) A worldwide bread wheat core collection arrayed in a 384-well plate. Theor Appl Genet 114:1265–1275. https://doi.org/10.1007/s00122-007-0517-1

Bandillo N, Jarquin D, Song Q, Nelson R, Cregan P, Specht J, Lorenz A (2015) A population structure and genome-wide association analysis on the USDA soybean germplasm collection. The plant. Genome 8(3):plantgenome2015-04

Bandillo NB, Lorenz AJ, Graef GL, Jarquin D, Hyten DL, Nelson RL, Specht JE (2017) Genome-wide association mapping of qualitatively inherited traits in a germplasm collection. The plant genome 10(2):plantgenome2016-06

Bari A, Amri A, Street K, Mackay M, De Pauw E, Sanders R et al (2014) Predicting resistance to stripe (yellow) rust (Puccinia striiformis) in wheat genetic resources using focused identification of germplasm strategy. J Agric Sci 152(6):906–916

Basigalup DH, Barnes DK, Stucker RE (1995) Development of a core collection for perennial Medicago plant introductions. Crop Sci 35:1163–1168

Benson DA, Cavanaugh M, Clark K, Karsch-Mizrachi I, Lipman DJ, Ostell J, Sayers EW (2017) GenBank. Nucleic Acids Res 45:D37–D42

Bertioli DJ, Clevenger J, Godoy IJ, Stalker HT, Wood S, Santos JF, Ballén-Taborda C, Abernathy B, Azevedo V, Campbell J, Chavarro C, Chu Y, Farmer AD, Fonceka D, Gao D, Grimwood J, Halpin N, Korani W, Michelotto MD et al (2021) Legacy genetics of Arachis cardenasii in the peanut crop shows the profound benefits of international seed exchange. Proc Natl Acad Sci U S A 118(38):e2104899118. https://doi.org/10.1073/pnas.2104899118

Bhattacharjee R, Khairwal IS, Bramel PJ (2007) Establishment of a pearl millet [Pennisetum glaucum (L.) R. Br.] core collection based on geographical distribution and quantitative traits. Euphytica 155:35–45. https://doi.org/10.1007/s10681-006-9298-x

Bhullar NK, Street K, Mackay M, Yahiaoui N, Keller B (2009) Unlocking wheat genetic resources for the molecular identification of previously undescribed functional alleles at the Pm3 resistance locus. Proc Natl Acad Sci 106(23):9519–9524

Bisht I, Mahajan R, Patel D (1998) The use of characterisation data to establish the Indian mungbean core collection and assessment of genetic diversity. Genet Resour Crop Evol 45: 127–133. https://doi.org/10.1023/A:1008670332570

Bouhssini ME, Street K, Joubi A, Ibrahim Z, Rihawi F (2009) Sources of wheat resistance to Sunn pest, Eurygaster integriceps Puton, in Syria. Genet Resour Crop Evol 56:1065–1069

Bouhssini ME, Street K, Amri A, Mackay M, Ogbonnaya FC, Omran A et al (2011) Sources of resistance in bread wheat to Russian wheat aphid (Diuraphis noxia) in Syria identified using the focused identification of germplasm strategy (FIGS). Plant Breed 130(1):96–97

Breese EL (1989) Regeneration and multiplication of germplasm resources in seed Genebanks: the scientific background. IPGRI, Rome

Brown AHD (1989) Core collections: a practical approach to genetic resources management. Genome 31(2):818–824

Childs KL, Miller FR, Cordonnier-Pratt MM (1997) The sorghum photoperiod sensitivity gene, Ma3, encodes a phytochrome B. Plant Physiol 113:611–619

Cochrane G, Karsch-Mizrachi I, Takagi T (2016) The international nucleotide sequence database collaboration. Nucleic Acids Res 44:D48–D50

De Beukelaer H, Davenport GF, Fack V (2018) Core hunter 3: flexible core subset selection. BMC bioinformatics 19:1–12

Diwan N, Bauchan GR, McIntosh MS (1995) Methods of developing a core collection of annual Medicago species. Theor Appl Genet 90:755–761

Dwivedi SL, Puppala N, Upadhyaya HD, Manivannan N, Singh S (2008) Developing a core collection of peanut specific to Valencia market type. Crop Sci 48:625–632. https://doi.org/10.2135/cropsci2007.04.0240

Dwivedi SL, Upadhyaya HD, Hegde DM (2005) Development of core collection using geographic information and morphological descriptors in safflower (Carthamus tinctorius L.) germplasm. Genet Resour Crop Evol 52:821–830

Ebana K, Kojima Y, Fukuoka S, Nagamine T, Kawase M (2008) Development of mini core collection of Japanese rice landrace. Breeding Sci 58:281–291

Ellis RH, Hong TD, Roberts EH (1985) Handbook of seed technology for genebanks, vol I. Principles and methodology. Handbooks for Genebanks No. 2. IPGRI, Rome

Endresen DTF, Street K, Mackay M, Bari A, De Pauw E (2011) Predictive association between biotic stress traits and eco-geographic data for wheat and barley landraces. Crop Sci 51(5):2036–2055

Endresen DTF, Street K, Mackay M, Bari A, Amri A, De Pauw E et al (2012) Sources of resistance to stem rust (Ug99) in bread wheat and durum wheat identified using focused identification of germplasm strategy. Crop Sci 52(2):764–773

Engelmann F, Takagi H (2000) Cryopreservation of tropical plant germplasm. In: Engelmann F, Takagi H (eds) Cryopreservation of tropical plant germplasm—current research Progress and applications. JIRCAS, Tsukuba, p 496

Engels JMM, Visser B (2003) A guide for effective germplasm collection management. Rome, Italy (in press), IPGRI

FAO (1998) The state of the World's plant genetic resources for food and agriculture. FAO, Rome

FAO (2002). International treaty on plant genetic resources for food and agriculture [Rome: food and agriculture Organization of the United Nations (FAO)]

FAO (2010) The second report on the state of the World's plant genetic resources for food and agriculture. FAO, Rome

FAO (2014) Genebank standards for plant genetic resources for food and agriculture. In *Commission on Genetic Resources for Food and Agriculture*. Food and Agriculture Organization of the United Nations Rome. ISBN 978–92–5-107855-6 (print) E-ISBN 978-92-5-107856-3 (PDF)

FAO (2019) Report on the Implementation and Operations of the Multilateral System. Eighth Session of the Governing Body, 11–16 November 2019, Rome [Rome: Food and Agriculture Organization of the United Nations (FAO)]

FAO–IPGRI (1994) Genebank Standards. FAO and IPGRI, Rome

Frankel OH (1984) Genetic perspective of germplasm conservation. In: Arber W, Limensee K, Peacock WJ, Stralinger P (eds) Genetic manipulations: impact on man and society. Cambridge University Press, Cambridge, pp 161–170

Gatto M, de Haan S, Laborte A, Bonierbale M, Labarta R, Hareau G (2021) Trends in varietal diversity of Main staple crops in Asia and Africa and implications for sustainable food systems. Frontiers in Sustainable Food Systems 5(February):1–15. https://doi.org/10.3389/fsufs.2021.626714

Gaur VS, Channappa G, Chakraborti M, Sharma TR, Mondal TK (2020) 'Green revolution' dwarf gene sd1 of rice has gigantic impact. Brief Funct Genomics 19(October):390–409. https://doi.org/10.1093/bfgp/elaa019

Gepts P (2006) Plant genetic resources conservation and utilization. Crop Sci 46:2278–2292. https://doi.org/10.2135/cropsci2006.03.0169gas

Gollin D, Smale M, Skovmand B (2000) Searching an ex-situ collection of wheat genetic resources. Am J Agric Econ 82(4):812–827

Gouesnard B, Bataillon TM, Decoux G, Rozale C, Schoen DJ, David JL (2001) MSTRAT: an algorithm for building germ plasm core collections by maximizing allelic or phenotypic richness. J Hered 92(1):93–94

Govindaraj M, Rai KN, Cherian B, Pfeiffer WH, Kanatti A, Shivade H (2019) Breeding biofortified pearl millet varieties and hybrids to enhance millet Markets for Human Nutrition. Agriculture 9: 106

Grenier C, Hamon P, Bramel-Cox PJ (2001) Core collection of sorghum: II. Comparison of three random sampling strategies. Crop Sci 41:241–246

Guarino L, Rao V.R, Reid R, eds. (1995). Collecting plant genetic diversity. Technical Guidelines. CAB International, Wallingford

Hannan RM, Kaiser WJ, Muehlbauer FJ (1994) Development and utilization of the USDA chickpea germplasm core collection. In: Agronomy abstracts. ASA, Madison, WI, p 217

Hanson J (1985) Procedures for handling seeds in genebanks, Practical manuals for Genebanks no. 1. IPGRI, Rome

Hilley J, Truong S, Olson S, Morishige D, Mullet J (2016) Identification of Dw1, a regulator of sorghum stem internode length. PLoS One 11:e0151271

Hilley JL, Weers BD, Truong SK, McCormick RF, Mattison AJ, McKinley BA, Morishige DT, Mullet JE (2017) Sorghum Dw2 encodes a protein kinase regulator of stem internode length. Sci Rep 7:4616–4628

Holbrook CC, Dong W (2005) Development and evaluation of a mini core collection for the US peanut germplasm collection. Crop Sci 2005(45):1540–1544

Holbrook CC, Anderson WF, Pittman RN (1993) Selection of core collection from the U.-S. germplasm collection of peanut. Crop Sci 33:859–861

Hong TD, Ellis RH (1996) A protocol to determine seed storage behaviour. In: Engels JMM, Toll J (eds) IPGRI Technical Bulletin No.1. IPGRI, Rome

Hong TD, Linnington S, Ellis RH (1996) Seed storage behaviour: a compendium, Handbooks for Genebanks No. 4. IPGRI, Rome

Jeong S, Kim JY, Jeong SC, Kang ST, Moon JK, Kim N (2017) GenoCore: a simple and fast algorithm for core subset selection from large genotype datasets. PLoS One 12(7):e0181420

Kang CW, Kim SY, Lee SW, Mathur PN, Hodgkin T, Zhou MD, Lee RJ (2006) Selection of a core collection of Korean sesame germplasm by a stepwise clustering method. Breeding Sci 56(1):85–91

Khazaei H, Street K, Santanen A, Bari A, Stoddard FL (2013) Do faba bean (Vicia faba L.) accessions from environments with contrasting seasonal moisture availabilities differ in stomatal characteristics and related traits? Genet Resour Crop Evol 60:2343–2357

Khoury CK, Bjorkmann AD, Dempewolf H, Ramirez-Villegas J, Guarino L, Jarvis A et al (2014) Increasing homogeneity in global food supplies and the implications for food security. Proc Natl Acad Sci U S A 111:4001–4006. https://doi.org/10.1073/pnas.1313490111

Kim KW, Chung HK, Cho GT, Ma KH, Chandrabalan D, Gwag JG et al (2007) PowerCore: a program applying the advanced M strategy with a heuristic search for establishing core sets. Bioinformatics 23(16):2155–2162

Klein RR, Klein PE, Mullet JE, Minx P, Rooney WL, Schertz KF (2005) Fertility restorer locus Rf1 of sorghum (Sorghum bicolor L.) encodes a pentatricopeptide repeat protein not present in the colinear region of rice chromosome 12. Theor Appl Genet 111:994–1012

Koo B, Wright BD (2000) The optimal timing of evaluation of genebank accessions and the effects of biotechnology. Am J Agric Econ 82(4):797–811

Lin ZW, Li XE, Shannon LM, Yeh CT, Wang ML, Bai GH, Peng Z, Li JR, Trick HN, Clemente TE, Doebley J, Schnable PS, Tuinstra MR, Tesso TT, White F, Yu JM (2012) Parallel domestication of the Shattering1 genes in cereals. Nat Genet 44:720–724

Lundqvist U, Franckowiak JD, Konishi T (1997) New and revised descriptions of barley genes. Barley Genet Newslett 1997(26):22–516

Madugula P, Uttam AG, Tonapi VA, Ragimasalawada M (2018) Fine mapping of Rf2, a major locus controlling pollen fertility restoration in sorghum A1 cytoplasm, encodes a PPR gene and its validation through expression analysis. Plant Breed 137:148–161

Mahajan RK, Bisht IS, Dhillon BS (2007) Establishment of a core collection of world sesame. (Sesamum indicum L.) germplasm accessions. Sabrao J Breed Genet 39:53–64

Mahalakshmi V, Ng Q, Atalobhor J (2007) Development of a west African yam Dioscorea spp. core collection. Genet Resour Crop Evol 54:1817–1825. https://doi.org/10.1007/s10722-006-9203-4

Malik SK, Chaudhury R, Pritchard HW (2012) Long-term, large scale banking of citrus species embryos: comparisons between cryopreservation and other seed banking temperatures. Cryo-Letters 33:453–464

Manzella D (2013) The design and mechanics of the multilateral system of access and benefit-sharing. In: Halewood M, López Noriega I, Louafi S (eds) Crop Genetic Resources as a Global Commons: challenges in international law and governance. Routledge, Abingdon, pp 150–163

Mashima J, Kodama Y, Fujisawa T, Katayama T, Okuda Y, Kaminuma E, Ogasawara O, Okubo K, Nakamura Y, Takagi T (2017) DNA Data Bank of Japan. Nucleic Acids Res 45:D25–D31

McIntyre CL, Drenth J, Gonzalez N, Henzell RG, Jordan DR (2008) Molecular characterization of the waxy locus in sorghum. Genome 51:524–533

Multani DS, Briggs SP, Chamberlin MA, Blakeslee JJ, Murphy AS, Johal GS (2003) Loss of an MDR transporter in compact stalks of maize br2 and sorghum dw3 mutants. Science 302:81–84

Murphy RL, Klein RR, Morishige DT, Brady JA, Rooney WL, Miller FR, Dugas DV, Klein PE, Mullet JE (2011) Coincident light and clock regulation of pseudoresponse regulator protein 37 (PRR37) controls photoperiodic flowering in sorghum. Proc Natl Acad Sci U S A 108: 16469–16474

Murphy RL, Morishige DT, Brady JA, Rooney WL, Yang S, Klein PE, Muller E (2014) Ghd7 (Ma6) represses sorghum flowering in long days: Ghd7 alleles enhance biomass accumulation and grain production. Plant Genome 7:1–10

Nida H, Girma G, Mekonen M, Lee S, Seyoum A, Dessalegn K, Tadesse T, Ayana G, Senbetay T, Tesso T, Ejeta G, Mengiste T (2019) Identification of sorghum grain mold resistance loci through genome wide association mapping. J Cereal Sci 85:295–304

Niino T (2006) Developments in plant genetic resources cryopreservation technologies. In: Proc. APEC Work. Eff. Gene Bank Manag. APEC Memb. Econ. Suwon Korea, pp 197–217

Niino T, Yamamoto SI, Fukui K, Martińez CRC, Valle Arizaga M, Matsumoto T, Engelmann F (2013) Dehydration improves cryopreservation of mat rush (Juncus decipiens Nakai) basal stem buds on cryo-plates. Cryo-Letters 34:549–560

Odong TL, Jansen J, Van Eeuwijk FA, van Hintum TJ (2013) Quality of core collections for effective utilisation of genetic resources review, discussion and interpretation. Theor Appl Genet 126:289–305

Oliveira MF, Nelson RL, Geraldi IO, Cruz CD, Toledo JFF (2010) Establishing a soybean germplasm core collection. Field Crops Res 119(2–3):277–289

Painting KA, Perry MC, Denning RA, Ayad WG (1995) Guidebook for genetic resources documentation. IPGRI, Rome

Peng J, Richards DE, Hartley NM (1999) 'Green revolution' genes encode mutant gibberellin response modulators. Nature 1999(400):256–261

Prasada Rao KE, Ramanatha Rao V (1995) Use of characterization data in developing a core collection of sorghum. In: Hodgkin T, Brown AHD, van Hinthum TJL, Morales EAV (eds) Core collection of plant genetic resources. John Wiley and Sons, pp 109–115

Praveen M, Anurag Uttam G, Suneetha N, Umakanth A, Patil JV, Madhusudhana R (2015) Inheritance and molecular mapping of *Rf6* locus with pollen fertility restoration ability on A1 and A2 cytoplasms in sorghum. Plant Sci 238:73–80

Rai KN, Patil HT, Yadav OP, Govindaraj M, Khairwal IS, Cherian B, Rajpurohit BS, Rao AS, Shivade H, Kulkarni MP (2014) Variety Dhanasakthi. Indian J Genet Plant Breeding 037:405–406

Reddy LJ, Upadhyaya HD, Gowda CLL, Singh S (2005) Development of core collection in pigeon pea [Cajanus cajan (L.) Millsp.] using geographical and qualitative morphological descriptors. Genet Res Crop Evol 52:1049–1056

Reed BM (2008) Cryopreservation—Practical considerations. In: Plant Cryopreservation: A Practical Guide. Springer, New York, NY. ISBN 978-0-387-72275-7

Roberts EH (1973) Predicting the viability of seeds. Seed Sci Technol 1:499–514

Rodiño A, Santalla M, De Ron A (2003) A core collection of common bean from the Iberian peninsula. Euphytica 131:165–175. https://doi.org/10.1023/A:1023973309788

Saballos A, Ejeta G, Sanchez E, Kang C, Vermerris W (2009) A genomewide analysis of the cinnamyl alcohol dehydrogenase family in sorghum [Sorghum bicolor (L.) Moench] identifies SbCAD2 as the Brown midrib6 gene. Genetics 181:783–795

Saballos A, Sattler SE, Sanchez E, Foster TP, Xin ZG, Kang C, Pedersen JF, Vermerris W (2012) Brown midrib2 (Bmr2) encodes the major 4-coumarate: coenzyme a ligase involved in lignin biosynthesis in sorghum (Sorghum bicolor (L.) Moench). Plant J 70:818–830

Sansaloni C, Franco J, Santos B, Percival-Alwyn L, Singh S et al (2020) Diversity analysis of 80,000 wheat accessions reveals consequences and opportunities of selection footprints. Nat Commun 11(1):4572. https://doi.org/10.1038/s41467-020-18404-w

Satish K, Srinivas G, Madhusudhana R, Padmaja PG, Nagaraja Reddy R, Murali Mohan S, Seetharama N (2009) Identification of quantitative trait loci for resistance to shoot fly in sorghum [Sorghum bicolor (L.) Moench]. Theo Appl Genet 119:1425–1439

Sattler SE, Singh J, Haas EJ, Guo LN, Sarath G, Pedersen JF (2009) Two distinct waxy alleles impact the granule-bound starch synthase in sorghum. Mol Breed 24:349–359

Sattler SE, Palmer NA, Saballos A, Greene AM, Xin ZG, Sarath G, Vermerris W, Pedersen JF (2012) Identification and characterization of four missense mutations in brown midrib 12 (bmr12), the Caffeic O-Methyltranferase (COMT) of sorghum. Bioenergy Res 5:855–865

Spielmeyer W, Ellis MH, Chandler PM (2002) Semi-dwarf (sd-1) "green revolution" rice contains a defective gibberellin 20-oxidase gene. Proc Natl Acad Sci U S A 2002(99):9043–9048

Studnicki, M., Debski, K., & Studnicki, M. M. (2012). Package 'ccChooser'

Tai PYP, Miller JD (2001) A core collection for Saccharum spontaneum L. from the world collection of sugarcane. Crop Sci 3:879–885

Tang H, Cuevas HE, Das S, Sezen UU, Zhou C, Guo H, Goff VH, Ge ZX, Clemente TE, Paterson AH (2013) Seed shattering in a wild sorghum is conferred by a locus unrelated to domestication. Proc Natl Acad Sci U S A 110:15824–15829

Tenhola-Roininen T, Tanhuanpää P (2010) Tagging the dwarfing gene Ddw1 in a rye population derived from doubled haploid parents. Euphytica 2010(172):303–312

The 3,000 rice genomes project (2014) The 3,000 rice genomes project. GigaSci **3**:7. https://doi.org/10.1186/2047-217X-3-7

Toribio AL, Alako B, Amid C, Cerdeño-Tarrága A, Clarke L, Cleland I, Fairley S, Gibson R, Goodgame N (2017) Nucleic Acids Res 45:D32–D36

Upadhyaya H, Ortiz R (2001) A mini core subset for capturing diversity and promoting utilization of chickpea genetic resources in crop improvement. Theor Appl Genet 102:1292–1298. https://doi.org/10.1007/s00122-001-0556-y

Upadhyaya HD, Bramel PJ, Singh S (2001) Development of a chickpea core subset using geographic distribution and quantitative traits. Crop Sci 2001(41):206–210

Upadhyaya HD, Bramel PJ, Ortiz R, Singh S (2002) Developing a mini core of peanut for utilization of genetic resources. Crop Sci 42:2150–2156

Upadhyaya HD, Ortiz R, Bramel PJ (2003) Development of a groundnut core collection using taxonomical, geographical and morphological descriptors. Genet Resour Crop Evol 50:139–148. https://doi.org/10.1023/A:1022945715628

Upadhyaya HD, Mallikarjuna Swamy BP, Goudar PVK, Kullaiswamy BY, Singh S (2005) Identification of diverse groundnut germplasm through multienvironment evaluation of a core collection for Asia. Field Crops Res. 93:293–299. https://doi.org/10.1016/j.fcr.2004.10.007

Upadhyaya HD, Reddy LJ, Gowda CLL, Reddy KN, Singh S (2006a) Development of a mini core subset for enhanced and diversified utilization of pigeon pea germplasm resources. Crop Sci 46: 2127–2132

Upadhyaya HD, Gowda CLL, Pundir RPS (2006b) Development of Core subset of finger millet germplasm using geographical origin and data on 14 quantitative traits. Genet Resour Crop Evol 53:679–685. https://doi.org/10.1007/s10722-004-3228-3

Upadhyaya HD, Gowda CLL, Reddy KN, Singh S (2009a) Augmenting the pearl millet core collection for enhancing germplasm utilization in crop improvement. Crop Sci. 49(2):573–580

Upadhyaya H, Pundir R, Dwivedi S, Gowda C, Reddy VG, Singh S (2009b) Developing a mini core collection of sorghum for diversified utilization of germplasm. Crop Sci 49:1769–1780. https://doi.org/10.2135/cropsci2009.01.0014

Upadhyaya H, Pundir R, Gowda C, Gopal Reddy V, Singh S (2009c) Establishing a core collection of foxtail millet to enhance the utilization of germplasm of an underutilized crop. Plant Genetic Resources 7(2):177–184. https://doi.org/10.1017/S1479262108178042

Upadhyaya H, Sarma N, Ravishankar C, Albrecht T, Narasimhudu Y, Singh S et al (2010) Developing a mini-core collection in finger millet using multilocation data. Crop Sci 50: 1924–1931. https://doi.org/10.2135/cropsci2009.11.0689

Upadhyaya HD, Shivali S, Gowda CLL, Gopal RV, Sube S (2011a) Developing proso millet (Panicum miliaceum L.) core collection using geographic and morpho-agronomic data. Crop and Pasture Science 62:383–389

Upadhyaya HD, Yadav D, Reddy KN, Gowda CLL, Singh S (2011b) Development of pearl millet minicore collection for enhanced utilization of germplasm. Crop Sci 51:217–223

Upadhyaya HD, Ravishankar CR, Narsimhudu Y, Sarma NDRK, Singh SK, Varshney SK, Reddy VG, Singh S, Parzies HK, Dwivedi SL, Nadaf HL, Sahrawat KL, Gowda CLL (2011c) Identification of trait-specific germplasm and developing a mini core collection for efficient use of foxtail millet genetic resources in crop improvement. Field Crops Res 124:459–467

Upadhyaya HD, Dronavalli N, Dwivedi SL, Kashiwagi J, Krishnamurthy L, Pande S et al (2013a) Mini core collection as a resource to identify new sources of variation. Crop Sci 53(6):2506–2517

Upadhyaya HD, Wang YH, Gowda CLL, Sharma S (2013b) Association mapping of maturity and plant height using SNP markers with the sorghum mini core collection. Theor Appl Genet 126: 2003–2015

Upadhyaya HD, Dwivedi SL, Vadez V, Hamidou F, Singh S, Varshney RK, Liao B (2014a) Multiple resistant and nutritionally dense germplasm identified from mini core collection in peanut. Crop Sci 54(2):679–693

Upadhyaya HD, Dwivedi SL, Singh SK (2014b) Forming core collections in barnyard, kodo, and little millets using morphoagronomic descriptors. Crop Sci 54:2673–2682. https://doi.org/10.2135/cropsci2014.03.0221

Upadhyaya HD, Vetriventhan M, Asiri AM, CR Azevedo V, Sharma HC, Sharma R et al (2019) Multitrait diverse germplasm sources from mini core collection for sorghum improvement. Agriculture 9(6):121

Van de Wouw M, Kik C, van Hintum T, van Treuren R, Visser B (2009) Genetic erosion in crops: concept, research results and challenges. Plant Genet Resour 8:1–15. https://doi.org/10.1017/S1479262109990062

Van Tassel DL, Tesdell O, Schlautman B, Rubin MJ, DeHaan LR, Crews TE, Streit Krug A (2020) New food crop domestication in the age of gene editing: genetic, agronomic and cultural change remain co-evolutionarily entangled. Front Plant Sci 11:789

Varshney RK, Bohra A, Yu J, Graner A, Zhang Q et al (2021) Designing future crops: genomics-assisted breeding comes of age. Trends Plant Sci 26(6):631–649. https://doi.org/10.1016/j.tplants.2021.03.010

Vikas VK, Kumar S, Archak S, Tyagi RK, Kumar J, Jacob S, Sivasamy M, Jayaprakash P, Saharan MS, Basandrai AK, Basandrai D, Srinivasan K, Radhamani J, Parimalan R, Tyagi S, Kumari J, Singh AK, Peter J, Nisha R, Yadav M, Kumari J, Dhillon HK, Chauhan D, Sharma S, Chaurasia S, Sharma RK, Dutta M, Singh GP, Bansal KC (2020) Screening of 19,460 genotypes of wheat species for resistance to powdery mildew and identification of potential candidates using focused identification of germplasm strategy (FIGS). Crop Sci 60:2857–2866

Visser B (2013) The moving scope of Annex 1: the list of crops covered under the multilateral system. In: Halewood M, López Noriega I, Louafi S (eds) Crop Genetic Resources as a Global Commons: challenges in international law and governance. Routledge, Abingdon, pp 265–282

von Wettberg E, Davis TM, Smýkal P (2020) Wild plants as source of new crops. Front Plant Sci 11:591554

Wang W, Mauleon R, Hu Z, Chebotarov D, Tai S, Wu Z et al (2018) Genomic variation in 3,010 diverse accessions of Asian cultivated rice. Nature 557(7703):43–49

Withers LA (1995) Collecting in vitro for genetic resources conservation. In: Guarino L, Rao VR, Reid R (eds) Collecting plant genetic diversity. Technical Guidelines, IPGRI/FAO/UNEP/IUCN and CAB International, pp 511–515

Xiurong Z, Yingzhong Z, Yong C (2000) Establishment of sesame germplasm core collection in China. Genet Resour Crop Evol 47:273–279. https://doi.org/10.1023/A:1008767307675

Yamamoto SI, Rafique T, Priyantha WS, Fukui K, Matsumoto T, Niino T (2011) Development of a cryopreservation procedure using aluminium cryo-plates. Cryo-Letters 32:256–265

Yan W, Rutger JN, Bryant RJ, Bockelman HE, Fjellstrom RG, Chen MH, Tai TH, McClung AM (2007) Development and evaluation of a core subset of the USDA rice germplasm collection. Crop Sci 2007(47):869–876

Zhang H, Zhang D, Wang M (2011) A core collection and mini core collection of Oryza sativa L. in China. Theor Appl Genet 122:49–61. https://doi.org/10.1007/s00122-010-1421-7

Chapter 3
Bioinformatics for Plant Genetics and Breeding Research

Yogesh Dashrath Naik, Chuanzhi Zhao, Sonal Channale, Spurthi N. Nayak, Karma L. Bhutia, Ashish Gautam, Rakesh Kumar, Vidya Niranjan, Trushar M. Shah, Richard Mott, Somashekhar Punnuri, Manish K. Pandey, Xingjun Wang, Rajeev K. Varshney, and Mahendar Thudi

Abstract Global food demand is expected to increase between 55 and 70% by 2050. Plant breeders and geneticists are constantly under pressure to develop high-yielding climate-resilient varieties using novel approaches. The quest for simplifying complex traits and efforts for developing high-yielding varieties during the twenty-first century led to a paradigm shift from phenotypic-based selection to genome-based breeding. On one hand, the development and utilization of diverse genetic

Y. D. Naik · K. L. Bhutia
Dr. Rajendra Prasad Central Agricultural University (RPCAU), Pusa, Bihar, India

C. Zhao · X. Wang
Shandong Academy of Agricultural Sciences (SAAS), Jinan, Shandong, China

S. Channale
University of Southern Queensland (USQ), Toowoomba, Queensland, Australia

S. N. Nayak
University of Agricultural Sciences, Dharwad, Karnataka, India

A. Gautam · R. Kumar
Central University of Karnataka, Kalaburagi, Karnataka, India

V. Niranjan
RV College of Engineering, Bengaluru, Karnataka, India

T. M. Shah
International Institute of Tropical Agriculture (IITA), Nairobi, Kenya

R. Mott
University College London, London, UK

S. Punnuri
College of Agriculture, Family Sciences and Technology, Dr. Fort Valley State University, Fort Valley, Geogia, USA

M. K. Pandey
Center of Excellence in Genomics and Systems Biology (CEGSB), International Crops Research Institute for the Semi-Arid Tropics (ICRISAT), Hyderabad, India

Shandong Academy of Agricultural Sciences (SAAS), Jinan, Shandong, China

M. K. Pandey et al. (eds.), *Frontier Technologies for Crop Improvement*, Sustainable Agriculture and Food Security, https://doi.org/10.1007/978-981-99-4673-0_3

resources, and advances in genomics on the other hand provided a kick start for the understanding the genetics of economically important complex traits at a faster pace. Further, the next-generation sequencing revolutionized our understanding of the genome architecture. As a result, there has been an increasing demand for statistical and bioinformatics tools to analyse and manage the enormous amount of data generated from sequencing of genomes, transcriptomes, proteome and metabolomes. In this chapter, we review the intervention of bioinformatics and computational tools for deploying the tremendous wealth of data for plant genetics and breeding research.

Keywords Bioinformatics · Next-generation sequencing · Database · Pangenome · Haplotype · Artificial intelligence

3.1 Introduction

Climate change and increasing population growth at an alarming rate poses the biggest challenges to food and nutritional security across the globe. By 2050, the global population is predicted to increase by 55 to 70%, as a result the proportion of people at risk of hunger may increase to around 8% (van Dijk et al. 2021a). With diminishing resources and limited arable land, sustainable production to cater the food and nutritional demands has been a daunting task. Plant breeders and geneticists are constantly under pressure to develop improved crop varieties that are climate-resilient and high-yielding to meet the food and nutritional demands. Low genetic diversity, prolonged breeding cycles, and limited access to high-quality seeds for cultivation have been serious obstacles to achieve greater genetic advancements (Varshney et al. 2020). Although conventional breeding programs contributed to the development of improved varieties, to achieve "zero hunger," the Sustainable Developmental Goal 2 adopted by United Nations Organization advocated the integration of modern breeding approaches in agriculture (Varshney et al. 2018).

Ever since the rediscovery of Mendelian laws, there has been a paradigm shift in understanding the phenotype-based trait genetics to the use of molecular markers, genomics, genomes and sequence-based trait dissection (Varshney et al. 2019; Thudi et al. 2023). During the last two decades, genomics and NGS (next-generation sequencing) technologies have not only revolutionized our understanding of

University of Southern Queensland (USQ), Toowoomba, Queensland, Australia

R. K. Varshney
Murdoch's Centre for Crop and Food Innovation, State Agricultural Biotechnology Centre, Food Futures Institute, Murdoch University, Murdoch, WA, Australia

M. Thudi (✉)
Dr. Rajendra Prasad Central Agricultural University (RPCAU), Pusa, Bihar, India

Shandong Academy of Agricultural Sciences (SAAS), Jinan, Shandong, China

University of Southern Queensland (USQ), Toowoomba, Queensland, Australia molecular basis of economically important traits, but also increased the rate of adoption of modern breeding approaches to develop climate-resilient crop varieties (Thudi et al. 2020; Varshney et al. 2021a). To date, draft genomes of more than 1000 plants representing 788 species are available in public domain (Sun et al. 2021). Not only draft genomes, gold standard reference genomes to platinum standard reference genomes are available in crops like rice (Zhou et al. 2020) and also in cetacean species (Morin et al. 2020). Efforts are also underway to sequence all the known eukaryotic species through "The Earth BioGenome Project" that provides insights into the biology of life (Lewin et al. 2018). Apart from draft genomes, several germplasm lines including wild species accessions have been sequenced in several crops including pearl millet (Varshney et al. 2017a), chickpea (Thudi et al. 2016; Varshney et al. 2021b), pigeon pea (Varshney et al. 2017b), rice (Wang et al. 2018; Stein et al. 2018). Development of pangenomes and super-pangenomes are underway in many crop species (Khan et al. 2020). With the rapid availability of biological data in public domain, rate-limiting factor in genomics research has shifted from sequencing to computer analysis (Kathiresan et al. 2017). The statistical, bioinformatics tools and algorithms developed earlier are becoming obsolete and computational tools and algorithms that handle "BIG data" are gaining importance (Edwards et al. 2009; Batley and Edwards 2016).

In this chapter, we review the NGS data analysis and available databases that are developed to store and retrieve biological information produced from different omics approaches. In addition, we also discuss the computational tools and approaches that enable development of pangenome, identification of haplotypes and editing genomes. Besides highlighting the challenges, we also highlight the scope of improving the bioinformatics approaches for effective use in crop improvement.

3.2 Understanding Genetic Diversity and Trait Mapping

Genetic diversity plays a major role for gaining greater insights and simplifying complex traits. Prior to advent of molecular markers, the phenotypic plasticity in a crop species was assessed using simple experimental analyses and programmes like XLstat or SPSS (Addinsoft 2021; IBM Corp Ibm 2017). In addition, statistical packages like INDOSTAT is being used to analyse variance, D^2 statistics, canonical roots, path analysis etc. (Khetan and Ameerpet 2015). The statistical tool for agricultural research (STAR) has modules for randomization and layout of crop research experimental designs, data management, and fundamental statistical analysis, including descriptive statistics, hypothesis testing, and ANOVA of designed experiments (Gulles et al. 2014). The stability of a crop over different locations and years is one of the crucial prospects in plant breeding. Software like GGE biplot, GEA-R, STABILITYSOFT, and AMMISOFT are used to analyse Genotype × Environment (G × E) interaction studies (Yan 2001; Pacheco et al. 2015; Pour-Aboughadareh et al. 2019; Gauch and Moran 2019). Stability and performance are examined simultaneously using these tools, allowing for a comprehensive

Table 3.1 List of commonly used software packages for plant breeding

Software/program	Key features	References
XLstat, SPSS	These are used for simple experimental analyses	Addinsoft (2021); IBM Corp Ibm (2017)
R and INDOSTAT	Used for analysis of variance, covariance matrices with ANOVA and ANCOVAS, D^2 statistics with Mahalanobis, stability model analysis, Diallel analysis, Heterosis, line × tester analysis, path analysis, joint scaling test (Cavilli), North Carolina design 1, North Carolina design 3, augmented design, double cross analysis, triple cross analysis and triple test cross	Ledesma (2008); Team (2013); Khetan and Ameerpet 2015
GGE biplot, GEA-R, STABILITYSOFT, and AMMISOFT	Analyses genotype × environment analysis for stability analysis	Yan (2001); Pacheco et al. (2015); Pour-Aboughadareh et al. (2019); Gauch and Moran (2019)
Mapmaker-QTL	It can perform only simple interval mapping	Lincoln et al. (1993)
QTL cartographer	Offers options for carrying out the majority of the documented QTL mapping methods	Basten et al. (2002)
Win-QTL cartographer	It maps quantitative trait loci (QTL) in cross populations from inbred lines	Wang (2005)
PLABQTL	Its primary goal is to identify and describe QTL in populations resulting from a biparental cross by selfing or the creation of doubled haploids. A rapid multiple regression approach achieves simple and composite interval mapping	Utz and Melchinger (1996)
MapQTL	It analyses composite interval mapping, interval mapping, nonparametric mapping, automatic cofactor selection, and permutation test for interval mapping	Van Ooijen and Maliepaard (1999)
STRUCTURE	Used for determining population structure	Pritchard et al. (2000)
TASSEL	It is used for evaluation of trait associations, evolutionary patterns, LD statistics, GLM, MLM, CMLM, P3D: Genomic selection; graphical interphase, PCA, and kinship analysis	Bradbury et al. (2007); Gupta et al. (2015)

understanding of the crop's behavior across different environments and conditions. (Table 3.1).

With the availability of molecular markers, efforts were made to map the genomic regions or genes responsible for the complex traits using both linkage mapping or QTL mapping and linkage disequilibrium-based mapping or association analysis. The most common software packages used for maapping genomic refions are Mapmaker-QTL, QTL Cartographer, Win-QTL Cartographer, PLABQTL, MapQTL are command-line software (Lincoln et al. 1993; Basten et al. 2002; Wang 2005; Utz and Melchinger 1996; Van Ooijen and Maliepaard 1999; Bradbury et al. 2007; Gupta et al. 2015. Mapmaker-QTL can only perform simple interval mapping (Lincoln et al. 1993). The most versatile QTL mapping software is QTL Cartographer. A range of software tools, including the widely used STRUCTURE, are available for determining population structure (Pritchard et al. 2000). Using this software, you can choose the number of subpopulations by using all marker data or a subset of unlinked markers from the marker collection. Alternatively, using the given marker data, principal component analysis (PCA) can be performed and the first few components used as variables to adjust for population structure. Association analysis can be done with TASSEL. Even without forming a core, one can test a population for its suitability as an association panel. Then it can be directly used for TASSEL analysis. However, some prerequisite analysis is required, like population structure, kinship analysis, and principal component analysis (PCA) (Bradbury et al. 2007). It uses marker data to calculate kinship, which helps to address family relatedness and population structure (Table 3.1) (Gupta et al. 2015).

3.3 Identification and Understanding Key Genes Using Multi-Omics Approaches

Interpretation of molecular complexity and variability at several levels, such as genome, transcriptome, proteome and metabolome, is necessary for comprehensive understanding of organism's entire metabolism. The data from various levels are together referred to as "multi-omics" data. Multi-omics data obtained from various approaches provide insights into the flow of biological information at various levels, can aid in figuring out the biological state of interests underlying mechanisms.

In the last decade, technological advancement in DNA sequencing (Le Nguyen et al. 2019), transcriptomics analysis via RNA-seq (Mashaki et al. 2018), SWATH-based proteomics (Zhu et al. 2020) and metabolomics via UPLC-MS and GC-MS (Balcke et al. 2012) has made a significant contribution in biological data. The first omics field to emerge is genomics that deals with study of complete genomes. Genomic studies like QTL/association mapping has been used to detect genomic regions associated with agronomically important traits (Varshney et al. 2014, 2021b; Bhatta et al. 2019; Thudi et al. 2021; Yoshida et al. 2022) and provide basic framework for other omics approaches. Additionally, differentially expressed genes under several biotic and abiotic stresses were identified using transcriptomics studies in several crop plants (Nayak et al. 2017; Channale et al. 2021; Chen et al. 2022; Pal et al. 2022). Gene expression atlas provides insights into the subsets of genes expressed during different growth stages for pigeon pea (Pazhamala et al. 2017), chickpea (Kudapa et al. 2018), groundnut (Sinha et al. 2020). The spatial transcriptomics method developed by Giacomello et al. (2017) enables

Table 3.2 Summary of widely used databases in plant genetics and breeding research

Databases	Key features	Link
AtMAD	Provide high-quality multi-omics data of *Arabidopsis thaliana*	http://www.megabionet.org/atmad
GoMapMan	Gene functional annotations in the plant sciences	http://www.gomapman.org/
HapRice	SNP-haplotype database for rice	http://qtaro.abr.affrc.go.jp/index.html
NPACT	Plant-derived natural compounds exhibiting anticancerous activity	https://webs.iiitd.edu.in/raghava/npact/faq.html
PGDD	Database for gene and genome duplication in plants	http://chibba.agtec.uga.edu/duplication/
PGDJ	DNA marker and linkage database	http://pgdbj.jp/plantdb/plantdb.html
Phytozome	Plant comparative genomics	https://phytozome-next.jgi.doe.gov/
PIECE	Plant intron exon comparison and evolution database	https://data.nal.usda.gov/dataset/piece-plant-intron-exon-comparison-and-evolution-database
Plant rDNA	Plant rDNA database	https://www.plantrdnadatabase.com/
PlantGDB	Plant genome browsers	https://www.plantgdb.org/prj/GenomeBrowser/
PlantRNA	Database for tRNAs of photo-synthetic eukaryotes	http://seve.ibmp.unistra.fr/plantrna/
PlnTFDB	Plant transcription factor prediction	http://planttfdb.gao-lab.org/
PLUTO	Contains information on plant varieties	http://www.upov.int/pluto/en/
PMRD	Plant microRNA database	http://bioinformatics.cau.edu.cn/PMRD/
PTGBase	Plant tandem duplicated genes database	http://ocri-genomics.org/PTGBase/
SALAD	Comparison of proteome data among the species	https://salad.dna.affrc.go.jp/salad/en/

high-throughput and spatially resolved transcriptomics in plant tissues using a combination of histological imaging and RNA sequencing. Functional analysis of translated regions of the genome is understood using proteomics, while metabolomics serves as a diagnostic tool for assessing the plant performance under different stimuli (Villate et al. 2021). A number of repositories were developed to organise data generated from different experiments and sequencing studies. The repositories include DNA, RNA and protein sequence databases, as well as specialized databases for specific information (Lai et al. 2012; Thudi et al. 2020). Based on different types of omics data, databases can be classified into four classes: (1) genomics databases contain nucleotide sequence or genomic sequence, (2) transcriptomics databases include functional RNA sequences, (3) proteomics databases contain information related to amino acid sequence and protein structure, and (4) metabolomics databases contain information about metabolites and metabolic pathways (Table 3.2, Fig. 3.1a).

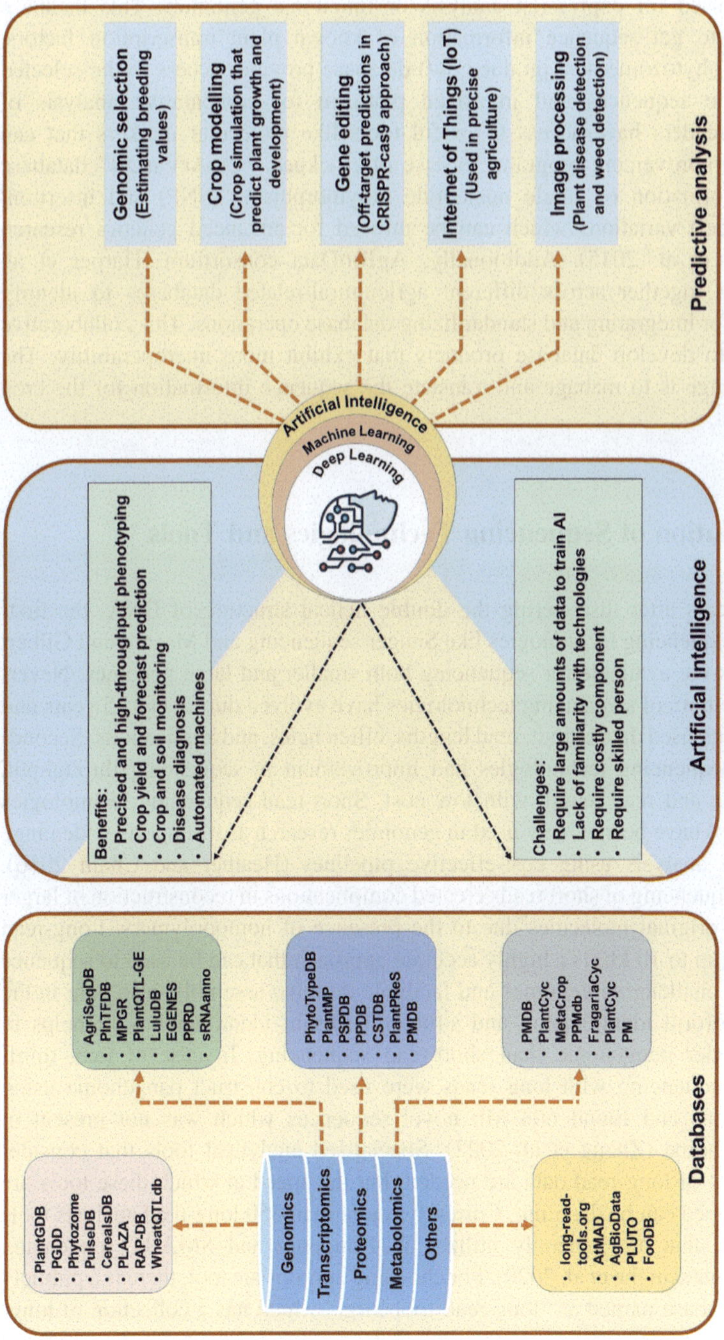

Fig. 3.1 Summary of databases and applications of artificial intelligence in agriculture: (**a**) Represent databases developed to store and retrieve the biological information produced from various omics approaches, includes Genomics, Transcriptomics, Proteomics and Metabolomics, (**b**, **c**) Showed different kinds of predictive analysis based on AI

Databases include PlnTFDB (planttfdb.gao-lab.org/) for plant transcription factor, widely used for expression analysis or functional genomics. This database allows user to get sequence information of known plant transcription factors. Phytozome (phytozome-next.jgi.doe.gov/) database provides access to the selected plant genome sequences and improved platform for comparative analysis of genomes. Breeders have access to useful tools like molecular markers that can speed crop improvement program. In case of chickpea, "CicArVarDB" database provides information of single nucleotide polymorphisms (SNP) and insertion/ deletion (Indel) variations which can be utilized for advanced genetics research (Doddamani et al. 2015). Additionally, AgBioData consortium (Harper et al. 2018) works together across different agricultural-related databases to identify approaches for integrating and standardizing database operations. This collaborative effort aims to develop database products that exhibit more interoperability. The major challenge is to manage and translate the sequence information for the crop improvement.

3.4 Evolution of Sequencing Technologies and Tools

About 25 years after discovering the double helical structure of DNA, the first-generation sequencing technologies like Sanger sequencing and Maxam and Gilbert sequencing were available for sequencing both smaller and large genomes. Nevertheless, a plethora of sequencing technologies have evolved during last 15 years and there is an increased data output, read lengths, efficiencies, and applications. Second-generation sequencing technologies had improvement in sequencing throughput, required time and read length with low cost. Short-read sequencing technologies (up to 600 bp) have been widely used in genomics research as it supports wide range of statistical analysis using cost-effective pipelines (Heather and Chain 2016). However, sequencing of short reads created complications in reconstruction of larger fragment or original molecules due to the presence of homopolymers. Long-read sequencing (up to 10 kb) is a highly accurate approach that can be used to sequence traditionally challenging genomes and facilitate de novo assembly, also help in the transcript isoform identification and structural variant identifications. It helps to construct better pangenome than short-read sequencing. In case of rice, third-generation sequencing with long reads were used to construct pangenome using 105 accessions and found 604 Mb novel sequences which was not present in reference genome (Zhang et al. 2022). Specialised analytical tools that consider the properties of long-read data are needed, but the speed at which these tools are being developed can be daunting. Currently, more than 350 long-read analysis tools are available that are generally utilized in Nanopore and SMART sequencing platform (Amarasinghe et al. 2020). For choosing appropriate tool, there is a publicly available database named as "long-read-tools.org," which has a collection of long-read analysis tools and allows us to choose appropriate tools for analysis (Amarasinghe et al. 2021). In order to analyse and interpret the NGS data, there is a need of highly qualified and competent bioinformaticians. For accurate downstream analysis of sequencing data, appropriate analysis tools are essential and it involves conversion of raw signal data to sequence data.

Sequencing data analysis includes raw read quality control, sequence alignment, variant calling, genome assembly, genome annotation and other advanced analysis. Numerous bioinformatics tools have been developed and used in sequence analysis (Table 3.3). It is essential to evaluate the raw sequence data to ensure the quality for any subsequent analysis. It can give a broad overview of read counts and lengths, coverage reads, contaminating sequences and sequence duplication level. In the first stage, adapter sequences and low-quality sequences are separated from whole genome sequencing data through a quality assessment process. FastQC is the well-known bioinformatics tool for calculating quality control of sequencing reads (Andrews 2010). More recently, fastp tool is also utilized in quality control, base correction and filtering of sequencing reads. The fastp tool is two to five times faster than previous approach (Chen et al. 2018) and ensures the read quality as well as adapter trimming.

The second step is to align the sequences with reference genome, that is, read/sequence alignment. In the case of non-availability of reference genome, de novo genome assembly method is used to generate the contigs by aligning the overlapping regions together. This step is the most crucial and important in the entire workflow. The sequence reads are precisely and quickly aligned to the appropriate places of the reference genome using a variety of tools and algorithms. Many tools have been developed for sequence alignment; the popular aligners include BWA (Li and Durbin 2009), Bowtie2 (Langmead and Salzberg 2012), CUSHAW3 (Liu et al. 2014), MOSAIK (Lee et al. 2014), and Novoalign (http://novocraft.com/). MOSAIK is the mapping tool currently available that can align reads produced by all the major sequencing technologies. Minimap2 is a flexible pairwise nucleotide sequence aligner and mapper. It can be used with short reads, assembly contigs, long noisy genomic and RNA-seq reads (Li 2018). The lra tool requires less time and memory for alignment as compared to Minimap2 (Ren and Chaisson, 2021). The recently developed kngMap (*k*-mer neighbourhood graph-based mapper) tool is specifically designed to align long noisy reads to a reference genome (Wei et al. 2022).

The third step is variant calling. The variations in the output sequences compared to the reference sequence are called as variants. The presence of SNPs, INDEL, presence/absence variations (PAVs), copy number variations and haplotypes blocks are detected using variant calling tools. Tools used for variant calling includes SAM tools (Li et al. 2009), Genome Analysis Tool Kit Haplotype Caller (GATK-HC) (McKenna et al. 2010), Freebayes (Garrison and Marth 2012), SNPSVM (O'Fallon et al. 2013), varScan (Koboldt et al. 2013), DeepVariant (Poplin et al. 2018), Torrent Variant Caller (TVC) (Life Technologies, Rockville, MD), etc. Numerous automated workflows have been developed to streamline the variant calling process. These workflows integrate various aligners and variant calling tools with other upstream and downstream tools to provide an end-to-end solution (Kanzi et al. 2020). Tools available like ToTem and Appreci8 (Tom et al. 2018; Sandmann et al. 2018) are completely automated variant calling pipelines. ToTem is becoming

Table 3.3 Bioinformatics tools used for NGS data analysis

Approach	Tool	Key feature	Link
Quality check	FastQC	Quality control checks on raw sequence data coming from high-throughput sequencing pipelines	https://www.bioinformatics.babraham.ac.uk/projects/fastqc/
	fastp	It can perform quality control, adapter trimming, quality filtering	https://github.com/OpenGene/fastp
Sequence alignment	BWA	Mapping low-divergent sequences against a large reference genome	http://bio-bwa.sourceforge.net/
	Bowtie2	Bowtie2 is an ultrafast and memory efficient tool for aligning sequencing reads	http://bowtie-bio.sourceforge.net/bowtie2/index.shtml
	CUSHAW3	Mapping with high computational efficiency	http://cushaw3.sourceforge.net/homepage.htm#latest
	kngMap	Align long reads to a reference sequence	https://github.com/zhang134/kngMap
	MOSAIK	Mapping second- and third-generation sequencing reads	https://github.com/wanpinglee/MOSAIK
	Novoalign	Mapping of short reads onto a reference genome from different NGS platforms	http://www.novocraft.com/products/novoalign/
	SOAP3-dp	SOAP3 is the first short-read alignment tool that leverages the multiprocessors in a graphic processing unit (GPU) to achieve a drastic improvement in speed. SOAP3 is the first short-read alignment tool that leverages The multiprocessors in a graphic processing unit (GPU) to achieve a drastic improvement in speed. SOAP3 is the first short-read alignment tool that leverages the multiprocessors in a graphic processing unit (GPU) to achieve a drastic improvement in speed. Consider alignment with Indels in addition to mismatches.	http://soap.genomics.org.cn/
	MAQ	Builds assembly by mapping short reads to reference sequences	http://maq.sourceforge.net/
	Minimap2	It is accurate and efficient for long noisy genomic and RNA sequences	https://github.com/lh3/minimap2
Variant calling	GATK	Set of bioinformatics tools for analysing high-throughput sequencing and variant call format data	https://software.broadinstitute.org/gatk/
	Freebayes	It is a haplotype-based variant detector and is a great tool for calling variants from a population	https://github.com/ekg/freebayes
	DeepVariant	It is an analysis pipeline that uses a deep neural network to call genetic variants	https://github.com/google/deepvariant
	Platypus	It is a haplotype-based variant caller	

(continued)

Table 3.3 (continued)

Approach	Tool	Key feature	Link
			http://www.well.ox.ac.uk/platypus
	VarScan	An open source tool for variant detection that is compatible with several short-read aligners	http://dkoboldt.github.io/varscan/
	ToTem	Primary role is to automatically generate, execute and benchmark different variant calling pipeline settings	https://totem.software/
	Appreci8	That combines and filters the variant calling results of eight different tools	https://hub.docker.com/r/wwuimi/appreci8/
Data visualization	IGV	It is a high-performance visualization tool for interactive exploration of large, integrated genomic datasets	https://igv.org/
	VISTA	Based on global alignment strategies and a curve-based visualization technique and it also used for comparative analysis	https://genome.lbl.gov/vista/index.shtml
	R software	Gosling: It is a grammar for scalable and interactive genomics data visualizations	http://gosling-lang.org/

a popular tool because it has automated pipeline optimization and efficient analysis management. Appreci8 gives an accurate variant calling as it uses eight different tools to perform the same task that filters and combines the outputs for appropriate calling. Final step is data visualization; there are various tools available for visualization depending on the experiments and the research objectives. One of the popular choices of visualization tool for reference genomes is integrated genome viewer (Thorvaldsdottir et al. 2012). VISTA is also visualization tool which can be used for comparing difference between two genomic sequences. To aid the biologists with no or little knowledge of using perl/python languages, desktop solutions for a wide range of genomic analysis needs, including transcriptomics, variant calling, epigenomics, metagenomics, comparative genomics, are available like Qiagen CLC Genomics Workbench, geWorkbench, Partek Genomics Suite, JMP Genomics, DNA Baser-NextGen Sequence Workbench, etc.

During NGS analysis, numerous intermediate analysis and result files are generated that require large storage. It is difficult to interpret these complicated NGS data files in terms of converting data into knowledge for important traits, especially for aggregated vast volumes of variants or heterogeneous sequencing data require a high-performance computational resource. The NGS data after analysis could be effectively interpreted using machine learning-based techniques.

3.5 Approaches for Development of Genome and Pangenome Assemblies

The wild relatives have a large genetic diversity and ability to survive under various biotic and abiotic stresses. Crop domestication and evolution have significantly decreased the genetic diversity in cultivated species, which has led to the loss of key loci that govern crucial traits. The traditional crop improvement approaches include selection of superior traits from either cultivated varieties or the wild relatives and utilizing them in the breeding programs (Dempewolf et al. 2017). During the process of selection, the crops became more susceptible to different stresses due to impact of climate change and evolution of pathogens and pests. To address these limitations, it is necessary to utilize crop wild relatives, which are known to have genes for several biotic/abiotic stress tolerance traits that have been lost during domestication or breeding procedures. As a result of advancement in sequencing technologies, reference genome sequences for a number of crops have been accessible, serving as the foundation for efforts to boost crop improvement programme (Varshney et al. 2017a, 2017b). In addition to cultivated crop genome, de novo assembled genomes of a number of wild relatives have also been made available. In addition, the idea of pangenomes is being adopted more widely due to the growing recognition that a single reference genome cannot capture the diversity contained within a species.

Pangenome is the collection of genes or DNA sequence in a species to provide useful sources for functional genomics, evolutionary studies that can be used for crop improvement. Pangenomic studies have been conducted in various model and crop plants including *Arabidopsis*, stiff brome, wheat, cabbage, tomato, soybean, rice, rapeseed, barley, chickpea and sorghum (Hurgobin et al. 2018; Gao et al. 2019; Jayakodi et al. 2020; Barchi et al. 2021; Ruperao et al. 2021; Varshney et al. 2021b; Jha et al. 2022) (Table 3.4). Genome assembly is the process of arranging nucleotides in the proper order. Sequence read lengths are currently far shorter than most of genomes or even most of the genes; therefore, it is important to assemble reads and construct genome or pangenome. In plants or other eukaryotic organisms, genes are found in the same physical place on the chromosome, but the frequency of copies and repeating sequences can vary, making assembly more difficult. Pangenomes have been constructed via de novo, iterative, and graph-based assembly techniques. The de novo assembly is straightforward and simplest approach for development of pangenome. This approach includes assembly using overlapping regions and does not require reference genome. It requires high depth sequencing of all the targeted accessions, then creates unique de novo assemblies for each accession. The comparison of the resulting individual assemblies identifies conserved and variable genomic regions across the genomes. Advancement in long-read sequencing technologies and complementary strategies like creation of Hi-C and BioNano maps make it possible to obtain high-quality plant genomes at the chromosomal level (Miga 2020). Comparative analysis is used to identify all types of variations and characterized genes found in core and dispensable regions (Mahmoud et al. 2019).

Table 3.4 Summary of important tools in various plant genetics and genomics approaches

Approach	Tool	Key feature	Link
Pangenome	EUPAN	Large-scale eukaryotic pangenome analyses and detection of gene PAVs at a relatively low sequencing depth	https://cgm.sjtu.edu.cn/eupan/index.html
	GET_HOMOLOGUES-EST	Highly customized and automated pipeline especially designed for people with non-bioinformatics background	https://github.com/eead-csic-compbio/get_homologues/releases
	PAN2HGENE	Computational tool that allows identification of gene products missing from the original genome sequence	https://sourceforge.net/projects/pan2hgene-software/
	Panakeia	Providing a detailed view of the pangenome structure which can efficiently be utilised for discovery, or further in-depth analysis	https://github.com/BioSina/Panakeia
	Pantools	A versatile tool for mapping the metagenomic and genomic reads in both prokaryotes and eukaryotes	https://git.wur.nl/bioinformatics/pantools
	PanViz	An interactive visualization tool to compare the individual genomes to the pangenome	https://github.com/thomasp85/PanViz/blob/master/package.json
	PATO	It performs common tasks of pangenome analysis and also integrates all the necessary functions for the complete analysis with high speed	https://github.com/irycisBioinfo/PATO
	PGAP	Perform pangenome profiling, gene cluster analysis, species evolution analysis, gene enrichment, and genetic variation analysis	https://sourceforge.net/projects/pgap/
	PGAP-X	Analyse pangenome profile curve, gene distribution analysis, genomic region variations, and comparative analysis of genome structure	http://pgapx.ybzhao.com/
	ppsPCP	Detect presence/absence variations and assembled comprehensive pangenome	http://cbi.hzau.edu.cn/ppsPCP/
	RPAN	Rich source for rice genomic research and breeding	https://cgm.sjtu.edu.cn/3kricedb/
Haplotype	Falcon phase	Groups long-read contigs into two separate haplotypes based on hi-C data	https://github.com/phasegenomics/FALCON-Phase

(continued)

Table 3.4 (continued)

Approach	Tool	Key feature	Link
	Hap10	Novel algorithm for haplotype assembly of polyploid genomes using linked reads	https://sourceforge.net/projects/sdhap/
	HapCut2	Robust and accurate haplotype assembly for diverse sequencing technologies	https://github.com/vibansal/HapCUT2
	HaploConduct	Package designed for reconstruction of individual haplotypes	https://github.com/HaploConduct/HaploConduct
	HaplotypeTools	Analysing hybrid or recombinant diploid or polyploid genomes and identifying parental ancestry for sub-genomic regions	https://github.com/rhysf/HaplotypeTools
	HAPLOVIEW	Analysis and visualization of LD and haplotype maps	https://www.broadinstitute.org/haploview/haploview
	HAPPE	Facilitates informative displays wherein data in plots are easy to read and access	https://github.com/fengcong3/HAPPE
	HapTree	Provide polyploid haplotype assembly tool based on a statistical framework.	http://cb.csail.mit.edu/cb/haptree/
	Hifiasm	Fast haplotype-resolved de novo assembler for PacBio HiFi reads	https://github.com/chhylp123/hifiasm
	SDip	Graph-based approach to haplotype-aware assembly	https://github.com/shilpagarg/sdip
	WhatsHap	Reconstruct the haplotypes and then write out the input VCF augmented with phasing information	https://whatshap.readthedocs.io/en/latest/
k-mer	BFCounter	Program for counting *k*-mers in DNA sequence data	http://pritch.bsd.uchicago.edu/bfcounter.html
	iMOKA	Utilized fast and accurate feature reduction step	https://github.com/RitchieLabIGH/iMOKA
	KAT	Multi-purpose software toolkit for reference-free quality control (QC) of WGS reads and de novo genome assemblies	https://github.com/TGAC/KAT
	KITSUNE	Identifying optimal *k*-mer length for alignment free phylogenomic analysis	https://github.com/natapol/kitsune
	KmerGO	Identify group-specific sequences using *k*-mers	

(continued)

Table 3.4 (continued)

Approach	Tool	Key feature	Link
			https://github.com/ChnMasterOG/KmerGO
Genome editing	CHOPCHOP	Web-based tool to select target sites for CRISPR/Cas9- or TALEN-directed mutagenesis	https://chopchop.cbu.uib.no/
	CLD	Suitable for the design of libraries using modified CRISPR enzymes and targeting non-coding regions	https://github.com/boutroslab/cld
	CRISPETa	Design optimal pairs of sgRNAs for deletion of desired genomic regions	http://crispeta.crg.eu/
	CRISPOR	Finds guide RNAs in an input sequence and ranks them according to different scores	http://crispor.tefor.net/
	CRISPR-FOCUS	Web-based platform to search and prioritize sgRNAs for CRISPR screen experiments	http://cistrome.org/crispr-focus/
	CROPSR	Highly effective and efficient to design gRNA in crop plants	https://github.com/H2muller/CROPSR
	E-CRISP	Computational tool to design and evaluate guide RNAs for use with CRISPR/Cas9	http://www.e-crisp.org/E-CRISP/

Several bioinformatics tools have been developed for assembling the prokaryotic pangenome and having the ability to handle less complex genomic content (Khan et al. 2020). For constructing eukaryotic pangenomes, some tools have been developed (Table 3.4) that include EUPAN (Hu et al. 2017), GET_HOMOLOGUES (Contreras-Moreira and Vinuesa 2013), PanTools (Sheikhizadeh et al. 2016), etc. One of the first attempts to examine eukaryotic pangenomes was EUPAN, which supported genome assembly, identification of core and dispensable gene databases using read coverage, and gene annotation of the pan-genomic dataset. GET_HOMOLOGUES can be used in eukaryotic pangenome development and it is written in Perl and R language platform. Additionally, Panconda tool (Warren et al. 2017) is used to compare whole genome multiple sequence and representing relations between sequence as graph and it is the initial step for the de Bruijn graph which can be used for pangenome construction. PanTools is also used to construct and visualize pangenome, the representation of pangenome depending on the de Bruijn graphs. PAN2HGENE (Silva de Oliveira et al. 2021) recently developed computational tools for pangenome analysis, which can do automated comparison analysis for both full and draft genomes and identifies gene that are missing from the original genome sequence.

3.6 Bioinformatics Tools Used in *K*-Mer Analysis

The importance of supporting sequencing technologies has been highlighted by our growing understanding of biological information and its implications for the vast volume of DNA data. Counting *k*-mers is an essential component for many bioinformatics techniques, such as nucleotides assembly, metagenomic sequencing and sequencing error correction (Melsted and Pritchard 2011). A *k*-mer is unique sub-sequence of nucleotide sequence. The distribution of statistically significant *k*-mers in a genomes and other regulatory subregions has been described in a number of recent studies (Hashim and Abdullah 2015; Cserhati et al. 2018). It has also been also employed in comparative studies (Cserhati et al. 2019), and major advantages of alignment-free approaches based on *k*-mer are their speed and ability to remove biases. Most of the association mapping studies has been done using SNPs. However, this approach has some limitations (Rahman et al. 2018). A *k*-mer-based analysis is alternative method to address some limitations of SNP-based analysis.

At its most basic, *k*-mer count analysis simply considers two parameters: the length of the *k*-mer and whether the orientation of the DNA strand is known. *k* is normally selected to be at least 20 and frequently falls between 20 and 31. Too small *k* will give redundant count information because the probability that a *k*-mer is unique to a genome is reduced. However, as *k* increases the probability that a *k*-mer contains an error increases. There are a number of bioinformatics tools developed to analyse the *k*-mer and further utilization of *k*-mers. BFCounter is a program that is used for counting *k*-mers in DNA sequence data (Melsted and Pritchard 2011) (Table 3.4). KAT (*k*-mer Analysis Toolkit) is a multipurpose tool for reference-free quality control and de novo assembly (Mapleson et al. 2017). iMOKA (interactive multi-objective *k*-mer analysis) is bioinformatical tool/software that enables comprehensive analysis of large collections of sequencing data based on *k*-mer. It uses efficient and effective steps that combines Naive Bayes classifier augmented by an adaptive entropy as well as graph-based filter to reduce search time (Lorenzi et al. 2020). KmerGO software is utilized to identify group-specific nucleotide sequences between two different groups. Furthermore, it is also used to check association between nucleotide sequence and quantitative traits (Wang et al. 2020). KITSUNE is a tool to identify the empirically optimal *k*-mer length for phylogenetic analysis and provides alternative alignment tool for comparative studies (Pornputtapong et al. 2020).

3.7 Artificial Intelligence

Artificial intelligence (AI) is the simulation of human intelligence processes by computer systems and it holds marvellous promise for better utilization of the available dataset to appropriate prediction and better understanding of genetic complexity (Fig. 3.1a, b). The three cognitive skills that make up AI encoding are learning (acquiring data and then developing algorithms to transform it into usable information), reasoning (selecting the appropriate algorithm to arrive at a desired result), and self-correction (constantly adjusting designed algorithms to ensure that they deliver the most accurate results) (Gharaei et al. 2019). Breeders have access to an ever-growing suite of high-throughput sensors and imaging techniques for a wide range of traits and situations in the field. In addition, novel genomic assays are constantly being developed that can reveal missing heritability (Harfouche et al. 2019). Nowadays, a major challenge in the advancement of technologies is the management and utilization of big data. The utilization of data with AI technologies can accelerate the breeding program to increase productivity and development of climate-resilient crop by phenotyping, efficient and effective diagnosis of disease and precise selection of individual for breeding (Fig. 3.1c). AI can also help breeders to quickly determine which plants grow the quickest in a specific climate, which genes support plant growth and adaptation, produce the best gene combination for a given location and choosing traits that increase yield and fend off the effects of a changing climate.

One of the important elements in AI is machine learning (ML), which helps to use data more efficiently and that uses statistical and mathematical approaches for appropriate predictions (Ayed and Hanana 2021). The ML has ability of ML to distinguish between various types of genomic regions, for instance, distinguishing active genes and pseudogenes, using feature like DNA methylation (Sartor et al. 2019). Additionally, ML was utilised to foresee the locations of DNA crossover (Demirci et al. 2018). Single-cell RNA sequencing is fascinating the new area in which ML is essential (Speranza et al. 2021; van Dijk et al. 2021b). This method makes it possible to examine cellular development and responses to environmental stimuli in diverse tissues. Digital plant phenotyping has been an active study area to accelerate plant science studies. Different imaging systems can be used to study the various macroscopic levels, for example, real-time stomata phenotyping using microscopic observation (Toda et al. 2021). Numerous sensors have been employed to accurate phenotyping, and it includes spectral sensor, lidar/laser sensor, fluorescence sensor, ultrasonic sensor and thermography (Qiu et al. 2018).

AI systems currently in use neural networks (NNs) and extreme gradient boosting (XGboost), both of which are popular machine learning models employed for a variety of tasks including regression and classification (Chen and Guestrin 2016). Deep learning techniques are based on neural networks, sometimes referred to as artificial/ simulated neural networks, which are a subset of machine learning. Leveraging AI in agriculture shows impressive results in image-based disease identification using deep learning model. It uses publicly available image datasets for disease identification (Mohanty et al. 2016). However, the supervised branch of machine learning includes the tree-based method known as XGboost. In maize, different models were used to predict yield using AI and found better results using XGBoost (Nyeki et al. 2019). These AI systems internal working and decision-making procedures are mysterious. It is possible to see the results, but it is not clear why a particular choice was picked. As a result, the introduction of new explainable AI algorithms that not only have a prediction model but also gives the appropriate reasons for choice is needed. It is the first stage in the development of next-generation AI (Harfouche et al. 2019).

3.8 Identification of Superior Haplotype for Crop Improvement

Second-generation molecular markers have been successfully used in plant breeding for development of improved varieties and also utilised in genome mapping, but gives low resolution of QTLs (Zargar et al. 2015). Advancement in the NGS technologies provide sequence-based markers (SNPs) having wide coverage with high density (Gouda et al. 2021), and have wide applications in plant breeding. These markers help to increase the resolution of genome mapping and the accuracy of genomic selection (Yadav et al. 2019). However, identified SNPs have some limitations which includes bi-allelic nature, difficult to identify rare alleles, less polymorphic, linkage drag problem and giving false positive results (Voss-Fels and Snowdon 2016; Bhat et al. 2021). In this context, the haplotype-based approaches are a successful strategy to get over SNPs limitations and boost the resolution of genomic regions (Qian et al. 2017). Haplotype is combination of nucleotide or markers that inherit together from polymorphic sites in the same or different chromosome having strong linkage disequilibrium between them (Bhat et al. 2021). Number of studies have demonstrated that a haplotype-based association study can find variants that would not be detected by a typical SNP-based investigation (Zakharov et al. 2013). Additionally, a recent study also identified several important genes, that can be utilized as important molecular markers for the purpose of genetic manipulation to design and develop robust and resistant crop cultivars (Pal et al. 2022).

The detection of haplotypes and their use in genetic investigations is significantly impacted by the availability of high-throughput sequencing technologies. Second-generation sequencing technologies generate 150 base pairs short reads. Therefore, the haplotypes identification is difficult and requires powerful statistical tools (Delaneau et al. 2019). On the other hand, third-generation sequencing technologies, such as Oxford Nanopore and Pacific Biosciences, generates long reads from which the haplotypes can be constructed directly (Maestri et al. 2020). The haplotype mining can be used to dissect complex traits by using approaches like haplotype-based breeding, haplotype-GWAS, haplotype-assisted genomic selection (Table 3.4).

Haplotype identification, characterization and visualization are important for utilization of haplotype for crop improvement. Many tools have been developed to estimate and visualize haplotypes. Haplotype identification/estimation also called as "phasing," is a process of estimation or construction of the haplotype sequences from genotypic data and it is utilized for understanding sequence-specific variation. Haplotype-based GWAS analysis is complicated as compared to SNP-based analysis to identify the associations, because it involves three major steps: phasing/haplotype estimation, block determination and statistical analysis. Estimation of haplotypes required pooled information of all individuals present in sample. Number of unrelated individuals is an important factor that can influence the estimation of haplotypes, and more individuals can give better results. However, related individuals can be phased by considering haplotypes shared by members of families which are descended from one another (Browning and Browning 2011). Numerous phasing techniques that enable the construction of haplotypes from long-read sequencing data have recently been established, such as reference-based phasing, de novo genome assembly and strain-resolved metagenome assembly (Garg 2021; Bhat et al. 2021). Choice of appropriate phasing, block determination algorithms and their interaction are important factors that can influence accuracy of phasing the haplotype blocks (Bkhetan et al. 2019). Various haplotype analysis approach combined with different computational tools such as DESMAN, Falcon phase, HapCut2, HapTree, Hifiasm, MetaMaps, POLYTE, SDip, and WhatsHap are extensively reviewed by Garg (2021). The combination of different analysis approaches and computational tools with long-reads sequencing technologies has allowed us to fully utilise the potential of these sequencing methodologies for haplotype construction. SNPViz v2.0 (Zeng et al. 2020) is a web-based tool that enhances the identification of large-scale haplotype blocks. HaplotypeTools (Farrer 2021) is tool to phase variant, based on detecting the reads overlapping ≥ 2 heterozygous positions and then extent of the reads; it is also a powerful tool for analysing hybrid and polyploid genomic regions. Recently, python coded tool HAPPE (Feng et al. 2022) was developed to construct and visualize the haplotypes easily (Table 3.4). Additionally, Practical Haplotype Graph is a powerful tool for storage, retrieval and imputation of haplotypes that can be used for genomic studies (Bradbury et al. 2022).

3.9 Genome Editing

CRISPR/Cas9 is the potent genetic modification technique that is a great example of genome editing technologies. This technology is proved to be extremely effective tool not only in the field of basic science but also in the plant breeding. The development of genome editing technologies (ZFN, TALEN, CRISPR/Cas9, etc.) drawn a lot of attention, because they eliminate the restrictions of traditional breeding approaches (Matres et al. 2021). These methods enable precise and effective targeted genome modifications, greatly shortening the time needed to obtain plants with desired traits for the development of new crop varieties. Sequence-specific nucleases and small guide RNA are the key components of CRISPR-based gene editing approach to generate precise modification. The CRISPR/Cas system is still evolving, but there are two significant obstacles: off-target effects and on-target efficiency (Xu et al. 2015; Zhang et al. 2015; Liu et al. 2020). To overcome these issues, optimizing small guide RNA by effective computer methods assist in silico gRNA design that plays an important role (Doench et al. 2016; Hassan et al. 2021). One of the key factors affecting gRNA effectiveness is the nucleotide content of a target sequence. The PAM (Protospacer Adjacent Motif) sequence and its nearby nucleotide is significantly important for the better efficiency (Liu et al. 2020). Guanines are favoured at first and second nucleotide position before the PAM sequence while thymines are not preferred within four nucleotides upstream/downstream of PAM sequence. Furthermore, sequences upstream of PAMs have no discernible influence, although sequences downstream can affect gRNA efficiency (Doench et al. 2014). At cleavage site, cytosine is preferred and GC content at downstream of the PAM sequence that increases high efficiency to gRNA. Numerous efficiency prediction models are available built using this important information. Various tools have been developed based on these models to design gRNA either by alignment-based, hypothesis-driven and/or learning-based models (Konstantakos et al. 2022). Hypothesis-driven and learning model-based tools perform better than alignment-based models. Several tools have been developed to predict gRNA with high target efficiency includes E-CRISP, CHOPCHOP, CRISPR-FOCUS, PROTOSPACER, CLD, CRISPOR, and CRISPETa (Table 3.4). WheatCRISPR is a web-based bioinformatics tool which is generally used for constructing target-specific gRNA in wheat (Cram et al. 2019). Additionally, CROPSR is the first open source bioinformatics tool to help design genome-wide guide RNA for CRISPR-based genome editing with high speed that reduces the challenges of complex crop genome (Paul et al. 2022).

3.10 Major Challenges in Bioinformatics

NGS technologies have made genomic revolution by generating enormous amount of data quickly and affordably. The use of bioinformatics in life science research is becoming more and more essential at the moment. Data analysis is frequently the main bottleneck because of the exponential growth in amount and complexity of life science data over the past two decades. Handling, analysing, and storing information has become a new barrier for biologists. Efficient data processing is necessary and there are many algorithms available for these specific tasks. To increase efficiency and accuracy, it needs combination of tools and enough resources for smooth operation. Another challenge for the biologists is to learn the languages like python, Perl or R for efficient handling of the data and lack of training in the field by the expert bioinformatician who knows biological problems and associated complexities. Genome assembly has gained more and more attention as advance sequencing technology are developed. Despite the abundance of genome assembly tools available, de novo genome assembly using next-generation reads still faces four significant obstacles: sequencing errors, sequencing bias, topological complexity of repetitive regions and huge computational resource consumption (Liao et al. 2019). The accuracy of results can have a big impact on downstream analysis of sequencing data. False positives and inaccurate findings may result from the errors during data processing. On the other side, poorly chosen approaches or tools may produce false negatives, which would result in the loss of genuine variants. Therefore, finding a suitable balance between accuracy of results and sensitivity is thus another big problem for data analysis. The application of ML in plant research is also an important issue. Traditionally, statistical techniques have been used to predict genotype-phenotype relationships. These techniques have been very effective and successful throughout the past century. Decision-making for researchers and practitioners typically involves the use of confidence measures and model interpretation. Further, data-driven flexibility of ML offers a range of advantages over stringent statistical approaches that make it a powerful tool for solving complex problems and extracting valuable insights from diverse and dynamic datasets.

3.11 Future Prospective and Conclusions

Bioinformatics has been emerging and cross-cutting different fields of agricultural sciences for enhancing our understanding of the complex mechanism underlying different traits in different crop plants in crop improvement (Fig. 3.2). A paradigm shift in the field of life sciences has been brought by NGS and has transformed genomics research. In addition to being crucial for fundamental genomic and molecular biology research, bioinformatics also has a significant influence on many fields of agricultural and medical sciences. Suitable computational tools and the right resources are essential for identifying biological information that adds value and offers novel insights into biological systems. The rise in omics-based research needs education in the relevant technologies and bioinformatics in order to correctly translate experimental and computational efforts. AI-based solutions are help to increase efficiency and regulate a number of factors, including crop yield, soil profile, crop irrigation, weeding, and crop monitoring (Bhardwaj et al. 2022). The possibility of using AI in agriculture will increase as the field of AI matures and more trained algorithms are added. Recently, the development of genetic algorithm-based Internet of precision agricultural things (IopaT) and becoming famous in rural areas to solve the real-time problems. Genetic algorithmic system is developed to predict water requirement (Roy and De 2020). This kind of system will also help in decision-making in agriculture, like crop patterns and water management at particular place (Xu et al. 2022a). Future applications of AI/ML in plant research include predicting which regions of the genome should be modified to produce a particular phenotype and providing the best possible local growing conditions by monitoring crop performance in vivo in the greenhouse or on the field. We are still very early in the genomics era, and undoubtedly, a long way from accomplishing the ambitious objective. In fact, efforts are still required for in-depth and appropriate analyses of genome, transcriptome, and metagenome data to identify link between organization and functionality. Moreover, chemical genomics approaches aid in the comprehension of overcoming stress conditions and improving crop yield and productivity (Pa et al. 2022; Adhinarayanreddy et al. 2022). Utilizing integrated multi-omics data, big data technology, and artificial intelligence proposed the new term called

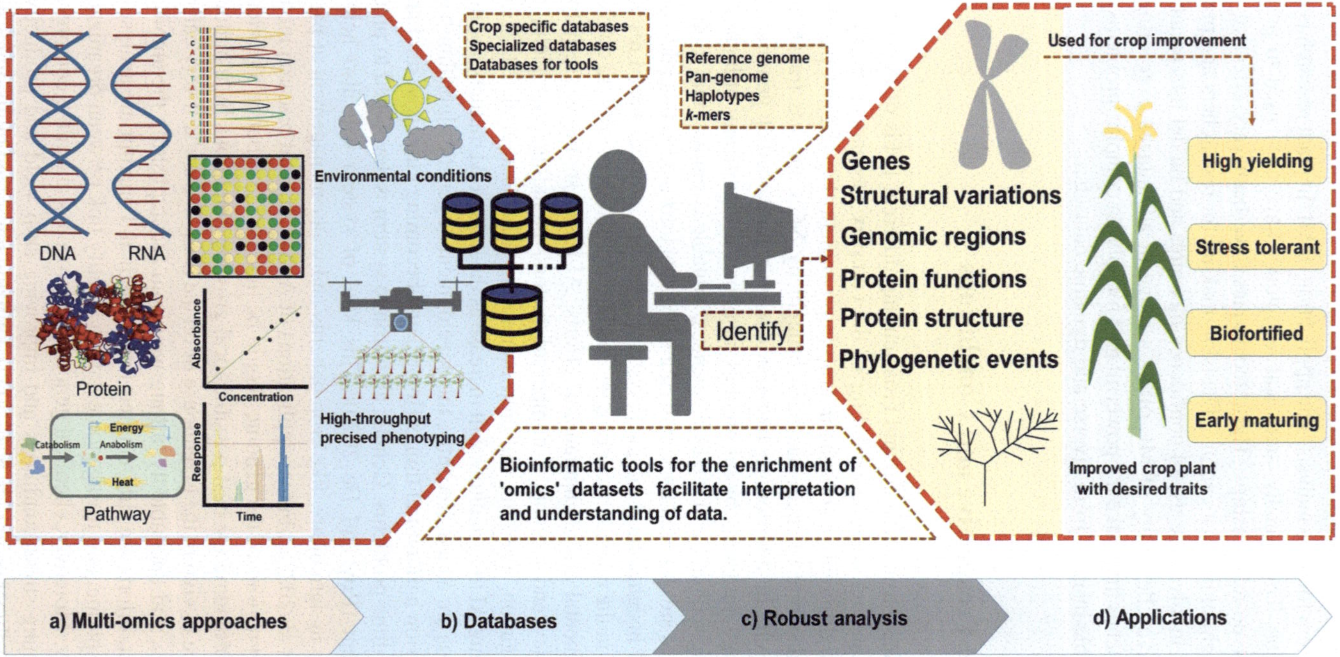

Fig. 3.2 Role of bioinformatics in genetics and plant breeding research for developing climate-resilient crops and sustainable food production. (**a**) Generation of biological data from various omics approaches as well as phenotyping data from multiple environments; (**b**) Storage and processing of different omics data generated. (**c**) Robust analysis of the raw data and transforming to useful information using bioinformatical tools for appropriate interpretation; (**d**) Application of bioinformatics in agricultural research

integrated genomic-enviromic prediction (Xu et al. 2022b), as an extension of genomic prediction will provide accelerating breeding programs. With the use of big data, AI and robust bioinformatical analysis, plant breeding in the future will become increasingly smart. The establishment of integrative plant breeding platforms and open-source breeding initiatives can help translate smart breeding efforts into genetic gains.

References

Addinsoft (2021) XLSTAT statistical and data analysis solution. New York, USA

Adhinarayanreddy V, Vijayaraghavareddy P, Vargheese A, Sujitha DA, Uttarkar A, Niranjan V, Anuradha CV, Sheshshayee MS, Vemanna R (2022) A simple and rapid oxidative stress screening method of small molecules for functional studies of transcription factor. Rice Sci 2022:3

Amarasinghe SL, Ritchie ME, Gouil Q (2021) Long-read-tools. Org: an interactive catalogue of analysis methods for long-read sequencing data. GigaScience 10(2):1–7

Amarasinghe SL, Su S, Dong X, Zappia L, Ritchie ME, Gouil Q (2020) Opportunities and challenges in long-read sequencing data analysis. Genome Biol 21(1):1–16

Andrews S (2010) FastQC: a quality control tool for high throughput sequence data

Ayed BR, Hanana M (2021) Artificial intelligence to improve the food and agriculture sector. J Food Qual 2021:1–7

Balcke GU, Handrick V, Bergau N, Fichtner M, Henning A, Stellmach H, Tissier A, Hause B, Frolov A (2012) An UPLC-MS/MS method for highly sensitive high-throughput analysis of phytohormones in plant tissues. Plant Methods 8(1):1–11

Barchi L, Rabanus-Wallace MT, Prohens J, Toppino L, Padmarasu S, Portis E, Rotino GL, Stein N, Lanteri S, Giuliano G (2021) Improved genome assembly and pan-genome provide key insights into eggplant domestication and breeding. Plant J 107(2):579–596

Basten CJ, Weir BS, Zeng ZB (2002) QTL cartographer, version 1.17. Department of Statistics, North Carolina State University, Raleigh, NC

Batley J, Edwards D (2016) The application of genomics and bioinformatics to accelerate crop improvement in a changing climate. Curr Opin Plant Biol 30(2):78–81

Bhardwaj A, Kishore S, Pandey DK (2022) Artificial Intelligence in Biological Sciences. Life 12: 1430

Bhat JA, Yu D, Bohra A, Ganie SA, Varshney RK (2021) Features and applications of haplotypes in crop breeding. Communications Biology 4(1):1–12

Bhatta M, Morgounov A, Belamkar V, Wegulo SN, Dababat AA, Erginbas-Orakci G, Bouhssini ME, Gautam P, Poland J, Akci N, Demir L (2019) Genome-wide association study for multiple biotic stress resistance in synthetic hexaploid wheat. Int J Mol Sci 20(15):3667

Bkhetan AZ, Zobel J, Kowalczyk A, Verspoor K, Goudey B (2019) Exploring effective approaches for haplotype block phasing. BMC Bioinform 20(1):1–14

Bradbury PJ, Casstevens T, Jensen SE, Johnson LC, Miller ZR, Monier B, Romay MC, Song B, Buckler ES (2022) The practical haplotype graph, a platform for storing and using pangenomes for imputation. Bioinform 38(15):3698–3702

Bradbury PJ, Zhang Z, Kroon DE, Casstevens TM, Ramdoss Y, Buckler ES (2007) TASSEL: software for association mapping of complex traits in diverse samples. Bioinform 23(19):2633–2635

Browning SR, Browning BL (2011) Haplotype phasing: existing methods and new developments. Nat Rev Genet 12(10):703–714

Channale S, Kalavikatte D, Thompson JP, Kudapa H, Bajaj P, Varshney RK, Zwart RS, Thudi M (2021) Transcriptome analysis reveals key genes associated with root-lesion nematode *Pratylenchus thornei* resistance in chickpea. Sci Rep 11(1):1–11

Chen T, Guestrin C (2016) Xgboost: a scalable tree boosting system. In proceedings of the 22nd acm sigkdd international conference on knowledge discovery and data mining 785–794

Chen C, Shang X, Sun M, Tang S, Khan A, Zhang D, Yan H, Jiang Y, Yu F, Wu Y, Xie Q (2022) Comparative transcriptome analysis of two sweet sorghum genotypes with different salt tolerance abilities to reveal the mechanism of salt tolerance. Int J Mol Sci 23(4):2272

Chen S, Zhou Y, Chen Y, Gu J (2018) Fastp: an ultra-fast all-in-one FASTQ preprocessor. Bioinform 34(17):884–890

Contreras-Moreira B, Vinuesa P (2013) GET_HOMOLOGUES, a versatile software package for scalable and robust microbial pangenome analysis. Appl Environ Microbiol 79(24):7696–7701

Cram D, Kulkarni M, Buchwaldt M, Rajagopalan N, Bhowmik P, Rozwadowski K, Parkin IA, Sharpe AG, Kagale S (2019) WheatCRISPR: a web-based guide RNA design tool for CRISPR/Cas9-mediated genome editing in wheat. BMC Plant Biol 19(1):1–8

Cserhati M, Xiao P, Guda C (2019) K-mer-based motif analysis in insect species across anopheles, drosophila, and Glossina genera and its application to species classification. Computational and mathematical methods in medicine 1–16

Cserhati MF, Mooter ME, Peterson L, Wicks B, Xiao P, Pauley M, Guda C (2018) Motifome comparison between modern human. Neanderthal and Denisovan BMC Genomics 19(1):1–9

Delaneau O, Zagury JF, Robinson MR, Marchini JL, Dermitzakis ET (2019) Accurate, scalable and integrative haplotype estimation. Nat Commun 10(3):1–10

Demirci S, Peters SA, de Ridder D, van Dijk AD (2018) DNA sequence and shape are predictive for meiotic crossovers throughout the plant kingdom. Plant J 95(4):686–699

Dempewolf H, Baute G, Anderson J, Kilian B, Smith C, Guarino L (2017) Past and future use of wild relatives in crop breeding. Crop Sci 57(3):1070–1082

Doddamani D, Khan AW, Katta MA, Agarwal G, Thudi M, Ruperao P, Edwards D, Varshney RK (2015) CicArVarDB: SNP and InDel database for advancing genetics research and breeding applications in chickpea. Database 2015:1–7

Doench JG, Fusi N, Sullender M, Hegde M, Vaimberg EW, Donovan KF, Smith I, Tothova Z, Wilen C, Orchard R, Virgin HW (2016) Optimized sgRNA design to maximize activity and minimize off-target effects of CRISPR-Cas9. Nat Biotechnol 34(2):184–191

Doench JG, Hartenian E, Graham DB, Tothova Z, Hegde M, Smith I, Sullender M, Ebert BL, Xavier RJ, Root DE (2014) Rational design of highly active sgRNAs for CRISPR-Cas9–mediated gene inactivation. Nat Biotechnol 32(12):1262–1267

Edwards D, Stajich J, Hansen D (eds) (2009) Bioinformatics: tools and applications. Springer, New York

Farrer RA (2021) HaplotypeTools: a toolkit for accurately identifying recombination and recombinant genotypes. BMC Bioinform 22(1):1–15

Feng C, Wang X, Wu S, Ning W, Song B, Yan J, Cheng S (2022) HAPPE: a tool for population haplotype analysis and visualization in editable excel tables. Front Plant Sci 13:1–7

Gao L, Gonda I, Sun H, Ma Q, Bao K, Tieman DM, Burzynski-Chang EA, Fish TL, Stromberg KA, Sacks GL, Thannhauser TW (2019) The tomato pan-genome uncovers new genes and a rare allele regulating fruit flavor. Nat Genet 51(6):1044–1051

Garg S (2021) Computational methods for chromosome-scale haplotype reconstruction. Genome Biol 22(1):1–24

Garrison E, Marth G (2012) Haplotype-based variant detection from short-read sequencing. arXiv 1207:3907

Gauch HG, Moran DR (2019) AMMISOFT for AMMI analysis with best practices. BioRxiv 538454

Gharaei A, Karimi M, Shekarabi SAH (2019) An integrated multi-product, multi-buyer supply chain under penalty, green, and quality control polices and a vendor managed inventory with consignment stock agreement: the outer approximation with equality relaxation and augmented penalty algorithm. Appl Math Model 69:223–254

Giacomello S, Salmen F, Terebieniec BK, Vickovic S, Navarro JF, Alexeyenko A, Reimegard J, McKee LS, Mannapperuma C, Bulone V, Stahl PL (2017) Spatially resolved transcriptome profiling in model plant species. Nat Plants 3(6):1–11

Gouda AC, Warburton ML, Djedatin GL, Kpeki SB, Wambugu PW, Gnikoua K, Ndjiondjop MN (2021) Development and validation of diagnostic SNP markers for quality control genotyping in a collection of four rice (*Oryza*) species. Sci Rep 11(1):1–11

Gupta AK, Zhang X, Andrews JG (2015) Potential throughput in 3D ultradense cellular networks. In 49th Asilomar conference on signals, systems and computers,1026–1030. IEEE

Gulles AA, Bartolome VI, Morantte RI, Nora LA, Relente CE, Talay DT, Caneda AA, Ye G (2014) Randomization and analysis of data using STAR [Statistical Tool for Agricultural Research]. Philippine J Crop Sci 39:137

Harfouche AL, Jacobson DA, Kainer D, Romero JC, Harfouche AH, Mugnozza GS, Moshelion M, Tuskan GA, Keurentjes JJ, Altman A (2019) Accelerating climate resilient plant breeding by applying next-generation artificial intelligence. Trends Biotechnol 37(11):1217–1235

Harper L, Campbell J, Cannon EK, Jung S, Poelchau M, Walls R, Andorf C, Arnaud E, Berardini TZ, Birkett C, Cannon S et al (2018) AgBioData consortium recommendations for sustainable genomics and genetics databases for agriculture. Database 2018:1–32

Hashim EK, Abdullah R (2015) Rare k-mer DNA: identification of sequence motifs and prediction of CpG Island and promoter. J Theor Biol 387:88–100

Hassan MM, Chowdhury AK, Islam T (2021) In silico analysis of gRNA secondary structure to predict its efficacy for plant genome editing. In: Islam, Molla (eds) CRISPR-Cas methods, New York, NY, pp 15–22

Heather JM, Chain B (2016) The sequence of sequencers: the history of sequencing DNA. Genomics 107(1):1–8

Hu Z, Sun C, Lu KC, Chu X, Zhao Y, Lu J, Shi J, Wei C (2017) EUPAN enables pan-genome studies of a large number of eukaryotic genomes. Bioinform 33(15):2408–2409

Hurgobin B, Golicz AA, Bayer PE, Chan CKK, Tirnaz S, Dolatabadian A, Schiessl SV, Samans B, Montenegro JD, Parkin IA, Pires JC (2018) Homoeologous exchange is a major cause of gene presence/absence variation in the amphidiploid Brassica napus. Plant Biotechnol J 16(7): 1265–1274

IBM Corp Ibm, SPSS (2017) Statistics for windows, version 25.0. IBM Corp, Armonk, NY

Jayakodi M, Padmarasu S, Haberer G, Bonthala VS, Gundlach H, Monat C, Lux T, Kamal N, Lang D, Himmelbach A, Ens J (2020) The barley pan-genome reveals the hidden legacy of mutation breeding. Nat 588(7837):284–289

Jha UC, Nayyar H, von Wettberg EJ, Naik YD, Thudi M, Siddique KH (2022) Legume Pangenome: status and scope for crop improvement. Plan Theory 22:3041

Kanzi AM, San JE, Chimukangara B, Wilkinson E, Fish M, Ramsuran V, De Oliveira T (2020) Next generation sequencing and bioinformatics analysis of family genetic inheritance. Front Genet 11:e544162

Kathiresan N, Temanni R, Almabrazi H, Syed N, Jithesh PV, Al-Ali R (2017) Accelerating next generation sequencing data analysis with system level optimizations. Sci Rep 7(1):1–11

Khan AW, Garg V, Roorkiwal M, Golicz AA, Edwards D, Varshney RK (2020) Super-pangenome by integrating the wild side of a species for accelerated crop improvement. Trends Plant Sci 25(2):148–158

Khetan M, Ameerpet M (2015) Indostat package for data analysis. Windostat version 9.3 from indostat services, Hyderabad

Koboldt DC, Larson DE, Wilson RK (2013) Using VarScan 2 for germline variant calling and somatic mutation detection. Curr Protoc Bioinform 44(1):15–14

Konstantakos V, Nentidis A, Krithara A, Paliouras G (2022) CRISPR–Cas9 gRNA efficiency prediction: an overview of predictive tools and the role of deep learning. Nucleic Acids Res 50(7):3616–3637

Kudapa H, Garg V, Chitikineni A, Varshney RK (2018) The RNA-Seq-based high resolution gene expression atlas of chickpea (*Cicer arietinum* L.) reveals dynamic spatio-temporal changes associated with growth and development. Plant Cell Environ 41(9):2209–2225

Lai K, Lorenc MT, Edwards D (2012) Genomic databases for crop improvement. Agron 2(1):62–73

Langmead B, Salzberg SL (2012) Fast gapped-read alignment with bowtie 2. Nat Methods 9(4): 357–359

Le Nguyen K, Grondin A, Courtois B, Gantet P (2019) Next-generation sequencing accelerates crop gene discovery. Trends Plant Sci 24(3):263–274

Ledesma R (2008) Software de análisis de correspondencias múltiples: una revisión comparativa. Metodología de encuestas 10(1):59–75

Lee WP, Stromberg MP, Ward A, Stewart C, Garrison EP, Marth GT (2014) MOSAIK: a hash-based algorithm for accurate next-generation sequencing short-read mapping. PLoS One 9(3): e90581

Lewin HA, Robinson GE, Kress WJ, Baker WJ, Coddington J, Crandall KA, Durbin R, Edwards SV, Forest F, Gilbert MTP, Goldstein MM (2018) Earth BioGenome project: sequencing life for the future of life. Proc Natl Acad Sci 115(17):4325–4333

Li H (2018) Minimap2: pairwise alignment for nucleotide sequences. Bioinform 34(18):3094–3100

Li H, Durbin R (2009) Fast and accurate short read alignment with Burrows–Wheeler transform. Bioinform 25(14):1754–1760

Li H, Handsaker B, Wysoker A, Fennell T, Ruan J, Homer N, Marth G, Abecasis G, Durbin R (2009) The sequence alignment/map format and SAMtools. Bioinform 25(16):2078–2079

Liao X, Li M, Zou Y, Wu FX, Wang J (2019) Current challenges and solutions of *de novo* assembly. Quantitat Biol 7(2):90–109

Lincoln SE, Daly MJ, Lander ES (1993) Constructing genetic linkage maps with MAPMAKER/EXP Version 3.0: a tutorial and reference manual. A whitehead institute for biomedical research technical report, 3

Liu G, Zhang Y, Zhang T (2020) Computational approaches for effective CRISPR guide RNA design and evaluation. Comput Struct Biotechnol J 18(2):35–44

Liu Y, Popp B, Schmidt B (2014) CUSHAW3: sensitive and accurate base-space and color-space short-read alignment with hybrid seeding. PLoS One 9(1):e86869

Lorenzi C, Barriere S, Villemin JP, Dejardin Bretones L, Mancheron A, Ritchie W (2020) iMOKA: k-mer based software to analyze large collections of sequencing data. Genome Biol 21(1):1–19

Maestri S, Maturo MG, Cosentino E, Marcolungo L, Iadarola B, Fortunati E, Rossato M, Delledonne M (2020) A long-read sequencing approach for direct haplotype phasing in clinical settings. Int J Mol Sci 21(23):9177

Mahmoud M, Gobet N, Cruz-Dávalos DI, Mounier N, Dessimoz C, Sedlazeck FJ (2019) Structural variant calling: the long and the short of it. Genome Bology 20(1):1–14

Mapleson D, Garcia Accinelli G, Kettleborough G, Wright J, Clavijo BJ (2017) KAT: a k-mer analysis toolkit to quality control NGS datasets and genome assemblies. Bioinform 33(4):574–576

Mashaki MK, Garg V, Nasrollahnezhad Ghomi AA, Kudapa H, Chitikineni A, Zaynali Nezhad K, Yamchi A, Soltanloo H, Varshney RK, Thudi M (2018) RNA-Seq analysis revealed genes associated with drought stress response in kabuli chickpea (*Cicer arietinum* L.). PLoS One 13(6):e0199774

Matres JM, Hilscher J, Datta A, Armario-Nájera V, Baysal C, He W, Huang X, Zhu C, Valizadeh-Kamran R, Trijatmiko KR, Capell T (2021) Genome editing in cereal crops: an overview. Transgenic Res 30(4):461–498

McKenna A, Hanna M, Banks E, Sivachenko A, Cibulskis K, Kernytsky A, Garimella K, Altshuler D, Gabriel S, Daly M, DePristo MA (2010) The genome analysis toolkit: a MapReduce framework for analyzing next-generation DNA sequencing data. Genome Res 20(9):1297–1303

Melsted P, Pritchard JK (2011) Efficient counting of k-mers in DNA sequences using a bloom filter. BMC Bioinform 12(1):1–7

Miga KH (2020) Centromere studies in the era of 'telomere-to-telomere'genomics. Exp Cell Res 394(2):e112127

Mohanty SP, Hughes DP, Salathe M (2016) Using deep learning for image-based plant disease detection. Front Plant Sci 7:1419

Morin PA, Alexander A, Blaxter M, Caballero S, Fedrigo O, Fontaine MC, Foote AD, Kuraku S, Maloney B, Mccarthy M, Mcgowen M (2020) Building genomic infrastructure: sequencing platinum-standard reference-quality genomes of all cetacean species. Mar Mamm Sci 36:1356–1366

Nayak SN, Agarwal G, Pandey MK, Sudini HK, Jayale AS, Purohit S, Desai A, Wan L, Guo B, Liao B, Varshney RK (2017) *Aspergillus flavus* infection triggered immune responses and host-pathogen cross-talks in groundnut during in-vitro seed colonization. Sci Rep 7(1):1–14

Nyeki AE, Kerepesi C, Daroczy BZ, Benczúr A, Milics G, Kovacs AJ, Nemenyi M (2019) Maize yield prediction based on artificial intelligence using spatio-temporal data precision agriculture '19, eds: John V Stafford, 1011–1017

O'Fallon BD, Wooderchak-Donahue W, Crockett DK (2013) A support vector machine for identification of single-nucleotide polymorphisms from next-generation sequencing data. Bioinform 29(11):1361–1366

Pa V, Vijayaraghavareddy P, Uttarkar A, Dawane A, KC B, Niranjan V, MS S, CV A, Makarla U, Vemanna RS (2022) Novel small molecules targeting bZIP23 TF improve stomatal conductance and photosynthesis under mild drought stress by regulating ABA. FEBS J 289(19):6058–6077

Pacheco A, Vargas M, Alvarado G, Rodríguez F, Crossa J, Burgueño J (2015) GEA-R (genotype x environment analysis with R for windows) version 4.1. hdl 11529(10203):16

Pal G, Bakade R, Deshpande S, Sureshkumar V, Patil SS, Dawane A, Agarwal S, Niranjan V, Prasanna MK, Vemanna RS (2022) Transcriptomic responses under combined bacterial blight and drought stress in rice reveal potential genes to improve multi-stress tolerance. BMC Plant Biol 22(1):1–20

Paul MH, Istanto DD, Heldenbrand J, Hudson ME (2022) CROPSR: an automated platform for complex genome-wide CRISPR gRNA design and validation. BMC Bioinform 23(1):1–19

Pazhamala LT, Purohit S, Saxena RK, Garg V, Krishnamurthy L, Verdier J, Varshney RK (2017) Gene expression atlas of pigeonpea and its application to gain insights into genes associated with pollen fertility implicated in seed formation. J Exp Bot 68(8):2037–2054

Poplin R, Chang PC, Alexander D, Schwartz S, Colthurst T, Ku A, Newburger D, Dijamco J, Nguyen N, Afshar PT, Gross SS (2018) A universal SNP and small-indel variant caller using deep neural networks. Nat Biotechnol 36(10):983–987

Pornputtapong N, Acheampong DA, Patumcharoenpol P, Jenjaroenpun P, Wongsurawat T, Jun SR, Yongkiettrakul S, Chokesajjawatee N, Nookaew I (2020) KITSUNE: a tool for identifying empirically optimal k-mer length for alignment-free phylogenomic analysis. Front Bioeng Biotechnol 23(8):556413

Pour-Aboughadareh A, Yousefian M, Moradkhani H, Poczai P, Siddique KH (2019) STABILITYSOFT: a new online program to calculate parametric and non-parametric stability statistics for crop traits. Appl Plant Sci 7(1):e01211

Pritchard JK, Stephens M, Donnelly P (2000) Inference of population structure using multilocus genotype data. Genetics 155:945–959

Qian L, Hickey LT, Stahl A, Werner CR, Hayes B, Snowdon RJ, Voss-Fels KP (2017) Exploring and harnessing haplotype diversity to improve yield stability in crops. Front Plant Sci 8(1):1–11

Qiu R, Wei S, Zhang M, Li H, Sun H, Liu G, Li M (2018) Sensors for measuring plant phenotyping: a review. International Journal of Agricultural and Biological Engineering 11(2):1–17

Rahman A, Hallgrimsdottir I, Eisen M, Pachter L (2018) Association mapping from sequencing reads using k-mers. elife 13(7):e32920

Ren J, Chaisson MJ (2021) lra: a long read aligner for sequences and contigs. PLoS Comput. Biol 17(6):e1009078

Roy SK, De D (2020) Genetic algorithm based internet of precision agricultural things (IopaT) for agriculture 4.0. Internet of Things 18:100201

Ruperao P, Thirunavukkarasu N, Gandham P, Selvanayagam S, Govindaraj M, Nebie B, Manyasa E, Gupta R, Das RR, Odeny DA, Gandhi H (2021) Sorghum pan-genome explores the functional utility for genomic-assisted breeding to accelerate the genetic gain. Front Plant Sci 12(1):963–980

Sandmann S, Karimi M, de Graaf AO, Rohde C, Gollner S, Varghese J, Ernsting J, Walldin G, van der Reijden BA, Müller-Tidow C, Malcovati L (2018) appreci8: a pipeline for precise variant calling integrating 8 tools. Bioinform 34(24):4205–4212

Sartor RC, Noshay J, Springer NM, Briggs SP (2019) Identification of the expressome by machine learning on omics data. Proc Natl Acad Sci 116(36):18119–18125

Sheikhizadeh S, Schranz ME, Akdel M, de Ridder D, Smit S (2016) PanTools: representation, storage and exploration of pangenomic data. Bioinform 32(17):487–493

Silva de Oliveira M, Thyeska Castro Alves J, Henrique Caracciolo Gomes de Sa P, Veras AADO (2021) PAN2HGENE–tool for comparative analysis and identifying new gene products. PLoS One 16(5):e0252414

Sinha P, Bajaj P, Pazhamala LT, Nayak SN, Pandey MK, Chitikineni A, Huai D, Khan AW, Desai A, Jiang H, Zhuang W (2020) *Arachis hypogaea* gene expression atlas for fastigiata subspecies of cultivated groundnut to accelerate functional and translational genomics applications. Plant Biotechnol J 18(11):2187–2200

Speranza E, Williamson BN, Feldmann F, Sturdevant GL, Pérez-Pérez L, Meade-White K, Smith BJ, Lovaglio J, Martens C, Munster VJ, Okumura A (2021) Single-cell RNA sequencing reveals SARS-CoV-2 infection dynamics in lungs of African green monkeys. Sci Transl Med 13(578): e8146

Stein JC, Yu Y, Copetti D, Zwickl DJ, Zhang L, Zhang C, Chougule K, Gao D, Iwata A, Goicoechea JL, Wei S (2018) Genomes of 13 domesticated and wild rice relatives highlight genetic conservation, turnover and innovation across the genus *Oryza*. Nat Genet 50(2): 285–296

Sun Y, Shang L, Zhu QH, Fan L, Guo L (2021) Twenty years of plant genome sequencing: achievements and challenges. Trends Plant Sci 27(4:391–401

Team RC (2013) R: a language and environment for statistical computing. R foundation for statistical computing, Vienna, Austria. http://www. R-project. org/

Thorvaldsdottir H, Robinson JT, Mesirov JP (2012) Integrative Genomics Viewer (IGV): High-performance genomics data visualization and exploration. Briefings in Bioinformatics 14 (2):178–192

Thudi M, Chen Y, Pang J, Kalavikatte D, Bajaj P, Roorkiwal M, Chitikineni A, Ryan MH, Lambers H, Siddique KH, Varshney RK (2021) Novel genes and genetic loci associated with root morphological traits, phosphorus-acquisition efficiency and phosphorus-use efficiency in chickpea. Front Plant Sci 1001

Thudi M, Khan AW, Kumar V, Gaur PM, Katta K, Garg V, Roorkiwal M, Samineni S, Varshney RK (2016) Whole genome re-sequencing reveals genome-wide variations among parental lines of 16 mapping populations in chickpea (*Cicer arietinum* L.). BMC Plant Biol 16(1):53–64

Thudi M, Palakurthi R, Schnable JC, Chitikineni A, Dreisigacker S, Mace E, Srivastava RK, Satyavathi CT, Odeny D, Tiwari VK, Lam HM (2020) Genomic resources in plant breeding for sustainable agriculture. J Plant Physiol 257(1):e153351

Thudi M, Samineni S, Li W, Boer MP, Roorkiwal M, Yang Z, Ladejobi F, Zheng C, Chitikineni A, Nayak S, He Z, Valluri V, Bajaj P, Khan AW, Gaur PM, van Eeuwijk F, Mott R, Xin L, Varshney RK (2023) Whole genome resequencing and phenotyping of MAGIC population for high resolution mapping of drought tolerance in chickpea. Plant Genome 30:e20333. https://doi.org/10.1002/tpg2.20333

Toda Y, Tameshige T, Tomiyama M, Kinoshita T, Shimizu KK (2021) An affordable image-analysis platform to accelerate stomatal phenotyping during microscopic observation. Front Plant Sci 12:715309

Tom N, Tom O, Malcikova J, Pavlova S, Kubesova B, Rausch T, Kolarik M, Benes V, Bystry V, Pospisilova S (2018) ToTem: a tool for variant calling pipeline optimization. BMC Bioinform 19(1):1–9

Utz HF, Melchinger AE (1996) PLABQTL: a program for composite interval mapping of QTL. J Quant Trait Loci 2(1):1–5

van Dijk ADJ, Kootstra G, Kruijer W, de Ridder D (2021b) Machine learning in plant science and plant breeding. iScience 24(1):101890

van Dijk M, Morley T, Rau ML, Saghai Y (2021a) A meta-analysis of projected global food demand and population at risk of hunger for the period 2010–2050. Nat Food 2(7):494–501

Van Ooijen JW, Maliepaard CA (1999) MapQTL: version 3.0: Software for the calculation of QTL positions on genetic maps

Varshney RK, Bohra A, Yu J, Graner A, Zhang Q, Sorrells ME (2021a) Designing future crops: genomics-assisted breeding comes of age. Trends Plant Sci 26(6):631–649

Varshney RK, Roorkiwal M, Sun S, Bajaj P, Chitikineni A, Thudi M, Singh NP, Du X, Upadhyaya HD, Khan AW, Wang Y (2021b) A chickpea genetic variation map based on the sequencing of 3,366 genomes. Nat 599(7886):622–627

Varshney RK, Saxena RK, Upadhyaya HD, Khan AW, Yu Y, Kim C, Rathore A, Kim D, Kim J, An S, Kumar V (2017b) Whole-genome resequencing of 292 pigeonpea accessions identifies genomic regions associated with domestication and agronomic traits. Nat Genet 49(7): 1082–1088

Varshney RK, Shi C, Thudi M, Mariac C, Wallace J, Qi P, Zhang H, Zhao Y, Wang X, Rathore A, Srivastava RK (2017a) Pearl millet genome sequence provides a resource to improve agronomic traits in arid environments. Nat Biotechnol 35(10):969–976

Varshney RK, Sinha P, Singh VK, Kumar A, Zhang Q, Bennetzen JL (2020) 5Gs for crop genetic improvement. Curr Opin Plant Biol 56:190–196

Varshney RK, Thudi M, Nayak SN, Gaur PM, Kashiwagi J, Krishnamurthy L, Jaganathan D, Koppolu J, Bohra A, Tripathi S, Rathore A (2014) Genetic dissection of drought tolerance in chickpea (*Cicer arietinum* L.). Theor Appl Genet 127(2):445–462

Varshney RK, Thudi M, Pandey MK, Tardieu F, Ojiewo C, Vadez V, Whitbread AM, Siddique KH, Nguyen HT, Carberry PS, Bergvinson D (2018) Accelerating genetic gains in legumes for the development of prosperous smallholder agriculture: integrating genomics, phenotyping, systems modelling and agronomy. J Exp Bot 69(13):3293–3312

Varshney RK, Pandey MK, Bohra A, Singh VK, Thudi M, Saxena RK (2019) Toward the sequence-based breeding in legumes in the post-genome sequencing era. Theoretical and Applied Genetics 132(3):797–816

Villate A, San Nicolas M, Gallastegi M, Aulas PA, Olivares M, Usobiaga A, Etxebarria N, Aizpurua-Olaizola O (2021) Metabolomics as a prediction tool for plants performance under environmental stress. Plant Sci 303:110789

Voss-Fels K, Snowdon RJ (2016) Understanding and utilizing crop genome diversity via high-resolution genotyping. Plant Biotechnol J 14(4):1086–1094

Wang SCJB (2005) Windows QTL cartographer 2.5. http://statgen.Ncsu.Edu/qtlcart/WQTLCart.Htm

Wang W, Mauleon R, Hu Z, Chebotarov D, Tai S, Wu Z, Li M, Zheng T, Fuentes RR, Zhang F, Mansueto L (2018) Genomic variation in 3,010 diverse accessions of Asian cultivated rice. Nature 557(7703):43–49

Wang Y, Chen Q, Deng C, Zheng Y, Sun F (2020) KmerGO: a tool to identify group-specific sequences with k-mers. Front Microbiol 11:2067

Warren AS, Davis JJ, Wattam AR, Machi D, Setubal JC, Heath LS (2017) Panaconda: application of pan-synteny graph models to genome content analysis. bioRxiv 2:1–15

Wei ZG, Fan XG, Zhang H, Zhang XD, Liu F, Qian Y, Zhang SW (2022) kngMap: sensitive and fast mapping algorithm for noisy long reads based on the *k*-mer neighborhood graph. Front Genet 13:890651

Xu H, Xiao T, Chen CH, Li W, Meyer CA, Wu Q, Wu D, Cong L, Zhang F, Liu JS, Brown M (2015) Sequence determinants of improved CRISPR sgRNA design. Genome Res 25(8): 1147–1157

Xu J, Gu B, Tian G (2022a) Review of agricultural IoT technology. Artificial Intelligence in Agriculture 6:10–22

Xu Y, Zhang X, Li H, Zheng H, Zhang J, Olsen MS, Varshney RK, Prasanna BM, Qian Q (2022b) Smart breeding driven by big data, artificial intelligence and integrated genomic-enviromic prediction. Mol Plant:1–32

Yadav S, Sandhu N, Singh VK, Catolos M, Kumar A (2019) Genotyping-by-sequencing based QTL mapping for rice grain yield under reproductive stage drought stress tolerance. Sci Rep 9(1):1–12

Yan W (2001) GGE biplot-a windows application for graphical analysis of multienvironment trial data and other types of two-way data. Agron J 93(5):1111–1118

Yoshida H, Hirano K, Yano K, Wang F, Mori M, Kawamura M, Koketsu E, Hattori M, Ordonio RL, Huang P, Yamamoto E (2022) Genome-wide association study identifies a gene responsible for temperature-dependent rice germination. Nat Commun 13(1):1–13

Zakharov S, Wong TY, Aung T, Vithana EN, Khor CC, Salim A, Thalamuthu A (2013) Combined genotype and haplotype tests for region-based association studies. BMC Genomics 14(1):1–12

Zargar SM, Raatz B, Sonah H, Bhat JA, Dar ZA, Agrawal GK, Rakwal R (2015) Recent advances in molecular marker techniques: insight into QTL mapping, GWAS and genomic selection in plants. J Crop Sci Biotechnol 18(5):293–308

Zeng S, Skrabisova M, Lyu Z, Chan YO, Bilyeu K, Joshi T (2020) SNPViz v20: a web-based tool for enhanced haplotype analysis using large scale resequencing datasets and discovery of phenotypes causative gene using allelic variations. In: In 2020 IEEE international conference on bioinformatics and biomedicine, pp 1408–1415

Zhang F, Xue H, Dong X, Li M, Zheng X, Li Z, Xu J, Wang W, Wei C (2022) Long-read sequencing of 111 rice genomes reveals significantly larger pan-genomes. Genome Res 32(5): 853–863

Zhang XH, Tee LY, Wang XG, Huang QS, Yang SH (2015) Off-target effects in CRISPR/Cas9-mediated genome engineering. Molecular Therapy-Nucleic Acids 4:e264

Zhou Y, Chebotarov D, Kudrna D, Llaca V, Lee S, Rajasekar S, Mohammed N, Al-Bader N, Sobel-Sorenson C, Parakkal P, Arbelaez LJ (2020) A platinum standard pan-genome resource that represents the population structure of Asian rice. Scientific Data 7(1):1–11

Zhu FY, Song YC, Zhang KL, Chen X, Chen MX (2020) Quantifying plant dynamic proteomes by SWATH-based mass spectrometry. Trends Plant Sci 25(11):1171–1172

Chapter 4
Evolution in the Genotyping Platforms for Plant Breeding

Awais Rasheed, Xianchun Xia, and Zhonghu He

Abstract Improving the crop productivity, resilience to climate extremes, resistance to biotic stress and improving the quality are the main breeding objectives. Different tools, resources and strategies are used to precisely select the desirable cultivars in crop breeding. One of such tools is the genomics-assisted breeding (GAB), which improves selection accuracy during breeding cycles. However, practicing GAB depends on the availability of molecular markers for selecting the desired phenotypes. Once a marker is available for use in breeding, the efforts are then made to make it cost-effective and high-throughput to integrate its use in applied breeding. However, different breeding scenario like gene tagging, marker-assisted recurrent selection (MARS), background selection, diversity estimates, and genomic selection require different genotyping platforms, and there is no 'one-size-fits-all' solution. We provided an overview of the efforts around developing cost-effective, high-throughput and breeding-oriented crop genotyping platforms. A successful genotyping platform would have the features of high genome coverage, least ascertainment bias, high power in gene discovery studies, balance between throughput and flexibility, provide high prediction accuracy in genomic selection, and, above all, affordable for most of the crop breeding programs.

Keywords Single nucleotide polymorphisms · SNP arrays · Genotyping-by-sequencing · Kompetitive allele-specific PCR

A. Rasheed
Department of Plant Sciences, Quaid-i-Azam University, Islamabad, Pakistan

Chinese Academy of Agricultural Sciences (CAAS), and CIMMYT-China office, Beijing, China

X. Xia · Z. He (✉)
Chinese Academy of Agricultural Sciences (CAAS), and CIMMYT-China office, Beijing, China

M. K. Pandey et al. (eds.), *Frontier Technologies for Crop Improvement*, Sustainable Agriculture and Food Security, https://doi.org/10.1007/978-981-99-4673-0_4

4.1 Introduction

Genetic improvement in crop productivity, resilience and quality are huge challenges to feed the global population, mitigate the effects of climate change and fulfil the end-user quality preferences, respectively. The conventional crop breeding approaches are not able to deliver 70% increase in crop productivity by the end of 2050 (Tester and Langridge 2010). The innovation would be required in all breeding components including selection accuracy, selection intensity, deploying new genetic variations and shortening of the breeding cycles in developing cultivars (Li et al. 2018). The conventional plant breeding heavily relied on selection of key phenotypes related to key yield-related traits, for instance, harvest index in wheat (Lopes et al. 2012), and it seems impossible further to improve harvest index using conventional breeding. Secondly, the phenotypic-based selections are labour-intensive and time-consuming, and off-spring can only be selected at the certain homozygous generation at the lateral growth stages. The concept of genomics-assisted breeding (GAB) was proposed as an alternate to overcome the selection challenges associated with conventional breeding (Varshney et al. 2005b). The marker-assisted selection (MAS) component dominated in the breeding programs where the diagnostic markers for the genes with major phenotypic effects were developed and successfully used for selection (Liu et al. 2012). However, many complex traits like yield and adaptability to stressed environments are controlled by many genes with minor effects or quantitative trait loci (QTL), further interacting with environment. Their individual effects are too small to be efficiently captured by one or few markers (Bernardo and Yu 2007). Therefore, a transition from marker to genome-based breeding is indispensable to achieve the productivity targets (Rasheed and Xia 2019).

Molecular markers emerged as a useful breeding tool in the past two decades, due to their ability to capture the polymorphisms between individuals and populations. The initial use of markers was to map genes or QTL of interest, and markers which were available were mostly restriction fragment length polymorphisms (RFLPs), simple sequence repeats (SSRs) and amplified fragment length polymorphisms (AFLPs) (Tanksley et al. 1989; Varshney et al. 2005a). However, the main limitation in using these markers were: (1) they were not abundant, (2) the genome-wide coverage was not uniform, (3) were not high-throughput, (4) data exchange between different labs were not possible, and (5) were cost-ineffective. Ultimately, single nucleotide polymorphisms (SNPs) became marker of choice due to their abundance in genome, uniform distribution, availability of cost-effective genotyping platforms, and easy in data exchange and sharing (Kim et al. 2016; Rasheed et al. 2017).

The next-generation sequencing (NGS) has revolutionized plant genomics and resulted in development of techniques and resources amenable to plant breeding (Bevan et al. 2017). The ever-growing plant genomic resources have provided plethora of SNP information distributed throughout the plant genomes, which have made them marker of choice for a variety of research applications, especially in breeding and genetics research. Until now, the reference genome sequence is available for most of the crop species, while pangenome sequences are increasing with the rapid pace. Characterization of the pangenome can rapidly identify variations within the candidate genes, which have direct application in breeding. We discussed the different genotyping platforms available now and how the technological and scientific advancements are making them more breeder-friendly (Fig. 4.1 and Table 4.1).

4.2 Genotyping Scenarios in Plant Breeding

In a plant breeding program, several types of different objectives have different marker platform requirements in terms of throughput and flexibility. The genotyping requirements can be easily divided into three categories: a) few to many samples with few markers, b) many samples with many markers, and c) few samples with many markers. The first scenario is usually in case of gene tagging, marker-assisted backcrossing (MABC) and gene introgressions. In such scenarios, flexibility is much more important compared to the throughput. Therefore, uniplex marker systems such as Kompetitive Allele-Specific PCR (KASP) (Semagn et al. 2014), and Semi-Thermal Asymmetry Reverse PCR (STARP) (Long et al. 2016) are the excellent platforms for such genotyping requirements. The second scenario is usually in the case of association mapping or linkage mapping where the marker density requirement is very high. Therefore, SNP arrays or NGS-based platforms are ideal for such genotyping scenario (Rasheed et al. 2017; Scheben et al. 2016). In third scenario, usually in the case of the hybrid purity, quality control, genetic characterization and in some cases genomic selection needs very high number of markers for relatively smaller populations. We propose to use a semi-flexible NGS-based targeted sequencing platform for such objectives. However, it will be a great technological advancement if the platform can achieve the objectives of first scenario in the cost-effective manner. Several technological solutions are available for the different density requirement of the markers. These platforms vary from high cost to low cost and high to low demand on facilities (Rasheed et al. 2017; Yu et al. 2021). Another important consideration for choosing any genotyping platform is the data sharing and compatibility among different among different labs. Due to the high 'demand on' facilities, many of the genotyping platforms are not affordable to the many breeding programs; therefore, the development of centralized or commercial genotyping facilities at regional or national level should be realized to use markers in the breeding (Rossetto and Henry 2014).

4.3 Molecular Markers Systems in Crop Genetics and Breeding

MAS has been widely used in crop breeding since the discovery of molecular markers. Various types of molecular markers were extensively used in the past, especially the RFLP, SSRs and sequence tagged sites (STS), which are gel-based

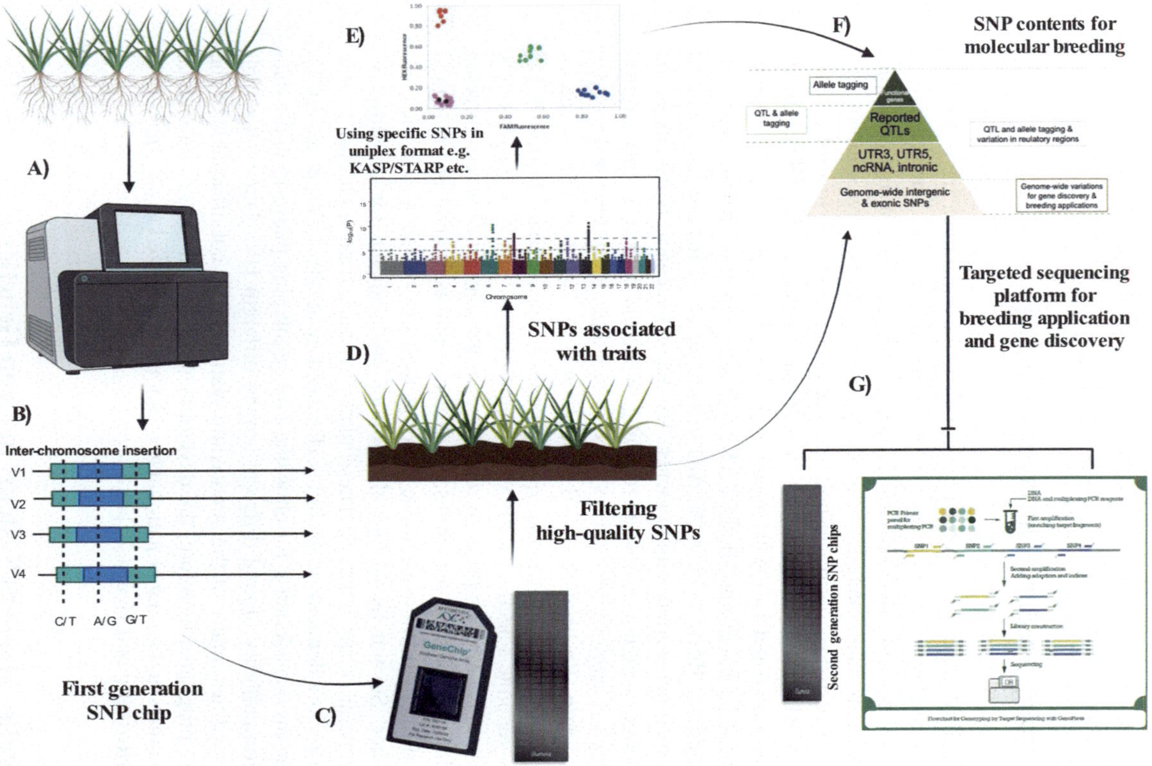

Fig. 4.1 An overview of different SNP genotyping platforms and their evolution over time to facilitate genotyping application in crop molecular breeding. In the initial phase, a small collection of diverse accession are sequenced (DNA or RNA) for collection of genome-wide SNPs (A, B), and then those qualifying specific criteria of being frequent can be fixed on SNP arrays (C), such arrays can be used for gene discovery studies like QTL mapping or genome-wide association studies to identify the SNPs associated with desired traits (D), those SNPs can be converted to KASP or STARP markers for their single-plex use in screening crop varieties for allele identification (E), the high-quality SNPs with uniform distribution in genic and intergenic regions along with QTL and causal gene SNPs can be identified and used to develop breeding-oriented SNP genotyping platforms (F), such platforms can be used either by genotyping-by-targeted

Table 4.1 Features of breeding-oriented genotyping platforms with low- to medium-density data generation and high-throughput turnaround

Platform	Name	Features	Crops	Reference
GT-seq	Genotyping-in-thousands by sequencing	Pooling PCR amplicons with dual barcodes; multiplex PCR	Capsicum	Jo et al. (2021)
MTA-seq	Multiplex PCR targeted amplicon sequencing	Low-density targeted amplicon sequencing	*Brachypodium distachyon*	Onda et al. (2018)
GMS	Genotyping by multiplexed sequencing	PCR-based multiplex sequencing	Wheat	Ruff et al. (2020)
TAS	Targeted amplicon sequencing	Two-step PCR target enrichment	Multiple species	Bybee et al. (2011)
GMAS	Genotyping by multiplexing amplicon sequencing	PCR-based multiplex sequencing	Wheat	Liu (2015)
GBTS	Genotyping-by-targeted sequencing	Genoplexes	Maize	Guo et al. (2019)
TGbyS	Targeted genotyping by sequencing	Oligonucleotide capture probes	Wheat	Burridge et al. (2017)
MRAseq	Multiplex restriction amplicon sequencing	Amplification of amplicons flanked by restriction sites; de novo	Wheat	Bernardo et al. (2019)
GBTS	Genotyping-by-targeted sequencing	Genoplexes	Cucumber	Zhang et al. (2020)
KASP	Kompetitive allele-specific PCR	High-throughput; non-gel-based genotyping; FRET-based plate reading or qPCR	Many crops	LGC, biosearch technology™
TaqMan	TaqMan	Minor groove binder (MGB) technology; high-throughput; non-gel-based genotyping; FRET-based plate reading or qPCR	Many crops	Thermo fisher scien-tific ™
STARP	Semi-thermal asymmetry reverse PCR	High-throughput; non-gel- or gel-based genotyping; FRET-based plate reading or qPCR can also be used	Many crops	Long et al. (2016)
Amplifluor	Ampl	High-throughput; non-gel-based genotyping; FRET-based plate reading or qPCR can also be used; however application is limited	Wheat, maize, Arabidopsis, sugar beet	Jatayev et al. (2017)

(continued)

Table 4.1 (continued)

Platform	Name	Features	Crops	Reference
ASQ	Allele-specific qPCR	High-throughput; non-gel-based genotyping; FRET-based plate reading or qPCR can also be used; can be multiplexed for 3–4 allelic variants or different SNPs	Many crops	Kalendar et al. (2022)

markers, and no automated, high-throughput and digital platform exist for their routine and large-scale use in plant breeding programs. In addition, the low-genome coverage was another disadvantage of gel-based markers, which ultimately paved the way to use SNP markers in variety of research applications. Furthermore, advances in NGS have enabled the use of SNP-based genotyping technologies at lower cost and offered opportunities to deliver high-throughput data to capture millions of variations in genome level. The availability of draft and reference genome sequences of many crop species, their wild relatives, and the availability of pangenome sequence availability have influenced the use of SNP markers for gene discovery and breeding applications (Della Coletta et al. 2021; Khan et al. 2020). The success of SNPs in dominating all other types of molecular markers is due to their abundance at genome-wide level, and biallelic and reproducible nature which make them the most desirable, precise and efficient tool for developing high-density genome scans that are amenable to different genotyping platforms (Gupta et al. 2008; Rasheed et al. 2017).

There are range of methods for detecting SNPs including NGS, SNP arrays, several uniplex methods like KASP and STARP. However, there are two factors to further opt for appropriate genotyping platform: low cost and high-throughput. Among the high-density SNP genotyping platforms, the array technology is cost-effective high-throughput genotyping platform, thus making it accessible to most of the researchers and breeding communities engaged in genetic studies and crop improvement applications. SNP arrays can be used for diversity studies, association mapping, genomic prediction and QTL mapping.

The other widely used method for SNP genotyping is based on NGS and is known as genotyping-by-sequencing (GBS). GBS can provide higher quantities of informative data by orders of magnitude. Although commercial SNP arrays still provide greater marker densities and are easier to analyse, they can be substantially more costly than GBS (Poland et al. 2012; Voss-Fels and Snowdon 2016). We previously reviewed 13 different GBS technologies with distinct features based on the cost, target genome-content and restriction enzyme (Rasheed et al. 2017). Apart from general GBS, there is possibility to capture targeted regions of chromosome by probes by genotyping-by-targeted sequencing (GBTS). The range of markers can be targeted by GBTS from 5 K (GenoPlexs) (Zhang et al. 2020) to 40 K through capture-in-solution (liquid chip) with regular PCR plates (GenoBaits) (Guo et al. 2019; Guo et al. 2021). The sequencing can be done using various currently available sequencing platforms. GBTS combines the advantages of solid chip-based technology (high stability and reliability) and GBS (high flexibility and cost-effectiveness). Its genotyping cost is significantly lower than that of chip-based genotyping when the same set of markers and samples are considered. It generates sharable and accumulative marker data with less bioinformatics support. With the same marker panel (e.g. 20 K maize SNPs), multiple panels with 1 K–20 K SNPs can be generated by sequencing at different depths (Guo et al. 2019).

4.4 Application of NGS for Developing Genotyping Platforms

4.4.1 First- and Second-Generation SNP Chips

SNP arrays have been extensively used in crop breeding and genetics research. The basic advantages of the SNP arrays include i) a range of species multiplex providing rapid high-density genome-wide scans, ii) robust allele-calling with high call rates, and iii) cost-effective per data point when genotyping large populations with large number of SNPs. However, main disadvantages remain the non-flexibility and ascertainment bias. The earlier versions of the SNP arrays were mostly based on the RNA-seq data and usually covered the exonic SNPs, for example, in wheat 9 K array (Cavanagh et al. 2013) and 90 K arrays (Wang et al. 2014). However, later availability of resequencing data in wide range of species diversity enabled to identify genome-wide intergenic SNPs, and the subsequent SNP arrays included relatively higher proportion of non-exonic SNPs like 280 K SNP array in wheat (Rimbert et al. 2018). Recently, a wheat barley 40 K SNP array was developed to genotype when DNA of both species are jointly hybridized (Keeble-Gagnere et al. 2021). In maize, several chip-based genotyping platforms have been established which include 1536 SNPs (Yan et al. 2010), 50 K SNP array (Xu et al. 2017), and 600 K SNP array (Unterseer et al. 2014). SNP arrays have been developed in other crop species including chickpea (Roorkiwal et al. 2018), pigeon pea (Saxena et al. 2018; Singh et al. 2020), barley (Bayer et al. 2017), soybean (Song et al. 2013) and many more. You et al. (2018) reviewed the SNP chip development in polyploidy species and several considerations in designing and analysis of SNP data from arrays in polyploidy species.

The first generation of SNP arrays has the common problem of ascertainment bias. Therefore, the SNP arrays were more carefully designed in the second phase which made them more useful for gene discovery studies and breeding application. For example, Rimbert et al. (2018) used the re-sequencing data for the mining of SNPs from genic and intergenic regions and developed a TaBW280K SNP array in wheat, where half of the SNPs showed a diploid-like clustering. Similarly, several SNP arrays have been developed in rice. A rice 50 K SNP array named RiceSNP50 was developed using illumina platform with 68% SNPs from genic regions (Chen et al. 2014). This SNP array was useful for varietal verification, GWAS, functional genetics studies and breeding applications. Recently, a 7 K SNP array named Cornell-IR LD Rice Array (C7AIR), which is a second-generation SNP array, was developed (Morales et al. 2020). The C7AIR SNP array has 7098 markers and has ability to detect genome-wide polymorphisms within and between subpopulations of *O. sativa*, as well as *O. glaberrima*, *O. rufipogon* and *O. nivara*. The high-quality SNP contents were also taken from the previously available SNP arrays, and some SNPs from the functional genes were also included. Therefore, the C7AIR array is a low-cost genotyping platform for genomic selection and other downstream applications in rice breeding.

The multi-species SNP arrays could be a next epoch for lowering the cost of genotyping when the target SNP number is low for a single species, and multiple species can be genotypes by pooling the DNA. For example, a 50 K multispecies Axiom SNP array was developed to defray the upfront cost of developing a SNP microarray (Grattapaglia et al. 2017). The contents of array were shared among five species including coffee, cashew, cassava, Brazilian pine, and eucalyptus. This SNP array has enabled a number of powerful genetic studies in breeding and conservation and represents the successful demonstration needed to build a larger array to accommodate SNPs for 12 species toward comprehensive large-scale genebanks genotyping. The 40 K wheat-barley SNP array is another example and has been described earlier. Such type of SNP arrays could serve both purposes, that is, research and breeding, because typically the high-density SNP contents are required for research purpose, while the number of SNPs required for breeding applications are typically low.

4.4.2 Sequencing-Based Second Generation of Crop Genotyping Platforms

The NGS-based genotyping platforms are exceptional tool for cost-effective genotyping, especially in crops where SNP arrays are not available or when available SNP arrays have high ascertainment bias. Another advantage of NGS-based genotyping platform is that sequencing and genotyping can be done simultaneously. The advantages and disadvantages of all the NGS-based genotyping platforms are described in detail (Rasheed et al. 2017; Scheben et al. 2016). Generally, the NGS-based platforms are classified into three formats. Whole genome sequencing (WGS) is a complete solution for identifying all sequence variability, but it is still very expensive for genotyping large breeding populations despite considerable decrease in sequencing costs over the past decade. The alternate strategy to the WGS is the reduced representation sequencing or GBS or exome capture platforms. The use of appropriate restriction enzymes is important to generate DNA markers to cover selected genomic regions. However, large-scale SNP discovery by this strategy needs a high-quality reference genome sequence and genotyping pipelines, and strong informatics support, to impute marker genotypes for some samples and loci (Glaubitz et al. 2014).

The third sequencing-based strategy is based on the capturing and targeted sequencing of predesigned probes of the loci. The strategy has been tested in plants and animals and is usually referred as genotyping by target sequencing (GBTS) or similar names (Guo et al. 2019). This was first used in humans to enrich a target of ~3.9 Mb and the SNP calling accuracy was 99% (Tewhey et al. 2009). The strategy was found highly effective to capture target regions, thus reducing sequencing costs and saving time (Mamanova et al. 2010). Burridge et al. (2018) used target enrichment strategy to genotype 3256 probes in wheat and found it highly reliable platform compared with Affymetrix SNP array and KASP markers. Another strategy called multiplex restriction amplicon sequencing (MRAseq) was used to generate thousands of SNPs in wheat and barley and platform was could be a promising tool for QTL mapping and other breeding applications (Bernardo et al. 2020).

Recently, GBTS was tested with a GenoBait technology to genotype 20 K SNPs in maize (Guo et al. 2019). Two germplasm panel consisting of 96 and 387 maize inbred lines were used to test and validate the markers. It was further tested that 20 K marker panel can be reduced to 10 K, 5 K, and 1 K SNP markers by sequencing the samples at the average sequencing depths of 20×, 7.5×, and 2.5×, respectively. The genotyping cost of GBTS was significantly lower than SNP arrays, when the same set of markers and samples are considered (Guo et al. 2019; Zhang et al. 2020). The other advantages of GBTS include the sharable data among different labs and less bioinformatics support needed for SNP calling. However, there is still need to improve the platform for cost-effectiveness and make it more breeding-oriented. Guo et al. (2021) further improved the maize GBTS from 20 K to 40 K through optimized procedures. Then a new protocol was developed to identify more than six SNPs from each individual amplicon and were termed as 'multiple single-nucleotide polymorphisms (mSNPs)'. Ultimately, the SNP number increased up to 251 K from the same set of designed SNP assays for the same cost, further reducing the cost of genotyping. Furthermore, after evaluation of the marker system and genotyping platform, a comparative analysis of three marker panels (40 K mSNPs, 251 K SNPs, and 159 K haplotypes with minor allele frequencies greater than 5%) was performed to evaluate their power for DNA variation detection and genome-wide association study (GWAS). This improved GBTS system has great potential for development and implementation in all organisms, including plants, animals, and microorganisms.

Another approach genotyping-in-thousands sequencing (GT-seq) is widely used in animals, fish, insects and recently been used on *Capsicum* spp. (Jo et al. 2021). GT-seq is based on pooling PCR amplicons with dual barcodes generated by multiplex PCR against multiple target loci and sequencing the resulting library in a single lane. This approach makes targeted sequencing cost-effective and flexible method to genotype multiple loci of interest by overcoming the limitation of sample size through dual barcoding (Campbell et al. 2015). GT-seq has been found to be very promising approach for MABC due to its cost-effectiveness. Conclusively, targeted sequencing platforms could facilitate medium-density semi-flexible approach in cost-effective manner which is applicable to variety of crop breeding applications.

4.4.3 Flexible Genotyping Systems for Gene Tagging

The high-density genotyping platforms like SNP arrays and GBS are useful for mapping and gene discovery and to identify the trait-associated markers, which are then used for gene tagging and gene pyramiding during crop breeding. The use of one or few markers is important for germplasm screening, and flexibility is extremely important for such markers as it gives choice in using any number of markers and any number of samples. The electrophoresis-based molecular markers are time-consuming and labour-intensive to use in breeding programs. In the last five years, various gel-free single-marker methods have been developed for SNP genotyping. These platforms include, but not limited to, KASP (Semagn et al. 2014), STARP (Long et al. 2016), TaqMan, Amplifluor and Amplifluor-like markers (Jatayev et al. 2017), and Allele-Specific qPCR (ASQ) (Baidyussen et al. 2021). The important factors in using such platforms remain high-throughput (the number of data points that can be generated in a short time period), ease of use, data quality (sensitivity, reliability, reproducibility, and accuracy), flexibility (genotyping few samples with many SNPs or many samples with few SNPs), assay development requirements, and genotyping cost per sample or data point. Among these platforms, KASP are amenable to different robotics or automated platforms like SNPLine™ and Nexar™, which increases sample throughput from several hundred per day to 153 K data points per day. The molecular markers important for diagnostics have been converted to KASP format in wheat (Rasheed et al. 2016) and rice (Pariasca-Tanaka et al. 2015).

KASP has been a leading uniplex SNP/InDel genotyping platform due to its ease of use, high-throughput and accessible technology. However, higher cost remained a major factor for KASP technology, and alternate genotyping method were kept emerging. Long et al. (2016) introduced a new SNP genotyping method known as STARP, and successfully validated it in rice, sunflower and *Aegilops tauschii*. STARP used specific PCR conditions to two universal priming element-adjustable primers (PEA primers) and one group of three locus-specific primers: two asymmetrically modified allele-specific primers (AMAS primers) and their common reverse primer. The two AMAS primers each were substituted one base in different positions at their 3′ regions to significantly increase the amplification specificity of the two alleles and tailed at 5′ ends to provide priming sites for PEA primers. The two PEA primers were developed for common use in all genotyping assays to stringently target the PCR fragments generated by the two AMAS primers with similar PCR efficiencies and for flexible detection using either gel-free fluorescence signals or gel-based size separation. The state-of-the-art primer design and unique PCR conditions endowed STARP with all the major advantages of high accuracy, flexible throughput, simple assay design, low operational costs, and platform compatibility.

In addition to SNPs, STARP can also be employed in genotyping of InDels. Contrary to the commercial KASP and TaqMan assays, which use the specific commercial master-mix, the STARP can be used with any commercial PCR master-mix. Recently, fluorescence resonance energy transfer (FRET)-based SNP genotyping method ASQ (allele-specific qPCR) was proposed which proved to be highly accurate for SNP genotyping and 2x to 10x cheaper than KASP markers. The ASQ was suitable not only for bi-allelic reactions but also for three- or four-allelic variants in a multiplex format in a range of applications.

4.5 Conclusion and Prospects

Development of appropriate genotyping platform which can fulfil most of the crop breeding requirement is now at center of efforts. Although a single platform cannot offer all the solutions, however efforts can be made to make them more breeding-oriented. Technological advancements have impacted all three types of high-, medium- and low-density genotyping platforms. SNP arrays and GBS remained the best strategy for high-density genotyping in crop genetics and breeding. However, there is significant progress in medium-density genotyping platforms and several types of 'targeted sequencing' platforms have been emerged in crop species. These platforms fulfil the needs of genetic mapping, germplasm characterization and gene tagging. Such genotyping-by-targeted sequencing (GBTS) approaches seems more appropriate because they are scalable, cost-effective and reliable. Similarly, there is huge technological surge in single-marker SNP genotyping platforms, and several alternates to KASP assays are available like STARP, Amplifluor and ASQ. However, significant effort is needed to convert the diagnostics or functional markers into these formats.

References

Bayer MM, Rapazote-Flores P, Ganal M, Hedley PE, Macaulay M, Plieske J, Ramsay L, Russell J, Shaw PD, Thomas W, Waugh R (2017) Development and evaluation of a barley 50k iSelect SNP Array. Frontiers. Plant Sci 8

Bernardo A, St. Amand P, Le HQ SZ, Bai G (2020) Multiplex restriction amplicon sequencing: a novel next-generation sequencing-based marker platform for high-throughput genotyping. Plant Biotechnol J 18:254–265

Bernardo R, Yu J (2007) Prospects for Genomewide selection for quantitative traits in maize. Crop Sci 47:1082–1090

Bevan MW, Uauy C, Wulff BB, Zhou J, Krasileva K, Clark MD (2017) Genomic innovation for crop improvement. Nature 543:346–354

Burridge AJ, Wilkinson PA, Winfield MO, Barker GLA, Allen AM, Coghill JA, Waterfall C, Edwards KJ (2018) Conversion of array-based single nucleotide polymorphic markers for use in targeted genotyping by sequencing in hexaploid wheat (Triticum aestivum). Plant Biotechnol J 16:867–876

Campbell NR, Harmon SA, Narum SR (2015) Genotyping-in-thousands by sequencing (GT-seq): a cost effective SNP genotyping method based on custom amplicon sequencing. Mol Ecol Resour 15:855–867

Cavanagh CR, Chao SM, Wang SC, Huang BE, Stephen S, Kiani S, Forrest K, Saintenac C, Brown-Guedira GL, Akhunova A, See D, Bai GH, Pumphrey M, Tomar L, Wong DB, Kong S, Reynolds M, da Silva ML, Bockelman H, Talbert L, Anderson JA, Dreisigacker S, Baenziger S, Carter A, Korzun V, Morrell PL, Dubcovsky J, Morell MK, Sorrells ME, Hayden MJ, Akhunov E (2013) Genome-wide comparative diversity uncovers multiple targets of selection for improvement in hexaploid wheat landraces and cultivars. Proc Natl Acad Sci U S A 110:8057–8062

Chen H, Xie W, He H, Yu H, Chen W, Li J, Yu R, Yao Y, Zhang W, He Y, Tang X, Zhou F, Deng XW, Zhang Q (2014) A high-density SNP genotyping array for rice biology and molecular breeding. Mol Plant 7:541–553

Della Coletta R, Qiu Y, Ou S, Hufford MB, Hirsch CN (2021) How the pan-genome is changing crop genomics and improvement. Genome Biol 22:3

Glaubitz JC, Casstevens TM, Lu F, Harriman J, Elshire RJ, Sun Q, Buckler ES (2014) TASSEL-GBS: a high capacity genotyping by Sequencing analysis pipeline. PLoS One 9:e90346

Guo Z, Wang H, Tao J, Ren Y, Xu C, Wu K, Zou C, Zhang J, Xu Y (2019) Development of multiple SNP marker panels affordable to breeders through genotyping by target sequencing (GBTS) in maize. Mol Breed 39:37

Guo Z, Yang Q, Huang F, Zheng H, Sang Z, Xu Y, Zhang C, Wu K, Tao J, Prasanna BM, Olsen MS, Wang Y, Zhang J, Xu Y (2021) Development of high-resolution multiple-SNP arrays for genetic analyses and molecular breeding through genotyping by target sequencing and liquid chip. Plant Communications 2:100230

Gupta PK, Rustgi S, Mir RR (2008) Array-based high-throughput DNA markers for crop improvement. Heredity (Edinb) 101:5–18

Jatayev S, Kurishbayev A, Zotova L, Khasanova G, Serikbay D, Zhubatkanov A, Botayeva M, Zhumalin A, Turbekova A, Soole K, Langridge P, Shavrukov Y (2017) Advantages of Amplifluor-like SNP markers over KASP in plant genotyping. BMC Plant Biol 17:254

Jo J, Kim Y, Kim GW, Kwon J-K, Kang B-C (2021) Development of a panel of genotyping-in-thousands by Sequencing in capsicum. Frontiers. Plant Sci 12

Khan AW, Garg V, Roorkiwal M, Golicz AA, Edwards D, Varshney RK (2020) Super-Pangenome by integrating the wild side of a species for accelerated crop improvement. Trends Plant Sci 25: 148–158

Kim C, Guo H, Kong W, Chandnani R, Shuang LS, Paterson AH (2016) Application of genotyping by sequencing technology to a variety of crop breeding programs. Plant Sci 242:14–22

Li H, Rasheed A, Hickey L, He Z (2018) Fast-forwarding genetic gain. Trends Plant Sci 23:184–186

Liu YN, He ZH, Appels R, Xia XC (2012) Functional markers in wheat: current status and future prospects. Theor Appl Genet 125:1–10

Long YM, Chao WS, Ma GJ, Xu SS, Qi LL (2016) An innovative SNP genotyping method adapting to multiple platforms and throughputs. Theor Appl Genet 130:597–607

Lopes MS, Reynolds MP, Manes Y, Singh RP, Crossa J, Braun HJ (2012) Genetic yield gains and changes in associated traits of CIMMYT spring bread wheat in a "historic" set representing 30 years of breeding. Crop Sci 52:1123–1131

Mamanova L, Coffey AJ, Scott CE, Kozarewa I, Turner EH, Kumar A, Howard E, Shendure J, Turner DJ (2010) Target-enrichment strategies for next-generation sequencing. Nat Methods 7: 111–118

Morales KY, Singh N, Perez FA, Ignacio JC, Thapa R, Arbelaez JD, Tabien RE, Famoso A, Wang DR, Septiningsih EM, Shi Y, Kretzschmar T, McCouch SR, Thomson MJ (2020) An improved 7K SNP array, the C7AIR, provides a wealth of validated SNP markers for rice breeding and genetics studies. PLoS One 15:e0232479

Pariasca-Tanaka J, Lorieux M, He C, McCouch S, Thomson MJ, Wissuwa M (2015) Development of a SNP genotyping panel for detecting polymorphisms in Oryza glaberrima/O. sativa interspecific crosses. Euphytica 201:67–78

Poland JA, Brown PJ, Sorrells ME, Jannink JL (2012) Development of high-density genetic maps for barley and wheat using a novel two-enzyme genotyping-by-sequencing approach. PLoS One 7:e32253

Rasheed A, Hao Y, Xia XC, Khan A, Xu Y, Varshney RK, He ZH (2017) Crop breeding chips and genotyping platforms: progress, challenges and perspectives. Mol Plant 10:1047–1064

Rasheed A, Xia X (2019) From markers to genome-based breeding in wheat. Theor Appl Genet 132:767–784

Rimbert H, Darrier B, Navarro J, Kitt J, Choulet F, Leveugle M, Duarte J, Rivière N, Eversole K, on behalf of The International Wheat Genome Sequencing C, Le Gouis J, on behalf The Breed Wheat C, Davassi A, Balfourier F, Le Paslier M-C, Berard A, Brunel D, Feuillet C, Poncet C, Sourdille P, Paux E (2018) High throughput SNP discovery and genotyping in hexaploid wheat. PLoS One 13:e0186329

Roorkiwal M, Jain A, Kale SM, Doddamani D, Chitikineni A, Thudi M, Varshney RK (2018) Development and evaluation of high-density axiom®CicerSNP Array for high-resolution genetic mapping and breeding applications in chickpea. Plant Biotechnol J 16:890–901

Rossetto M, Henry RJ (2014) Escape from the laboratory: new horizons for plant genetics. Trends Plant Sci 19:554–555

Saxena RK, Rathore A, Bohra A, Yadav P, Das RR, Khan AW, Singh VK, Chitikineni A, Singh IP, Kumar CVS, Saxena KB, Varshney RK (2018) Development and application of high-density axiom Cajanus SNP Array with 56K SNPs to understand the genome architecture of released cultivars and founder genotypes. The Plant Genome 11:180005

Scheben A, Batley J, Edwards D (2016) Genotyping-by-sequencing approaches to characterize crop genomes: choosing the right tool for the right application. Plant Biotechnol J 15:149–161

Semagn K, Babu R, Hearne S, Olsen M (2014) Single nucleotide polymorphism genotyping using Kompetitive allele specific PCR (KASP): overview of the technology and its application in crop improvement. Mol Breed 33:1–14

Singh S, Mahato AK, Jayaswal PK, Singh N, Dheer M, Goel P, Raje RS, Yasin JK, Sreevathsa R, Rai V, Gaikwad K, Singh NK (2020) A 62K genic-SNP chip array for genetic studies and breeding applications in pigeonpea (Cajanus cajan L. Millsp). Sci Rep 10:4960

Song Q, Hyten DL, Jia G, Quigley CV, Fickus EW, Nelson RL, Cregan PB (2013) Development and evaluation of SoySNP50K, a high-density genotyping array for soybean. PLoS One 8: e54985

Tanksley SD, Young ND, Paterson AH, Bonierbale MW (1989) RFLP mapping in plant breeding: new tools for an old science. Nat Biotechnol 7:257–264

Tester M, Langridge P (2010) Breeding technologies to increase crop production in a changing world. Science 327:818–822

Tewhey R, Nakano M, Wang X, Pabón-Peña C, Novak B, Giuffre A, Lin E, Happe S, Roberts DN, LeProust EM, Topol EJ, Harismendy O, Frazer KA (2009) Enrichment of sequencing targets from the human genome by solution hybridization. Genome Biol 10:R116

Unterseer S, Bauer E, Haberer G, Seidel M, Knaak C, Ouzunova M, Meitinger T, Strom TM, Fries R, Pausch H, Bertani C, Davassi A, Mayer KF, Schon CC (2014) A powerful tool for genome analysis in maize: development and evaluation of the high density 600 k SNP genotyping array. BMC Genomics 15:823

Varshney RK, Graner A, Sorrells ME (2005a) Genic microsatellite markers in plants: features and applications. Trends Biotechnol 23:48–55

Varshney RK, Graner A, Sorrells ME (2005b) Genomics-assisted breeding for crop improvement. Trends Plant Sci 10:621–630

Voss-Fels K, Snowdon RJ (2016) Understanding and utilizing crop genome diversity via high-resolution genotyping. Plant Biotechnol J 14:1086–1094

Wang SC, Wong DB, Forrest K, Allen A, Chao SM, Huang BE, Maccaferri M, Salvi S, Milner SG, Cattivelli L, Mastrangelo AM, Whan A, Stephen S, Barker G, Wieseke R, Plieske J, Lillemo M, Mather D, Appels R, Dolferus R, Brown-Guedira G, Korol A, Akhunova AR, Feuillet C, Salse J, Morgante M, Pozniak C, Luo MC, Dvorak J, Morell M, Dubcovsky J, Ganal M, Tuberosa R, Lawley C, Mikoulitch I, Cavanagh C, Edwards KJ, Hayden M, Akhunov E, Sequencing IWG (2014) Characterization of polyploid wheat genomic diversity using a high-density 90 000 single nucleotide polymorphism array. Plant Biotechnol J 12:787–796

Xu C, Ren Y, Jian Y, Guo Z, Zhang Y, Xie C, Fu J, Wang H, Wang G, Xu Y, Li P, Zou C (2017) Development of a maize 55 K SNP array with improved genome coverage for molecular breeding. Mol Breed 37:20

Yan J, Yang X, Shah T, Sánchez-Villeda H, Li J, Warburton M, Zhou Y, Crouch JH, Xu Y (2010) High-throughput SNP genotyping with the GoldenGate assay in maize. Mol Breed 25:441–451

You Q, Yang X, Peng Z, Xu L, Wang J (2018) Development and applications of a high throughput genotyping tool for Polyploid crops: single nucleotide polymorphism (SNP) Array. Frontiers. Plant Sci 9

Yu D, Song L, Gu W, Guan Y, Wang H, Shi B, Zhou Z, Zheng H, Jiang Y, Yao Y (2021) Genome-wide comparative analysis of genetic diversity of regular and specialty maize inbred lines through genotyping by target Sequencing (GBTS). Plant Mol Biol Report 40:221

Zhang J, Yang J, Zhang L, Luo J, Zhao H, Zhang J, Wen C (2020) A new SNP genotyping technology target SNP-seq and its application in genetic analysis of cucumber varieties. Sci Rep 10:5623

Chapter 5
Rapid Generation Advancement for Accelerated Plant Improvement

Aladdin Hamwieh, Naglaa Abdallah, Shiv Kumar, Michael Baum, Nourhan Fouad, Tawffiq Istanbuli, Sawsan Tawkaz, Tapan Kumar, Khaled Radwan, Fouad Maalouf, and Rajeev K. Varshney

Abstract In 2020, more than 800 million people suffered from hunger, and this number will continue to rise as the world's population increases, in addition to heightening the consequences of climate change and the probability of increasing the risk of wars. We cannot continue to use the conventional breeding techniques employed 50 years ago, which require 7–10 years to develop a high-yielding and stable variety. Several technologies, including shuttle breeding, off-season planting, tissue culture (embryo rescue), doubled haploid (DH), marker-assisted selection (MAS), high-throughput genotyping, genomic selection (GS), plant transformation, speed breeding, and genome editing, have been developed for rapid generation advancement (RGA). Utilizing these technologies can expedite the development of climate-resilient plant varieties with enhanced yield and resilience to biotic and abiotic challenges. This chapter goes deep into these technologies and approaches that have emerged in the last 10 years and could be used to accelerate crop improvement.

A. Hamwieh · N. Fouad
International Center for Agricultural Research in the Dry Areas (ICARDA), Cairo, Egypt

N. Abdallah
Faculty of Agriculture, Cairo University, Cairo, Egypt

S. Kumar · M. Baum
International Center for Agricultural Research in the Dry Areas (ICARDA), Rabat, Morocco

T. Istanbuli · S. Tawkaz · F. Maalouf
International Center for Agricultural Research in the Dry Areas (ICARDA), Terbol, Lebanon

T. Kumar
International Center for Agricultural Research in the Dry Areas (ICARDA), New Delhi, India

K. Radwan
Agricultural Genetic Engineer Research Institute (AGERI), Agricultural Research Center (ARC), Giza, Egypt

R. K. Varshney (✉)
Murdoch's Centre for Crop and Food Innovation, State Agricultural Biotechnology Centre, Food Futures Institute, Murdoch University, Murdoch, WA, Australia
e-mail: rajeev.varshney@murdoch.edu.au

M. K. Pandey et al. (eds.), *Frontier Technologies for Crop Improvement*, Sustainable Agriculture and Food Security, https://doi.org/10.1007/978-981-99-4673-0_5

Keywords Speed breeding · Shuttle breeding · Doubled haploid · Genomic selection · Genome editing · Rapid generation advancement · Marker-assisted selection

5.1 Introduction

Hunger is currently rising globally, and the proportion of starvation worldwide has been increased to 11% in 2016. According to the Food and Agricultural Organization (FAO), the number of hungry people in the world reached about 821 million in 2017 (FAO 2018). The Sustainable Development Goal (SDG) is aiming to reach hunger eradication by 2030, food scarcity is on rise, posing serious threat to global food security (FAO 2018). Furthermore, the climate change leading to floods, droughts, carbon dioxide increase, 2 °C temperature rise in the climate, increasing of the saline soil, pest, and diseases outbreaks, and more, are additional challenges to the plant breeders to develop tolerant/resistant varieties (Van Oort and Zwart 2018).

Agriculture is the mainstay for major food supply to the rising population which is expected to be 10 billion by 2057 (www.theworldcounts.com), that is, 25% population increase over the next 30 years (Hickey et al. 2019). This will cause a 'perfect storm' of food, energy, and water shortages as the demand for food will jump by 60%. In the past, conventional breeding methods have been successful in developing high-yielding varieties, adoption of which along with good agronomic practices has led to the green revolution in the 1960s. Breeding cycles are lengthy with 7–10 years to produce genetically pure varieties. Reduction in the time taken in breeding a variety can have a profound effect on annual genetic gain. For rapid generation advancement (RGA), several technologies have been developed such as shuttle breeding, off-season planting, tissue culture (embryo rescue), doubled haploid (DH), marker-assisted selection (MAS), high-throughput genotyping, genomic selection (GS), plant transformation, speed breeding and genome editing. Using such tools will enable the rapid development of climate-resilient varieties with improved yield and resistance to biotic and abiotic stresses.

The early rapid generation advancement method was to produce more than one generation in different locations by off-season planting. It was possible to apply such shuttle breeding for some crops to develop two generations per year. This was efficient enough to reduce 50% of the breeding time to release a new variety. The DH has also been considered as one of the most promising technologies for RGA and obtaining homozygous lines in one year. Marker-assisted selection (MAS) further holds great promise because it increases the accuracy of the selection, helps the breeders to focus on fewer promising plants, resulting in a higher efficiency and accuracy of selection (Collard and Mackill 2008; Wessel and Botes 2014; Humphreys and Knox 2015; Wiśniewska et al. 2019). The use of MAS can also reduce the selection time for the breeder, especially when used for backcrossing approach. The latter is called as marker-assisted backcrossing method and was used by Indian breeders to develop two drought-tolerant varieties, and one *Fusarium* wilt-resistant variety (Mannur et al. 2019) in chickpea. The development of high-throughput genotyping and sequencing technologies has contributed to the speeding up of the genetic gain across the target environments. The efficiency of association mapping, high-density genetic mapping, and quantitative traits loci (QTL) have been significantly increased to identify the right markers for selection. These new tools will help breeders and geneticists to have a better understanding of quantitative traits which are controlled by several major and minor genes. However, although MAS can reduce the breeding time needed for variety release, it is more complicated to use quantitative traits controlled by many genes with small effects (Desta and Ortiz 2014).

Genomic selection (GS) has become an established methodology in breeding (Meuwissen et al. 2001). Potentially it can accelerate breeding progress in a very cost-effective process because of early selection before phenotypes are measured. Even in some cases like in chickpea, DNA could be extracted from the population's seed coats in summer, which enables for early selection before the planting in winter. The advantage of GS is that it is not based on QTL conferring traits as in MAS. The concept is based on the use of molecular markers and the phenotypic data from a certain population (called training population) to predict the phenotypic values in the breeding population without field screening and enhancing the RGA.

Doubled haploid technology as an accelerated approach to crop improvement has become routine for some crops at many research centres leading to the release of about 300 varieties (Sood and Dwivedi 2015). Recent publications have utilized haploid induction editing technology (HI-Edit) to edit inbred maize lines (Kelliher et al. 2019). The maize pollen parent, transformed with Cas9, was used to pollinate wheat spikes, resulting in edited wheat embryos (Kelliher et al. 2019).

Novel sensor modules have been developed to improve the high-throughput phenotyping such as imaging cameras, drones, light-emitting diodes (LED), lightings and portable devices, for screening a wide range of traits (Fiorani and Schurr 2013). High-throughput phenotyping technologies are leading to large populations screening in shorter times than the conventional screening with better accuracy. Among these high-throughput phenotyping platforms utilized for improving genetic gain are the unmanned aerial vehicle for assessing grain yield in wheat (Hu et al. 2020), mobile field vehicle for flowering time (Wang et al. 2019), ArduCrop wireless IR thermometer and airborne thermography using manned helicopter for measuring the canopy temperature (Deery et al. 2019). LED light innovation was applied on plant growth in 1990, and this came after several attempts to grow the plants under artificial light period aiming to grow the plants outside the season. Speed breeding has been applied on wheat, barley, chickpea, pea, and canola where the plants were exposed to photoperiod 22 hours/day by using LEDs which reduce the operational cost of lighting and cooling. The results indicated a great potential for integrating speed breeding with other novel plant breeding technologies to accelerate the crop improvement process (Watson et al. 2018) and can shorten the life cycle of the crops up to 50% (Samineni et al. 2020).

5.2 Shuttle Breeding

The shuttle breeding methodology helps crop breeders to advance segregating populations at different locations (latitude, altitude, and rainfall) within the year, and this shortens the breeding cycle. Off-season has been used to advance the generation of breeding populations of crops and vegetables for a long time. It is very useful to speed up the breeding process, but it has a limit on the number of generations per year which are generally two, and in rare cases, three generations. However, it requires finding an appropriate location suitable for the off-season within the breeders' reach. Another difficulty facing off-season is shipping the seeds from one location to another through the political borders. This is a time-consuming process until the seeds are cleared from customs.

CIMMYT shuttle-breeding program was initiated when Norman Borlaug began growing two crops of wheat a year in contrasting growing conditions in the early 1960s. ICARDA has applied off-season planting by using two locations (Tal Hadya, North Aleppo, Syria, and Terbol station, Terbol, Lebanon) for cereal and legumes. Shuttle breeding in wheat was very effective in selecting photoperiod-insensitive genes *(Ppd1* and *Ppd2)* and dwarfing genes (*Rht1* and *Rht2)*. The results of shuttle breeding produced widely adapted, high-yielding, lodging-tolerant, and input-responsive semidwarf wheat varieties. For spring bread wheat at ICARDA, the modified pedigree-bulk selection method is applied when advancing F_2 to F_6 generations where about 2000 F_2 plants are selected individually. The shuttle breeding involves two cycles (winter-summer) at Terbol station (33°,48′34″, 35°59′24, elev. 894 m) in Lebanon, one cycle (winter) at Marchouch station (Morocco) and one cycle (early winter) at Sids station (28°54′21″, 30°57′01″ elev 32 m) in Egypt, and one cycle (summer) at Kulumsa station (8°01′09″, 39°09′12″ elev. 2175 m) in Ethiopia (Fig. 5.1). This method shortens the breeding cycle by half as two generations can be grown per year (Tadesse et al. 2016).

5.3 Doubled Haploid

Doubled haploid (DH) is a rapid generation technology to reduce the time of obtaining homozygous genotypes by using chemicals (such as colchicine) to double the number of the chromosomes from the haploid cells. This technology has been successfully developed and applied in several crops such as barley, rice, wheat and maize to shorten the time of releasing new varieties within 3–4 years (Brennan and Martin 2007). However, DH was not applied in crops like legumes because of the nature of embryo growth which requires complicated tissue culture protocols. Another disadvantage of DH is that some species do not respond to it.

DH has been applied in wheat by using two technologies, intergeneric hybridization by using maize pollen where chromosome elimination takes place after 2–3 days of hybridization. Chromosome elimination following fertilization was

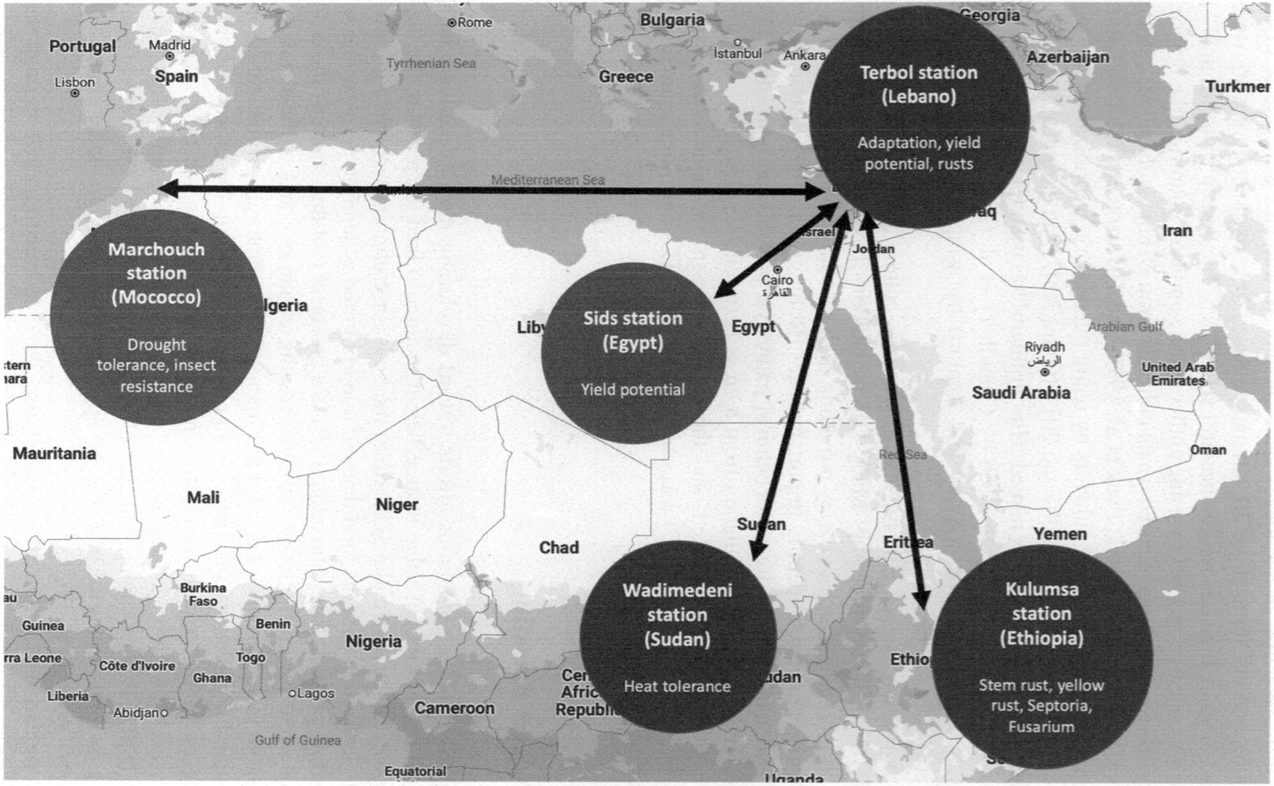

Fig. 5.1 Shuttle breeding sites of ICARDA’s wheat breeding program in West-Asia North and sub-Saharan Africa; A*rrow* directions indicate the direction of shipment of subsequent segregating populations and fixed germplasms

proposed based on the influence of the genome balance in reciprocal crosses between the diploid and autotetraploid levels. Another technology of DH is the culture of anthers or microspores. In Sudan, two heat-tolerant DH varieties were released (Ali et al. 2006), and in China, about 20 varieties released from tissue culture work were cultivated on more than one million hectares (Hu 1997). The first cultivar (Jinghua 1) in China was released in 1983 by using anther culture and later countries like France, Hungary and Sweden have released wheat cultivars from DH by using anther culture methods (De Buyser et al. 1987; Pauk et al. 1995; and Tuvesson et al. 2003). Later, Romania produced four wheat cultivars by using intergeneric hybridization following maize pollination between 2005 and 2011 (Giura 2007; Săulescu et al. 2012). In India, the first DH wheat variety named 'Him Pratham' has been released by utilizing the chromosome elimination approach by Chaudhary et al. (2015). In 2018, USA released two wheat varieties (Avery and Langin) by using maize hybridization (Haley et al. 2018 a, b).

The bulbosum, anther culture, and microspore culture have been applied to produce DH in barley breeding. Many DH populations have been developed for molecular marker research in barley providing high-resolution maps (chromosome-based) which enabled to identify important genes (Rollins et al. 2013). These genes can be associated with various biochemical pathways, proteins, and enzymes. For example, The S42IL DH population from a cross between *H. vulgare* ssp. *vulgare* German malting variety 'Scarlett' and the *H. vulgare* ssp. *spontaneum* accession 'ISR42–8' (Honsdorf et al. 2017) was used to study the QTL controlling grain filling under terminal drought stress.

5.4 Speed Breeding

Plant breeders make continuous efforts toward developing modern cultivars with high yield and nutritional value in an efficient manner to feed the growing population. However, the conventional breeding methods for crop improvement are insufficient to keep pace with the food demand due to limited generation per year during the breeding process. Speed breeding (SB) is a powerful tool that shortens the crop breeding cycle by manipulating various growing conditions and artificial intelligence tools, including photoperiod, light intensity, temperature, soil moisture, soil nutrition and high-density planting. SB significantly speeds up the process of crop breeding by inducing early flowering and seed set, reducing the time taken to generate each breeding generation. For the plant breeding program, single seed descent (SSD) is time-saving and cost-efficient as compared to the conventional breeding methods and is the best way for applying speed breeding under controlled conditions (Wang et al. 2021).

LED lights have been used for accelerating the crop breeding cycle through the manipulation of light (hours/day). The LED lights were first used for plant growth in 1990 (Fasol and Nakamura 1997; Bula et al. 1991). Interestingly, the National Aeronautics and Space Administration (NASA), USA, in association with Utah

State University utilized the LED technology to grow wheat in the space station; they found that LED reduced the life cycle of plants drastically (Bugbee and Koerner 1997). However, this technology has been refined for application in plant breeding, paralleled with cost reduction (where 40–60% of energy being converted into light rather than heat in the old light systems), and therefore, it is a cooler light system with low-energy consumption (Stutte 2015). All these features have made this technology an optimal system for use in crop breeding. In addition to photoperiod, two other factors are important to secure efficient speed breeding system: one is an air conditioner to control the temperature and the second is the humidity which should be optimized at 70–75%. The time of photoperiod should be 22-h/day and the temperature should be between 22–25 °C, the remaining 2 h of dark, temperature could be reduced to 17 °C (Watson et al. 2018).

Speed breeding techniques have been applied in various crops including the most strategic cereal and legume crops, allowing four to eight breeding generations per year (Table 5.1). In addition, SB has also been applied extensively for phenotyping of agronomic traits in wheat by using different selection methods (Li et al. 2018b) and for multi-trait phenotyping of quantitative traits in barley (Hickey et al. 2017). Speed breeding has also been applied for phenotyping of root traits, disease reaction and plant height in durum wheat (Alahmad et al. 2018). This approach has helped breeders to identify and combine disease resistance in a short time. For the screening of disease resistance in wheat, speed breeding has been applied for stripe rust disease (Hickey et al. 2011), leaf rust (Riaz et al. 2016) and *Fusarium* head blight (Watson et al. 2018), and in Australia, the first spring wheat variety released by using speed breeding method was in 2017 (Hickey et al. 2019). Speed breeding facilities have been applied under hydroponics system to evaluate the boron resistance in pea (Bennett et al. 2017), *Fusarium* wilt in chickpea, and salt tolerance in chickpea and wheat (Hamwieh et al., unpublished data). The advantage of screening for

Table 5.1 List of improved cereal and legume crops applied by speed breeding

Category	Crop	Field (generations/ per year)	Speed breeding generations/per year	Reference
Cereals	Wheat	1–2	4–6	(Yao et al. 2017)
	Barley	1–2	~6	(Hickey et al. 2017)
	Rice	1–2	4	(Rana et al. 2019)
Legumes	Chickpea	1–2	~6	(Watson et al. 2018; Gaur et al. 2019)
	Faba bean	1–2	7	(Mobini et al. 2015)
	Pea	1–2	6–7	(Ribalta et al. 2017; Samantara et al. 2022)
	Lentil	1–2	~8	(Idrissi 2020)

different traits under speed breeding facilities has the potential to improve the genetic gain in strategic crops (Li et al. 2018b).

Speed breeding is the next logical step to aid the wide-scale screening of the numerous plant genotypes to keep pace with the advances in plant science research. The use of speed breeding coupled with modern-day breeding and genetic engineering techniques will be useful for making tailor-made crops having high nutritional value, high yield, and tolerance to abiotic and biotic stresses. Speed breeding protocols have been developed for many pulse crops including lentil, chickpea, faba bean, grass pea, field pea, and pigeon pea to obtain four to six generations per year (Ghosh et al. 2018; Lulsdorf and Banniza 2018; Samineni et al. 2019; Saxena et al. 2019). In brief, speed breeding protocol is capable of improving flower induction by 25–28% of the field time in lentil (Mobini et al. 2016), The SB results have enhanced the per year generations three to four times (Idrissi 2020). In faba bean, the SB was able to reduce flower initiation time and increase grain-setting rates. In consequence, it produces up to seven generations per year (Mobini et al. 2015). In a recent study on pea, up to six generations per year have been optimized (Ribalta et al. 2017; Mobini and Warkentin 2016; Samantara et al. 2022).

Speed breeding technology has been applied in chickpea in India for advancing three generations per year under field and greenhouse conditions (Gaur et al. 2007). Later, the number of generations has increased to four generations per year in chickpea under modified glasshouses by using sodium vapor lamps (Watson et al. 2018), but this method is highly energy-consuming. The most cost-effective method in chickpea was applied in 2019, by using long photoperiod via LED light leading to the potential of harvesting five generations per year (Samineni et al. 2019). The team of Samineni has proved over 2 years experiment that by using six cultivated chickpea (early and late maturity) genotypes, it is possible to harvest seven generations when using the early maturity genotypes, and six with late maturity genotypes (Samineni et al. 2019). The results indicated that the days needed for one life cycle (from seed to seed) have been reduced to about 50–55 days under speed breeding (Table 5.2) (Samineni et al. 2020).

Table 5.2 Six chickpea accessions evaluated under speed breeding facilities and compared to the normal planting

Genotype	Maturity group	Days to harvest under normal conditions	Days to harvest under speed breeding	% Life reduction
JG 11	Early	97	54	55.7
JG 14	Early	97	52	53.6
ICCV 10	Medium	101	54	53.5
JG 16	Medium	104	56	53.8
C 235	Late	117	64	54.7
CDC-frontier	Late	123	61	49.6

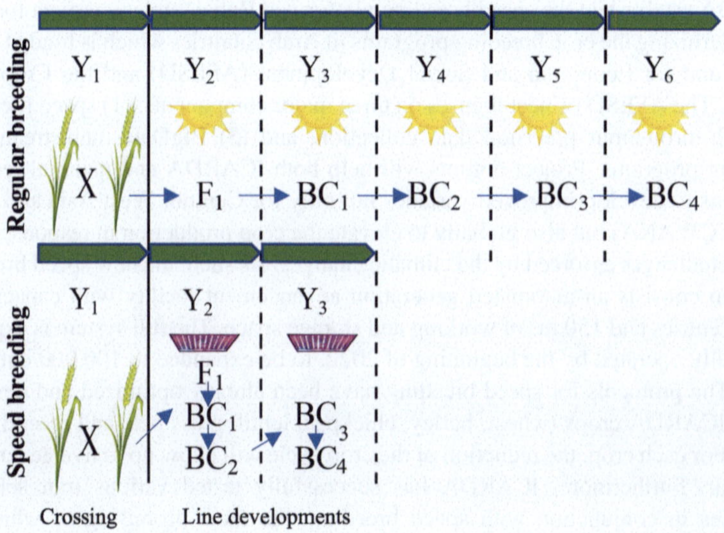

Fig. 5.2 Schematic representation of advantage of using SB techniques to enhance the crop improvement. BC: backcross, Y: year

In cereals, wheat (*Triticum aestivum*) and barley (*Hordeum vulgare*) were the first crops where the SB was tested allowing for a faster selection in breeding programs by using the single seed descent (SSD) method (Ghosh et al. 2018; Rai and Rai 2022). The time taken from seed to seed when developing RILs by SSD was successfully optimized to obtain six to eight generations per year, depending on the genotype (Yao et al. 2017; Cha et al. 2020). By introducing the desired genes into the elite lines in shorter time, the breeders could complete early selection under the speed breeding method (Fig. 5.2).

5.5 Implementing Speed Breeding in CGIAR

After the development of SB protocol for many crops, it is important to scale up the application of this technology at a large scale to support international breeding programs, especially in the international centres. As examples of speed breeding adoption as a promising technology, ICRISAT has successfully optimized a SB protocol in chickpea for having up to six generations of per year in a controlled greenhouse of full daylight (Watson et al. 2018; Gaur et al. 2019). ICARDA started testing and implementing speed breeding in 2017 with the establishment of a pilot speed breeding platform in Rabat (Morocco). This initial platform had the capacity for 3000 entries and was used to develop protocols for speed breeding for mandate cereals and legumes. In 2021, after validating the facilities and methodology,

ICARDA established the speed breeding platform at Rabat, under a project focusing on modernizing the crop breeding programs in Arab countries which is funded by the Arab Fund for Economic and Social Development (AFESD) and the Crop Trust Project. The AFESD project focuses on three major components: (1) speed breeding, (2) high-throughput precision data collection, and (3) BigData mainstreamed in breeding programs. Project outputs will help both ICARDA and national research programs to develop improved varieties not only for Central, West Asia and North Africa (CWANA) but also globally to elevate the crop production in response to the recent challenges enforced by the climatic changes. As such, the new speed breeding platform consists an automated generation advancement facility with capacity for 50,000 entries and 150 m^2 of working and storage space. The full system is expected to be fully operated by the beginning of 2022, to be expanded to 100,000 entries in 2023. The protocols for speed breeding have been already optimized and designed for all ICARDA crops (wheat, barley, chickpea, lentil, grass pea, and recently faba bean). For each crop, the reduction of the crop cycle will allow up to five generations per year. Furthermore, ICARDA has successfully tested various trait selection strategies in conjunction with speed breeding, the most promising of which are disease screening and non-destructive end-use quality.

Optimized protocol for accelerating breeding has been applied in chickpea and lentil breeding cycle in ICARDA. The seeds of five genotypes were grown in a growth chamber and greenhouse at a high density of supplemental lighting (using far-red-enriched LED and blue LED lights) under an extended photoperiod of 22 h light/25.5 °C and 2 h dark/15 °C. The protocol was undertaken within the controlled growth house facilities at the Agricultural Research Center (ARC) in Egypt. The plants were supplemented by blue and far-red LED lights. Chickpea and lentil seeds were planted in pots (0.5 L) filled with soil mix (1:1 peat moss and soil). The flowering started in 23–32 days after the sowing of lentil and chickpea, and the first pod set in 31–42 days. The matured pods were harvested 51–65 days from the planting date. The harvested seeds were dried at 30–35 °C for 2 days. These seeds then were planted using the same protocol to get the next generation (Fig. 5.3). This protocol of speed breeding is expected to allow four to five generations/year instead of one or two generations by the conventional greenhouse or field methods. This protocol will save both the field and labour cost. Furthermore, it enables to integrate with the high-throughput genotyping and genomic selection in chickpea and lentil.

5.6 MAS and Genomic Selection

Recently, the most common technology used for plant selection is through the usage of molecular markers, which has the advantage of time-saving and the selection precision in plant breeding programs over conventional breeding (Heffner et al. 2009). Molecular markers such as simple sequence repeats (SSR) markers, which are

Fig. 5.3 *Optimization of speed breeding protocol (from seed to seed in 60 days).* The experiment has been conducted in the greenhouse at the Agricultural Research Center (ARC), Egypt. The plants were supplemented by cool blue and far-red LED and at 22 h light/25 °C and 2 h dark/15 °C. Days were counted for the first flower, first pod, and full maturity after the date of sowing. (**a**) Speed breeding stages of chickpea growth. (**b**) Speed breeding stages of lentil growth (Unpublished data). (**c**) Speed breeding of chickpea under the hydroponic system for trait screening

commonly used in traditional methods for marker-assisted selection in local breeding programs.

Speed breeding requires more adaptable molecular techniques that can provide more control while also improving our understanding of plant-environment interactions (Panjabi et al. 2019). The high-density molecular markers development such as single nucleotide polymorphisms (SNPs) has enabled the plant breeders to innovate genomic selection (GS) as a promising tool in crop breeding (Meuwissen et al. 2001).

The GS is a predictive computational model developed from a set of genotypes (called as training population) for learning from the markers associated with the desired traits to predict the best selection according to the genetic value (Jannink et al. 2010). Still GS needs to be more developed to increase the selection accuracy for the complex quantitative traits such as yield. In chickpea, the GS indicated prediction accuracies from 0.138 for seed yield to 0.912 for 100-seed weight (Roorkiwal et al. 2016). In rice, the studies show the feasibility and great potential of GS where the average predictive for the flowering date reach up to 0.85, 0.6, 0.63 (Onogi et al. 2015; Isidro et al. 2015; Spindel et al. 2015, respectively), and for plant height up to 0.54, 0.58, 0.7, and 0.86, (Grenier et al. 2015; Cui et al. 2020; Isidro et al. 2015; Wang et al. 2015, respectively). However, the predictive ability value is still low for grain yield with an average of 0.35 obtained from six research studies, where the maximum prediction obtained was 0.54 by using 102,795 SNP markers and GBLUP statistical model (Spindel et al. 2015; Grenier et al. 2015; Wang et al. 2015; Cui et al. 2020). In wheat, the accuracy of the predictive value for the grain yield was ranged 0.2–0.4 by using 1726 DArT markers and GBLUP (Poland et al. 2012), but the value was higher (0.3–0.6, 0.48–0.61, and 0.5–0.6) when using the ridge regression best linear unbiased prediction (rrBLUP) statistical model (Zhao et al. 2013; Crossa et al. 2010; and Pérez-Rodríguez et al. 2012; respectively). These values could be improved with improving the predictive computational models and improving the accuracy of the phenotyping data. There are several packages for GS assessment which is available in the several software operated under R environment such as the Bayesian linear regression BLR (Pérez et al. 2010); genomic best linear unbiased prediction (GBLUP; Endelman 2011), the genome-wide regression and prediction BGLR (Pérez and de Los Campos 2014); genomic estimates of breeding values BWGS (Charmet et al. 2020), genomic prediction of hybrid performance Predhy (Xu et al. 2021), and Solving Mixed Model Equations in R SOMMER (Covarrubias-Pazaran et al. 2021).

5.7 Genome Editing

Targeted mutation is a technology that can generate desirable mutations in the genome and is called genome editing (GE). The change could be made by GE in targeted DNA sequences in the form of deletion, addition or even substitution in one or a few genetic bases. GE requires prior information about the gene and its function to design the location of the targeted mutation. Therefore, GE is considered as precision breeding that can increase the efficiency of the use of plant genetic resources for speeding up the release of new crop varieties. New breeding technology needs exhaustive knowledge of the genome of the target species to enable the development of a new crop with the target trait. Gene editing is an option to translate gene biology into practical breeding strategies, that is, several CRISPR variant technologies have already been developed and quickly deployed for basic and applied research, including plant breeding. (Li et al. 2013; Chavez et al. 2015; Piatek et al. 2015; Xie et al. 2015; Zetsche et al. 2015; Endo et al. 2016; Kleinstiver et al. 2016; Lin et al. 2016; Minkenberg et al. 2017; Shimatani et al. 2017; Hu et al. 2018; Nishimasu et al. 2018; Ren et al. 2019). Programmable nucleases can produce specific changes at a desired location within the genome (Zhang et al. 2019a, 2019b, 2019d). Targeted gene editing (GE) aims to introduce modified targeted change in genomes using site-directed nucleases (SDN), producing double-strand breaks (DSBs) in the DNA and using the normal existing cellular DNA repair machinery of the cell to introduce mutations.

Gene editing, started during the last decade (Fig. 5.4), is a fast-growing technique that has been utilized in many crops because of its acceptability, cost-effectiveness, less time taken and enhanced and focused targeting (Xu et al. 2016). In the early approaches, GE depended on protein binding motif to target desired genome sequence for introducing mutations and included meganucleases, Transcription Activator-Like Effector Nucleases (TALEN), and Zinc-Finger Nucleases (ZNF), while the new approach includes CRISPR/Cas system that depends on a small molecule of RNA to guide the nuclease to the target genome site (Abdallah et al. 2015; Bortesi et al. 2016). CRISPR/Cas techniques enable editing of genes with significant importance include CRISPR-associated protein system 9 (CRISPR/Cas9) derived from *Streptococcus pyogenes*, the CRISPR system from *Prevotella* and *Francisella1* (Cpf1), base editing (BE) and prime editing (PE) have proved powerful tools for the successful modification of genome sequence in a simple and precise way.

CRISPR/Cas9 system requires short guide sequence RNA (sgRNA) to direct Cas9 nuclease to cleave the double-stranded DNA target site complementary to the sgRNA. CRISPR-Cas9 is now commonly used system for genome editing in model crops (Wang et al. 2014). CRISPR/Cpf1 system has significant benefits of efficiency and accuracy in genome manipulation. The Cpf1 endonuclease is smaller than Cas9 and thus requires a shorter CRISPR RNA (crRNA). The base editing system allows conversion of nucleotides without the formation of DSBs in target DNA. Cytosine base editing allows conversion of cytosine (C) to thymine (T) gRNA that binds to dead Cas9 (dCas9) fused to C deaminase and uracil DNA glycosylase inhibitor. While adenine base editing has been developed to allow conversion of adenine (A) to guanine (G), the prime editing (PE) system allows the efficient replacement of stretches of genomic DNA with RNA-specified edits. It depends on combining Cas9 nickase (nCas9) with a reverse transcriptase (RT) and an extended gRNA, the prime editing gRNA (pegRNA), containing the genetic information to correct the target sequence. The 5'end of the pegRNA contains the gRNA for the nCas9 to bind the target genome sequence and perform a nick in the complementary strand. While the 3′ of the pegRNA contains a sequence that will serve as template for the RT system and the reverse transcription of the template will create an ssDNA that will serve in operating independently of the host cell's homology-directed repair (HDR)

Fig. 5.4 Timeline of forefront plants modified by genome editing in 10 years (2012–2021) by using different techniques (TALENs, ZFN and CRISPR/Cas9)

machinery. Epigenome editing (EE) depends on removing or inactivating the nuclease and adding an effector domain to provide a new function for manipulation of the epigenome causing alternation in gene expression in an inheritable fashion without changing the DNA sequence. EE could be achieved by targeting a specific gene promoter or histone and causing methylation or demethylation to control gene expression (Abdallah et al. 2021).

Each of these technologies seeks to induce a precise modification in the genome, creating desired novel alleles to develop and release new varieties in a short time and broaden the current narrow genetic pool with new genes. There are two basic requirements for the delivery of editing components into the genome: availability of the genome sequences and effective transformation methods. There are three types of GE that developed by site-directed nucleases (SDN); SDN1 (site-directed nuclease 1) involves double-strand breaks (DSB) that lead to a mutation causing gene silencing, gene knock-out or a change in the activity of a gene. SDN2 (site-directed nuclease 2) needs short single-stranded DNA a template guided to repair the targeted DSB allow the introduction of one or several mutations at the target site. SDN3 depends on using a large sequence donor of double-stranded DNA to repair targeted DSB allowing the introduction of the gene or genetic element(s) at the target site (Das et al. 2021; Menz et al. 2020). A number of review articles addressing gene editing and its importance in various plants have been published recently (Table 5.3).

Combining speed breeding, single nucleotide polymorphism-marker-assisted selection (SNP-MAS), with gene transformation or genome editing would open a new era of research to enhance the breeding capacity to develop plants tolerant to abiotic and biotic stresses in shorter time (Table 5.4). For example, SB methodology has been applied in rice breeding program to develop transgenic lines resistant to salinity derived from six successive generations (including three backcrosses and two self-pollinated) during only 1.4 years (Rana et al. 2019). Such a revolution is needed to meet the new challenges posed by climate change.

Table 5.3 An overview of the genome-edited most economic crops using CRISPR/CAS9

Species	Trait targeted	Gene(s) edited	References
Abiotic stress			
Maize	Herbicide tolerance	*Pat, aadl*	Ainley et al. (2013)
Cotton	Herbicide tolerance	*cry2Ae/bar*	D'Halluin et al. (2013)
Rice	Drought resistance	OsSAPK2	Lou et al. (2017)
Rice	Low cadmium content	*OsNramp5*	Tang et al. (2017a)
Tomato	Plant chilling tolerance	*SlCBF1*	Li et al. (2018c)
Wheat	Drought resistance	TaDREB2, TaERF3	Kim et al. (2018)
Watermelon	Herbicide resistant	*ALS*	Tian et al. (2018)
Rice	Salinity tolerance	OsRR22	Zhang et al. (2019a)
Rice	Decreased cd accumulation	OsLCT1, OsNramp5	Songmei et al. (2019)
Rice	Decreased cd accumulation	OsNRAMP5	Yang et al. (2019)
Rice	Thermo-sensitive genic male-sterile	TMS5	Barman et al. (2019)
Tomato	Reduced drought tolerance with increased stomatal aperture, higher electrolytic leakage, malondialdehyde	*NPR1*	Li et al. (2019)
Rice	Cold tolerance	OsMYB30	Zeng et al. (2020)
Rice	Drought adaptive	OsDST	Santosh Kumar et al. (2020)
Rice	Drought tolerance	OsEBP89	Zhang et al. (2020)
Rice	Root structure for saline soils	*DRO1*	Kitomi et al. (2020)
Rice	Herbicide tolerance	*OsALS*	Wang et al. (2021)
Biotic resistance			
Rice	Bacterial blight resistant	*Os11N3* (*OsSWEET14*)	Li et al. (2012)
Wheat	Fungi resistance	*TaMLO*	Wang et al. (2014)
Rice	Herbicide resistance	BEL	Xu et al. (2014)
Soybean	Herbicide resistance	*ALS1*	Li et al. (2015)
Corn	Herbicide resistance	*ALS1, ALS2, MoPAT*	Svitashev et al. (2015)
Cucumber	Virus resistance	*eIF4E*	Chandrasekaran et al. (2016)
Rice	Herbicide resistance	EPSPS	Li et al. (2016)
Rice	Rice leaf blast resistance	*OsERF922*	Wang et al. (2016)
Rice	Blast resistance	*OsERF922*	Wang et al. (2016)
Tomato	Fungi resistance	*Mlo*	Nekrasov et al. (2017)
Rice	Powdery mildew resistance	*TaEDR1*	Zhang et al. (2017)
Wheat	Powdery mildew	*TaEDR1*	Zhang et al. (2017)
Rice	Broad spectrum of biotic and abiotic stresses	*Bsr*-k1	Zhou et al. (2018)
Cocoa	Disease resistance	*TcNPR3*	Fister et al. (2018)
Grape	Fungi resistance	*VvWRKY52*	Wang et al. (2018a)
Rice	Resistance to rice tungro spherical virus	*eIF4G*	Macovei et al. (2018)

(continued)

Table 5.3 (continued)

Species	Trait targeted	Gene(s) edited	References
Rice	Thermo-sensitive genic male sterility (TGMS)	*TMS5*	Barman et al. (2019)
Barley	Virus resistance	WDV	Kis et al. (2019)
Barley	Virus resistance	WDV	Kis et al. (2019)
Barley	Wheat dwarf virus	MP, CP, rep/RepA, LIR	Kis et al. (2019)
Rice	Bacterial blight resistance	*SWEET11, 13,14*	Oliva et al. (2019)
Banana	Disease resistance	Aspartic protease	Tripathi et al. (2019)
Banana	Banana streak virus	eBSV	Tripathi et al. (2019)
Rice	Bacterial blight resistance	*Os8N3*	Kim et al. (2019)
Grape	Powdery mildew	*VvMLO3*	Wan et al. (2020
Wheat	Haploid	*TaMTL*	Liu et al. (2020)
Rice	Bacterial blight resistance	*Pi21*	Nawaz et al. (2020)
Yield and quality			
Soybean	Seed oil quality	*FAD2–1A*, *FAD2–1B*	Haun et al. (2014)
Tomato	Bigger seedlings	*PROCERA* (*PRO*)	Lor et al. (2014)
Potato	Quality	SSR2	Sawai et al. (2014)
Rice	Decrease amylose content	OsWaxy	Ma et al. (2015)
Rice	Grain weight	*GW2, GW5 and TGW6*	Xu et al. (2016)
Wheat	Plant height regulator, lipoxygenase, grain weight negative regulator	*TaGASR7, TaDEP1, TaNAC2, TaPIN1, TaLOX2, TdGASR7* and *TaGW2*	Zhang et al. (2016)
Wheat	Increase yield	TaGS5-3A	Ma et al. (2016)
Rice	Production of high amylose rice	*SBEIIb*	Sun et al. (2017)
Canola	Seed oil quality	*ALCATRAZ* (ALC)	Braatz et al. (2017)
Millet	Reduction of lignin	*4CL:Pv4CL1*, *Pv4CL2, Pv4CL3*	Park et al. (2017)
Tobacco	Auxin biosynthesis	*NtPIN4*	Xie et al. (2017)
Wheat	Grain size and weight	*TaGW2*	Wang et al. (2018c)
Wheat	Low gluten content	α-Gliadin	Sánchez-León et al. (2018)
Rice	Starch depletion	OsAPL2 a, OsAPS2b	Pérez et al. (2018)
Rice	Stabilize amylose content	OsMADS7	Zhang et al. (2018a)
Rice	Increased oleic acid	*OsFAD2–1*	Abe et al. (2018)
Sorghum	Improved protein quality and digestibility	*k1C*	Li et al. (2018a)
Wheat	Grain type	*TaGW2*	Wang et al. (2018c)
Lettuce	Yield increase	*LsNCED4*	Bertier et al. (2018)
Barley	Grain yield	*HvCKX1* or *HvCKX3*	Gasparis et al. (2019)
Rice	Lower phytic acid	*OsPLDα1*	Khan et al. (2019)
Rice	Decrease grain number, panicle and grain size, and plant shape and size	*OsNramp5*	Haun et al. (2014)

(continued)

Table 5.3 (continued)

Species	Trait targeted	Gene(s) edited	References
Wheat	Grain number	TaCKX2-D1	Zhang et al. (2019d)
Wheat	Grain yield	TaCKX2–1, TaGLW7, TaGW2, and TaGW8	Zhang et al. (2019a, b, c, d)
Kiwi	Growth performance	*AcCEN4*, *AcCEN*	Varkonyi-Gasic et al. (2019)
Cowpea	Nitrogen fixation	SYMRK	Ji et al. (2019)
Cabbage	Growth performance	*BoMS1*, *BoSRK3*	Ma et al. (2019)
Maize	Waxy corn	*Wx/Wx1*	Gao et al. (2020)
Rice	Controlling grain protein content	*OsAAP6, OsAAP10*	Wang et al. (2020)

Table 5.4 Estimated time to develop GM and GE annual crops combined with SB techniques to enhance crop improvement

New breeding technique	Stages	Time without SB	Time With SB
 GM wheat	Transformation	1 year	1 year
	$T_1 - T_4$ selection	4 years	1 year
	Total required time	5 years	2 years
 ExpressEdit using gene editing	Developing edited plants	1 year	3–6 months
	Selection of non-GM with homozygote edited plants	2 year	1 year
	Total required time	3 years	1.5 year

References

Abdallah NA, Hamwieh A, Radwan K, Fouad N, Prakash C (2021) Genome editing techniques in plants: a comprehensive review and future prospects toward zero hunger. GM Crops Food 12(2):601–615. https://doi.org/10.1080/21645698.2021.2021724. Epub 2022 Feb 9

Abdallah NA, Prakash CS, McHughen AG (2015) Genome editing for crop improvement: challenges and opportunities. GM Crops Food 6(4):183–205

Abe K, Araki E, Suzuki Y, Toki S, Saika H (2018) Production of high oleic/low linoleic rice by genome editing. Plant Physiol Biochem 131:58–62

Ainley WM, Sastry-Dent L, Welter ME, Murray MG, Zeitler B, Amora R, Corbin DR, Miles RR, Arnold NL, Strange TL, Simpson MA (2013) Trait stacking via targeted genome editing. Plant Biotechnol J 11(9):1126–1134

Alahmad S, Dinglasan E, Leung KM, Riaz A, Derbal N, Voss-Fels KP, Able JA, Bassi FM, Christopher J, Hickey LT (2018) Speed breeding for multiple quantitative traits in durum wheat. Plant Methods 14(1):1–15

Ali AM, Mustafa HM, Tahir IS, Elahmadi AB, Mohamed MS, Ali MA, Suliman AM, Baum M, Ibrahim AES (2006) Two doubled haploid bread wheat cultivars for irrigated heat-stressed environments. Sudan J Agric Res 6:35–42

Barman HN, Sheng Z, Fiaz S, Zhong M, Wu Y, Cai Y, Wang W, Jiao G, Tang S, Wei X, Hu P (2019) Generation of a new thermo-sensitive genic male sterile rice line by targeted mutagenesis of TMS5 gene through CRISPR/Cas9 system. BMC Plant Biol 19(1):1–9

Bennett RG, Ribalta FM, Pazos-Navarro M, Leonforte A, Croser JS (2017) Discrimination of boron tolerance in Pisum sativum L. genotypes using a rapid, high-throughput hydroponic screen and precociously germinated seed grown under far-red enriched light. Plant Methods 13(1):70

Bertier LD, Ron M, Huo H, Bradford KJ, Britt AB, Michelmore RW (2018) High-resolution analysis of the efficiency, heritability, and editing outcomes of CRISPR/Cas9-induced modifications of NCED4 in lettuce (*Lactuca sativa*). G3: Genes, Genomes. Genetics 8(5):1513–1521

Bortesi L, Zhu C, Zischewski J, Perez L, Bassié L, Nadi R, Forni G, Lade SB, Soto E, Jin X, Medina V (2016) Patterns of CRISPR/Cas9 activity in plants, animals and microbes. Plant Biotechnol J 14(12):2203–2216

Braatz J, Harloff HJ, Mascher M, Stein N, Himmelbach A, Jung C (2017) CRISPR-Cas9 targeted mutagenesis leads to simultaneous modification of different homoeologous gene copies in polyploid oilseed rape (*Brassica napus*). Plant Physiol 174(2):935–942

Brennan JP, Martin PJ (2007) Returns to investment in new breeding technologies. Euphytica 157(3):337–349

Bugbee B, Koerner G (1997) Yield comparisons and unique characteristics of the dwarf wheat cultivar 'USU-Apogee'. Adv Space Res 20(10):1891–1894

Bula RJ, Morrow RC, Tibbitts TW, Barta DJ, Ignatius RW, Martin TS (1991) Light-emitting diodes as a radiation source for plants. HortScience 26(2):203–205

Cha JK, Lee JH, Lee SM, Ko JM, Shin D (2020) Heading date and growth character of Korean wheat cultivars by controlling photoperiod for rapid generation advancement. Korean Society of Breeding Science 52(1):20–24

Chandrasekaran J, Brumin M, Wolf D, Leibman D, Klap C, Pearlsman M, Sherman A, Arazi T, Gal-On A (2016) Development of broad virus resistance in non-transgenic cucumber using CRISPR/Cas9 technology. Mol Plant Pathol 17(7):1140–1153

Charmet G, Tran LG, Auzanneau J, Rincent R, Bouchet S (2020) BWGS: AR package for genomic selection and its application to a wheat breeding programme. PLoS One 15(4):e0222733

Chaudhary HK, Badiyala A, Jamwal NS (2015) New frontiers in doubled haploidy breeding in wheat. Agric Res J 52(4):1–12

Chavez A, Scheiman J, Vora S, Pruitt BW, Tuttle M, Iyer EP, Lin S, Kiani S, Guzman CD, Wiegand DJ, Ter-Ovanesyan D (2015) Highly efficient Cas9-mediated transcriptional programming. Nat Methods 12(4):326–328

Collard BC, Mackill DJ (2008) Marker-assisted selection: an approach for precision plant breeding in the twenty-first century. Philosophical Transactions of the Royal Society B: Biological Sciences 363(1491):557–572

Covarrubias-Pazaran G, Martini JW, Quinn M, Atlin G (2021) Strengthening Public Breeding Pipelines by Emphasizing Quantitative Genetics Principles and Open-Source Data Management. Front Plant Sci 12

Crossa J, Campos, de los G, Pérez P, Gianola D, Burgueno J, Araus JL, Makumbi D, Singh RP, Dreisigacker S, Yan J (2010) Prediction of genetic values of quantitative traits in plant breeding using pedigree and molecular markers. Genetics 186(2):713–724

Cui Y, Li R, Li G, Zhang F, Zhu T, Zhang Q, Ali J, Li Z, Xu S (2020) Hybrid breeding of rice via genomic selection. Plant Biotechnol J 18(1):57–67

Das A, Ghana P, Rudrappa B, Gandhi R, Tavva VS, Mohanty A (2021) Genome Editing of Rice by CRISPR-Cas: End-to-End Pipeline for Crop Improvement. In: Rice Genome Engineering and Gene Editing. Humana, New York, NY, pp 115–134

De Buyser J, Henry Y, Lonnet P, Hertzog R, Hespel A (1987) 'Florin': a doubled haploid wheat variety developed by the anther culture method. Plant Breed 98(1):53–56

Deery DM, Rebetzke GJ, Jimenez-Berni JA, Bovill WD, James RA, Condon AG, Furbank RT, Chapman SC, Fischer RA (2019) Evaluation of the phenotypic repeatability of canopy temperature in wheat using continuous-terrestrial and airborne measurements. Front Plant Sci 10:875
Desta ZA, Ortiz R (2014) Genomic selection: genome-wide prediction in plant improvement. Trends Plant Sci 19(9):592–601
D'Halluin K, Vanderstraeten C, Van Hulle J, Rosolowska J, Van Den Brande I, Pennewaert A, D'Hont K, Bossut M, Jantz D, Ruiter R, Broadhvest J (2013) Targeted molecular trait stacking in cotton through targeted double-strand break induction. Plant Biotechnol J 11(8):933–941
Endelman JB (2011) Ridge regression and other kernels for genomic selection with R package rrBLUP. Plant genome 4(3):250–255
Endo A, Masafumi M, Kaya H, Toki S (2016) Efficient targeted mutagenesis of rice and tobacco genomes using Cpf1 from *Francisella novicida*. Sci Rep 6(1):1–9
FAO (2018) Food outlook – biannual report on the global food markets, Nov 2018. Rome. p104. License CC BY-NC-SA 3.0 IGO
Fasol G, Nakamura S (1997) The Blue Laser Diode
Fiorani F, Schurr U (2013) Future scenarios for plant phenotyping. Annu Rev Plant Biol 64:267–291
Fister AS, Landherr L, Maximova SN, Guiltinan MJ (2018) Transient expression of CRISPR/Cas9 machinery targeting TcNPR3 enhances defense response in *Theobroma cacao*. Front Plant Sci 9:268
Gao H, Gadlage MJ, Lafitte HR, Lenderts B, Yang M, Schroder M, Farrell J, Snopek K, Peterson D, Feigenbutz L, Jones S (2020) Superior field performance of waxy corn engineered using CRISPR–Cas9. Nat Biotechnol 38(5):579–581
Gasparis S, Przyborowski M, Kała M, Nadolska-Orczyk A (2019) Knockout of the HvCKX1 or HvCKX3 gene in barley (*Hordeum vulgare* L.) by RNA-Guided Cas9 Nuclease affects the regulation of cytokinin metabolism and root morphology. Cell 8(8):782
Gaur PM, Samineni S, Gowda CLL, Rao BV (2007) Rapid generation advancement in chickpea. J SAT Agric Res 3(1)
Gaur PM, Samineni S, Thudi M, Tripathi S, Sajja SB, Jayalakshmi V, Mannur DM, Vijayakumar AG, Ganga Rao NV, Ojiewo C, Fikre A (2019) Integrated breeding approaches for improving drought and heat adaptation in chickpea (*Cicer arietinum* L.). Plant Breed 138(4):389–400
Ghosh S, Watson A, Gonzalez-Navarro OE, Ramirez-Gonzalez RH, Yanes L, Mendoza-Suárez M, Simmonds J, Wells R, Rayner T, Green P, Hafeez A (2018) Speed breeding in growth chambers and glasshouses for crop breeding and model plant research. Nat Protocols 13(12):2944–2963
Giura A (2007) Haploids and doubled haploid lines production by Zea system in Triticum durum and Triticale. In: *Cercetări ştiinţifice*, Horticultură, Inginerie Genetică, vol XI. Ed. Agroprint, USAMVB, Timişoara, pp. 32–37
Grenier C, Cao TV, Ospina Y, Quintero C, Châtel MH, Tohme J, Courtois B, Ahmadi N (2015) Accuracy of genomic selection in a rice synthetic population developed for recurrent selection breeding. PLoS One 10(8):e0136594
Haley SD, Johnson JJ, Peairs FB, Stromberger JA, Hudson-Arns EE, Seifert SA, Anderson VA, Bai G, Chen X, Bowden RL, Jin Y (2018) Registration of 'Avery'hard red winter wheat. Journal of Plant Registrations 12(3):362–366
Haun W, Coffman A, Clasen BM, Demorest ZL, Lowy A, Ray E, Retterath A, Stoddard T, Juillerat A, Cedrone F, Mathis L (2014) Improved soybean oil quality by targeted mutagenesis of the fatty acid desaturase 2 gene family. Plant Biotechnol J 12(7):934–940
Heffner, E. L., Sorrells, M. E., and Jannink, J. L. (2009). Genomic selection for crop improvement
Hickey LT, Germán SE, Pereyra SA, Diaz JE, Ziems LA, Fowler RA, Platz GJ, Franckowiak JD, Dieters MJ (2017) Speed breeding for multiple disease resistance in barley. Euphytica 213(3):64
Hickey LT, Lawson W, Platz GJ, Dieters M, Arief VN, German S, Fletcher S, Park RF, Singh D, Pereyra S, Franckowiak J (2011) Mapping Rph20: a gene conferring adult plant resistance to Puccinia hordei in barley. Theor Appl Genet 123(1):55–68

Hickey LT, Hafeez AN, Robinson H, Jackson SA, Leal-Bertioli SCM, Tester M, Gao C, Godwin ID, Hayes BJ, Wulff BBH (2019) Breeding crops to feed 10 billion. Nat Biotechnol 37(7): 744–754

Honsdorf N, March TJ, Pillen K (2017) QTL controlling grain filling under terminal drought stress in a set of wild barley introgression lines. PLoS One 12(10):e0185983

Hu H (1997) In vitro induced haploids in wheat. In: In vitro haploid production in higher plants. Springer, Dordrecht, pp 73–97

Hu X, Meng X, Liu Q, Li J, Wang K (2018) Increasing the efficiency of CRISPR-Cas9-VQR precise genome editing in rice. Plant Biotechnol J 16(1):292–297

Hu Y, Knapp S, Schmidhalter U (2020) Advancing high-throughput phenotyping of wheat in early selection cycles. Remote Sens 12(3):574

Humphreys DG, Knox RE (2015) Doubled haploid breeding in cereals. In: Al-Khayri JM et al (eds) Advances in plant breeding strategies: breeding, biotechnology and molecular tools. Springer International Publishing, New York. https://doi.org/10.1007/978-3-319-22521-0_9

Idrissi O (2020) Application of extended photoperiod in lentil: Towards accelerated genetic gain in breeding for rapid improved variety development. Moroccan J Agric Sci 1(1)

Isidro J, Jannink JL, Akdemir D, Poland J, Heslot N, Sorrells ME (2015) Training set optimization under population structure in genomic selection. Theor Appl Genet 128(1):145–158

Jannink JL, Lorenz AJ, Iwata H (2010) Genomic selection in plant breeding: from theory to practice. Brief Funct Genomics 9(2):166–177

Ji J, Zhang C, Sun Z, Wang L, Duanmu D, Fan Q (2019) Genome editing in cowpea Vigna unguiculata using CRISPR-Cas9. Int J Mol Sci 20(10):2471

Kelliher T, Starr D, Su X, Tang G, Chen Z, Carter J, Wittich PE, Dong S, Green J, Burch E, McCuiston J (2019) One-step genome editing of elite crop germplasm during haploid induction. Nat Biotechnol 37(3):287–292

Khan MSS, Basnet R, Islam SA, Shu Q (2019) Mutational analysis of *OsPLDα1* reveals its involvement in phytic acid biosynthesis in rice grains. J Agric Food Chem 67(41):11436–11443

Kim D, Alptekin B, Budak H (2018) CRISPR/Cas9 genome editing in wheat. Funct Integr Genomics 18(1):31–41

Kim YA, Moon H, Park CJ (2019) CRISPR/Cas9-targeted mutagenesis of *Os8N3* in rice to confer resistance to *Xanthomonas oryzae* pv. *oryzae*. Rice 12(1):1–13

Kis A, Hamar É, Tholt G, Bán R, Havelda Z (2019) Creating highly efficient resistance against wheat dwarf virus in barley by employing CRISPR/Cas9 system. Plant Biotechnol J 17(6):1004

Kitomi Y, Hanzawa E, Kuya N, Inoue H, Hara N, Kawai S, Kanno N, Endo M, Sugimoto K, Yamazaki T, Sakamoto S (2020) Root angle modifications by the DRO1 homolog improve rice yields in saline paddy fields. Proc Natl Acad Sci 117(35):21242–21250

Kleinstiver BP, Pattanayak V, Prew MS, Tsai SQ, Nguyen NT, Zheng Z, Joung JK (2016) High-fidelity CRISPR–Cas9 nucleases with no detectable genome-wide off-target effects. Nature 529(7587):490–495

Li A, Jia S, Yobi A, Ge Z, Sato SJ, Zhang C, Angelovici R, Clemente TE, Holding DR (2018a) Editing of an alpha-kafirin gene family increases, digestibility and protein quality in sorghum. Plant Physiol 177(4):1425–1438

Li H, Rasheed A, Hickey LT, He Z (2018b) Fast-forwarding genetic gain. Trends Plant Sci 23(3): 184–186

Li J, Meng X, Zong Y, Chen K, Zhang H, Liu J, Li J, Gao C (2016) Gene replacements and insertions in rice by intron targeting using CRISPR–Cas9. Nat Plants 2:16139

Li JF, Norville JE, Aach J, McCormack M, Zhang D, Bush J, Church GM, Sheen J (2013) Multiplex and homologous recombination–mediated genome editing in Arabidopsis and *Nicotiana benthamiana* using guide RNA and Cas9. Nat Biotechnol 31(8):688–691

Li R, Liu C, Zhao R, Wang L, Chen L, Yu W, Zhang S, Sheng J, Shen L (2019) CRISPR/Cas9-Mediated SlNPR1 mutagenesis reduces tomato plant drought tolerance. BMC Plant Biol 19(1): 1–13

Li R, Zhang L, Wang L, Chen L, Zhao R, Sheng J, Shen L (2018c) Reduction of tomato-plant chilling tolerance by CRISPR–Cas9-mediated SlCBF1 mutagenesis. J Agric Food Chem 66(34):9042–9051

Li T, Liu B, Spalding MH, Weeks DP, Yang B (2012) High-efficiency TALEN-based gene editing produces disease-resistant rice. Nat Biotechnol 30(5):390–392

Li Z, Liu ZB, Xing A, Moon BP, Koellhoffer JP, Huang L, Ward RT, Clifton E, Falco SC, Cigan AM (2015) Cas9-guide RNA directed genome editing in soybean. Plant Physiol 169:960–970

Lin C, Li H, Hao M, Xiong D, Luo Y, Huang C, Yuan Q, Zhang J, Xia N (2016) Increasing the efficiency of CRISPR/Cas9-mediated precise genome editing of HSV-1 virus in human cells. Sci Rep 6(1):1–13

Liu H, Wang K, Jia Z, Gong Q, Lin Z, Du L, Pei X, Ye X (2020) Efficient induction of haploid plants in wheat by editing of TaMTL using an optimized Agrobacterium-mediated CRISPR system. J Exp Bot 71(4):1337–1349

Lor VS, Starker CG, Voytas DF, Weiss D, Olszewski NE (2014) Targeted mutagenesis of the tomato PROCERA gene using transcription activator-like effector nucleases. Plant Physiol 166(3):1288–1291

Lou D, Wang H, Liang G, Yu D (2017) OsSAPK2 confers abscisic acid sensitivity and tolerance to drought stress in rice. Front Plant Sci 8:993

Lulsdorf MM, Banniza S (2018) Rapid generation cycling of an F2 population derived from a cross between *Lens culinaris* Medik. and Lens ervoides (Brign.) Grande after aphanomyces root rot selection. Plant Breed 137(4):486–491

Ma C, Zhu C, Zheng M, Liu M, Zhang D, Liu B, Li Q, Si J, Ren X, Song H (2019) CRISPR/Cas9-mediated multiple gene editing in Brassica oleracea var. capitata using the endogenous tRNA-processing system. Horticul Res 6(1):1–15

Ma L, Li T, Hao C, Wang Y, Chen X, Zhang X (2016) Ta GS 5-3A, a grain size gene selected during wheat improvement for larger kernel and yield. Plant Biotechnol J 14(5):1269–1280

Ma X, Zhang Q, Zhu Q, Liu W, Chen Y, Qiu R, Wang B, Yang Z, Li H, Lin Y, Xie Y (2015) A robust CRISPR/Cas9 system for convenient, high-efficiency multiplex genome editing in monocot and dicot plants. Mol Plant 8(8):1274–1284

Macovei A, Sevilla NR, Cantos C, Jonson GB, Slamet-Loedin I, Čermák T, Voytas DF, Choi IR, Chadha-Mohanty P (2018) Novel alleles of rice eIF4G generated by CRISPR/Cas9-targeted mutagenesis confer resistance to Rice tungro spherical virus. Plant Biotechnol J 16(11): 1918–1927

Mannur DM, Babbar A, Thudi M, Sabbavarapu MM, Roorkiwal M, Sharanabasappa BY, Bansal VP, Jayalakshmi SK, Yadav SS, Rathore A, Chamarthi SK (2019) Super Annigeri 1 and improved JG 74: two Fusarium wilt-resistant introgression lines developed using marker-assisted backcrossing approach in chickpea (*Cicer arietinum* L.). Mol Breed 39(1):1–13

Menz J, Modrzejewski D, Hartung F, Wilhelm R, Sprink T (2020) Genome edited crops touch the market: a view on the global development and regulatory environment. Front Plant Sci 11

Meuwissen TH, Hayes BJ, Goddard ME (2001) Prediction of total genetic value using genome-wide dense marker maps. Genetics 157(4):1819–1829

Minkenberg B, Wheatley M, Yang Y (2017) CRISPR/Cas9-enabled multiplex genome editing and its application. Prog Mol Biol Transl Sci 149:111–132

Mobini SH, Lulsdorf M, Warkentin TD, Vandenberg A (2015) Plant growth regulators improve in vitro flowering and rapid generation advancement in lentil and faba bean. In Vitro Cell Develop Biol-Plant 51(1):71–79

Mobini SH, Lulsdorf M, Warkentin TD, Vandenberg A (2016) Low red: far-red light ratio causes faster in vitro flowering in lentil. Can J Plant Sci 96(5):908–918

Mobini SH, Warkentin TD (2016) A simple and efficient method of in vivo rapid generation technology in pea (*Pisum sativum* L.). In Vitro Cell Develop Biol-Plant 52(5):530–536

Nawaz G, Usman B, Peng H, Zhao N, Yuan R, Liu Y, Li R (2020) Knockout of pi21 by crispr/cas9 and itraq-based proteomic analysis of mutants revealed new insights into M. oryzae resistance in elite rice line. Genes 11(7):735

Nekrasov V, Wang C, Win J, Lanz C, Weigel D, Kamoun S (2017) Rapid generation of a transgene-free powdery mildew resistant tomato by genome deletion. Sci Rep 7(1):1–6

Nishimasu H, Shi X, Ishiguro S, Gao L, Hirano S, Okazaki S, Noda T, Abudayyeh OO, Gootenberg JS, Mori H, Oura S (2018) Engineered CRISPR-Cas9 nuclease with expanded targeting space. Science 361(6408):1259–1262

Oliva R, Ji C, Atienza-Grande G, Huguet-Tapia JC, Perez-Quintero A, Li T, Eom JS, Li C, Nguyen H, Liu B, Auguy F (2019) Broad-spectrum resistance to bacterial blight in rice using genome editing. Nat Biotechnol 37(11):1344–1350

Onogi A, Ideta O, Inoshita Y, Ebana K, Yoshioka T, Yamasaki M, Iwata H (2015) Exploring the areas of applicability of whole-genome prediction methods for Asian rice (*Oryza sativa* L.). Theor Appl Genet 128(1):41–53

Panjabi P, Yadava SK, Kumar N, Bangkim R, Ramchiary N (2019) Breeding Brassica juncea and B. rapa for sustainable oilseed production in the changing climate: progress and prospects. In: Genomic Designing of Climate-Smart Oilseed Crops. Springer, Cham, pp 275–369

Park JJ, Yoo CG, Flanagan A, Pu Y, Debnath S, Ge Y, Ragauskas AJ, Wang ZY (2017) Defined tetra-allelic gene disruption of the 4-coumarate: coenzyme A ligase 1 (Pv4CL1) gene by CRISPR/Cas9 in switchgrass results in lignin reduction and improved sugar release. Biotechnol Biofuels 10(1):1–11

Pauk J, Kertész Z, Beke B, Bóna L, Csösz M, Matuz J (1995) New winter wheat variety: 'GK Délibáb' developed via combining conventional breeding and in vitro androgenesis. Cereal Res Commun 23:251–256

Pérez L, Soto E, Villorbina G, Bassie L, Medina V, Muñoz P, Capell T, Zhu C, Christou P, Farré G (2018) CRISPR/Cas9-induced monoallelic mutations in the cytosolic AGPase large subunit gene APL2 induce the ectopic expression of APL2 and the corresponding small subunit gene APS2b in rice leaves. Transgenic Res 27(5):423–439

Pérez P, de Los Campos G (2014) Genome-wide regression and prediction with the BGLR statistical package. Genetics 198(2):483–495

Pérez P, de Los Campos G, Crossa J, Gianola D (2010) Genomic-enabled prediction based on molecular markers and pedigree using the Bayesian linear regression package in R. Plant Genome 3(2):106

Pérez-Rodríguez P, Gianola D, González-Camacho JM, Crossa J, Manès Y, Dreisigacker S (2012) Comparison between linear and non-parametric regression models for genome-enabled prediction in wheat. G3: Genesl Genomesl. Genetics 2(12):1595–1605

Piatek A, Ali Z, Baazim H, Li L, Abulfaraj A, Al-Shareef S, Aouida M, Mahfouz MM (2015) RNA-guided transcriptional regulation in planta via synthetic dC as9-based transcription factors. Plant Biotechnol J 13(4):578–589

Poland JA, Endelman J, Dawson J, Rutkoski J, Wu S, Manes Y, Dreisigacker S, Crossa J, Sánchez-Villeda H, Sorrells M, Jannink JL (2012) Genomic selection in wheat breeding using genotyping-by-sequencing. Plant Genome 5(3):103–113

Rai AC, Rai KK (2022) Speed breeding for rapid cycling of crops for stress management and global food security. In: Microbial Biocontrol: Food Security and Post Harvest Management. Springer, Cham, pp 23–37

Rana MM, Takamatsu T, Baslam M, Kaneko K, Itoh K, Harada N, Sugiyama T, Ohnishi T, Kinoshita T, Takagi H, Mitsui T (2019) Salt tolerance improvement in rice through efficient SNP marker-assisted selection coupled with speed-breeding. Int J Mol Sci 20(10):2585

Ren B, Liu L, Li S, Kuang Y, Wang J, Zhang D, Zhou X, Lin H, Zhou H (2019) Cas9-NG greatly expands the targeting scope of the genome-editing toolkit by recognizing NG and other atypical PAMs in rice. Mol Plant 12(7):1015–1026

Riaz A, Periyannan S, Aitken E, Hickey L (2016) A rapid phenotyping method for adult plant resistance to leaf rust in wheat. Plant Methods 12(1):1–10

Ribalta FM, Pazos-Navarro M, Nelson K, Edwards K, Ross JJ, Bennett RG, Munday C, Erskine W, Ochatt SJ, Croser JS (2017) Precocious floral initiation and identification of exact timing of embryo physiological maturity facilitate germination of immature seeds to truncate the lifecycle of pea. Plant Growth Regul 81(2):345–353

Rollins JA, Drosse B, Mulki MA, Grando S, Baum M, Singh M, Ceccarelli S, Von Korff M (2013) Variation at the vernalisation genes Vrn-H1 and Vrn-H2 determines growth and yield stability in barley (*Hordeum vulgare*) grown under dryland conditions in Syria. Theor Appl Genet 126(11): 2803–2824

Roorkiwal M, Rathore A, Das RR, Singh MK, Jain A, Srinivasan S, Gaur PM, Chellapilla B, Tripathi S, Li Y (2016) Genome-enabled prediction models for yield related traits in chickpea. Front Plant Sci 7:1666

Samantara K, Bohra A, Mohapatra SR, Prihatini R, Asibe F, Singh L, Reyes VP, Tiwari A, Maurya AK, Croser JS, Wani SH (2022) Breeding More Crops in Less Time: A Perspective on Speed Breeding. Biology 11(2):275

Samineni S, Sen M, Sajja S, Gaur P (2019) Rapid generation advance (RGA) in chickpea to produce up to seven generations per year and enable speed breeding. Crop J 8:164–169

Samineni S, Sen M, Sajja SB, Gaur PM (2020) Rapid generation advance (RGA) in chickpea to produce up to seven generations per year and enable speed breeding. Crop J 8(1):164–169

Sánchez-León S, Gil-Humanes J, Ozuna CV, Giménez MJ, Sousa C, Voytas DF, Barro F (2018) Low-gluten, non-transgenic wheat engineered with CRISPR/Cas9. Plant Biotechnol J 16(4): 902–910

Santosh Kumar VV, Verma RK, Yadav SK, Yadav P, Watts A, Rao MV et al (2020) CRISPR-Cas9 mediated genome editing of drought and salt tolerance (OsDST) gene in indica mega rice cultivar MTU1010. Physiol Mol Biol Plants 26:1099–1110

Săulescu NN, Ittu G, Giura A, Mustătea P, Ittu M (2012) Results of using Zea method for doubled haploid production in wheat breeding at Nardi Fundulea – Romania. Rom Agric Res 29:3–8

Sawai S, Ohyama K, Yasumoto S, Seki H, Sakuma T, Yamamoto T, Takebayashi Y, Kojima M, Sakakibara H, Aoki T, Muranaka T (2014) Sterol side chain reductase 2 is a key enzyme in the biosynthesis of cholesterol, the common precursor of toxic steroidal glycoalkaloids in potato. Plant Cell 26(9):3763–3774

Saxena KB, Saxena RK, Hickey LT, Varshney RK (2019) Can a speed breeding approach accelerate genetic gain in pigeonpea? Euphytica 215(12):1–7

Shimatani Z, Kashojiya S, Takayama M, Terada R, Arazoe T, Ishii H, Teramura H, Yamamoto T, Komatsu H, Miura K, Ezura H (2017) Targeted base editing in rice and tomato using a CRISPR-Cas9 cytidine deaminase fusion. Nat Biotechnol 35(5):441–443

Songmei LIU, Jie JIANG, Yang LIU, Jun MENG, Shouling XU, Yuanyuan TAN, Youfa LI, Qingyao SHU, Jianzhong HUANG (2019) Characterization and evaluation of OsLCT1 and OsNramp5 mutants generated through CRISPR/Cas9-mediated mutagenesis for breeding low Cd rice. Rice Sci 26(2):88–97

Sood S, Dwivedi S (2015) Doubled haploid platform: an accelerated breeding approach for crop improvement. In: Plant biology and biotechnology. Springer, New Delhi, pp 89–111

Spindel J, Be Spindel J, Begum H, Akdemir D, Virk P, Collard B, Redona E, Atlin G, Jannink JL, McCouch SR (2015) Genomic selection and association mapping in rice (*Oryza sativa*): effect of trait genetic architecture, training population composition, marker number and statistical model on accuracy of rice genomic selection in elite, tropical rice breeding lines. PLoS Genet 11(2):e1004982

Stutte GW (2015) Commercial transition to LEDs: A pathway to high-value products. HortScience 50(9):1297–1300

Sun Y, Jiao G, Liu Z, Zhang X, Li J, Guo X, Du W, Du J, Francis F, Zhao Y, Xia L (2017) Generation of high-amylose rice through CRISPR/Cas9-mediated targeted mutagenesis of starch branching enzymes. Front Plant Sci 8:298

Svitashev S, Young JK, Schwartz C, Gao H, Falco SC, Cigan AM (2015) Targeted mutagenesis, precise gene editing, and site-specific gene insertion in maize using Cas9 and guide RNA. Plant Physiol 169:931–945

Tadesse W, Amri A, Ogbonnaya FC, Sanchez-Garcia M, Sohail Q, Baum M (2016) Wheat. In: Genetic and Genomic Resources for Grain Cereals Improvement. Academic Press, pp 81–124

Tang L, Mao B, Li Y, Lv Q, Zhang L, Chen C, He H, Wang W, Zeng X, Shao Y, Pan Y (2017a) Knockout of OsNramp5 using the CRISPR/Cas9 system produces low Cd-accumulating indica rice without compromising yield. Sci Rep 7(1):1–12

Tang X, Lowder LG, Zhang T, Malzahn AA, Zheng X, Voytas DF, Zhong Z, Chen Y, Ren Q, Li Q, Kirkland ER (2017b) A CRISPR–Cpf1 system for efficient genome editing and transcriptional repression in plants. Nat Plants 3(3):1–5

Tian S, Jiang L, Cui X, Zhang J, Guo S, Li M, Zhang H, Ren Y, Gong G, Zong M, Liu F (2018) Engineering herbicide-resistant watermelon variety through CRISPR/Cas9-mediated base-editing. Plant Cell Rep 37(9):1353–1356

Tripathi JN, Ntui VO, Ron M, Muiruri SK, Britt A, Tripathi L (2019) CRISPR/Cas9 editing of endogenous banana streak virus in the B genome of Musa spp. overcomes a major challenge in banana breeding. Commun Biol 2(1):1–11

Tuvesson S, Von Post R, Ljungberg A (2003) Wheat anther culture. In: Doubled Haploid Production in Crop Plants. Springer, Dordrecht, pp 71–76

Van Oort PA, Zwart SJ (2018) Impacts of climate change on rice production in Africa and causes of simulated yield changes. Glob Chang Biol 24(3):1029–1045

Varkonyi-Gasic E, Wang T, Voogd C, Jeon S, Drummond RS, Gleave AP, Allan AC (2019) Mutagenesis of kiwifruit CENTRORADIALIS-like genes transforms a climbing woody perennial with long juvenility and axillary flowering into a compact plant with rapid terminal flowering. Plant Biotechnol J 17(5):869–880

Wan DY, Guo Y, Cheng Y, Hu Y, Xiao S, Wang Y, Wen YQ (2020) CRISPR/Cas9-mediated mutagenesis of VvMLO3 results in enhanced resistance to powdery mildew in grapevine (*Vitis vinifera*). Horticul Res 7(1):1–14

Wang F, Wang C, Liu P, Lei C, Hao W, Gao Y, Liu YG, Zhao K (2016) Enhanced rice blast resistance by CRISPR/Cas9-targeted mutagenesis of the ERF transcription factor gene OsERF922. PLoS One 11(4):e0154027

Wang F, Xu Y, Li W, Chen Z, Wang J, Fan F, Tao Y, Jiang Y, Zhu QH, Yang J (2021) Creating a novel herbicide-tolerance OsALS allele using CRISPR/Cas9-mediated gene editing. Crop J 9(2):305–312

Wang S, Yang Y, Guo M, Zhong C, Yan C, Sun S (2020) Targeted mutagenesis of amino acid transporter genes for rice quality improvement using the CRISPR/Cas9 system. Crop J 8(3): 457–464

Wang W, Pan Q, He F, Akhunova A, Chao S, Trick H, Akhunov E (2018b) Transgenerational CRISPR-Cas9 activity facilitates multiplex gene editing in allopolyploid wheat. CRISPR J 1(1): 65–74

Wang W, Simmonds J, Pan Q, Davidson D, He F, Battal A, Akhunova A, Trick HN, Uauy C, Akhunov E (2018c) Gene editing and mutagenesis reveal inter-cultivar differences and additivity in the contribution of TaGW2 homoeologues to grain size and weight in wheat. Theor Appl Genet 131(11):2463–2475

Wang X, Tu M, Wang D, Liu J, Li Y, Li Z, Wang Y, Wang X (2018a) CRISPR/Cas9-mediated efficient targeted mutagenesis in grape in the first generation. Plant Biotechnol J 16:844–855

Wang X, Xuan H, Evers B, Shrestha S, Pless R, Poland J (2019) High-throughput phenotyping with deep learning gives insight into the genetic architecture of flowering time in wheat. GigaScience 8(11):giz120

Wang Y, Cheng X, Shan Q, Zhang Y, Liu J, Gao C, Qiu JL (2014) Simultaneous editing of three homoeoalleles in hexaploid bread wheat confers heritable resistance to powdery mildew. Nat Biotechnol 32(9):947–951

Wang Y, Du Y, Yang Z, Chen L, Condon AG, Hu YG (2015) Comparing the effects of GA-responsive dwarfing genes Rht13 and Rht8 on plant height and some agronomic traits in common wheat. Field Crop Res 179:35–43

Watson A, Ghosh S, Williams MJ, Cuddy WS, Simmonds J, Rey MD, Hatta MAM, Hinchliffe A, Steed A, Reynolds D, Adamski NM (2018) Speed breeding is a powerful tool to accelerate crop research and breeding. Nat Plants 4(1):23–29

Wessel E, Botes WC (2014) Accelerating resistance breeding in wheat by integrating marker assisted selection and doubled haploid technology. South Afr J Plant Soil 31:35–43

Wiśniewska H, Majka M, Kwiatek M, Gawlowska M, Surma M, Adamski T, Kaczmarek Z, Drzazga T, Lugowska B, Korbas M, Belter J (2019) Production of wheat doubled haploids resistant to eyespot supported by marker-assisted selection. Electron J Biotechnol 37:11–17

Xie K, Minkenberg B, Yang Y (2015) Boosting CRISPR/Cas9 multiplex editing capability with the endogenous tRNA-processing system. Proc Natl Acad Sci 112(11):3570–3575

Xie X, Qin G, Si P, Luo Z, Gao J, Chen X, Zhang J, Wei P, Xia Q, Lin F, Yang J (2017) Analysis of Nicotiana tabacum PIN genes identifies NtPIN4 as a key regulator of axillary bud growth. Physiol Plant 160(2):222–239

Xu R, Li H, Qin R, Wang L, Li L, Wei P, Yang J (2014) Gene targeting using the Agrobacterium tumefaciens-mediated CRISPR-Cas system in rice. Rice 7(1):1–4

Xu R, Yang Y, Qin R, Li H, Qiu C, Li L, Wei P, Yang J (2016) Rapid improvement of grain weight via highly efficient CRISPR/Cas9-mediated multiplex genome editing in rice. J Genet Genomics 43(8):529

Xu Y, Ma K, Zhao Y, Wang X, Zhou K, Yu G, Li C, Li P, Yang Z, Xu C, Xu S (2021) Genomic selection: A breakthrough technology in rice breeding. Crop J 9:669–677

Yang CH, Zhang Y, Huang CF (2019) Reduction in cadmium accumulation in japonica rice grains by CRISPR/Cas9-mediated editing of OsNRAMP5. J Integr Agric 18(3):688–697

Yao Y, Zhang P, Liu H, Lu Z, Yan G (2017) A fully in vitro protocol towards large scale production of recombinant inbred lines in wheat (*Triticum aestivum* L.). Plant Cell, Tissue and Organ Culture (PCTOC) 128(3):655–661

Zeng Y, Wen J, Zhao W, Wang Q, Huang W (2020) Rational improvement of Rice yield and cold tolerance by editing the three genes OsPIN5b, GS3, and OsMYB30 with the CRISPR–Cas9 system. Front Plant Sci 10:1663

Zetsche B, Gootenberg JS, Abudayyeh OO, Slaymaker IM, Makarova KS, Essletzbichler P, Volz SE, Joung J, Van Der Oost J, Regev A, Koonin EV (2015) Cpf1 is a single RNA-guided endonuclease of a class 2 CRISPR-Cas system. Cell 163(3):759–771

Zhang A, Liu Y, Wang F, Li T, Chen Z, Kong D, Bi J, Zhang F, Luo X, Wang J, Tang J (2019a) Enhanced rice salinity tolerance via CRISPR/Cas9-targeted mutagenesis of the OsRR22 gene. Mol Breed 39(3):1–10

Zhang H, Xu H, Feng M, Zhu Y (2018a) Suppression of OsMADS7 in rice endosperm stabilizes amylose content under high temperature stress. Plant Biotechnol J 16(1):18–26

Zhang HX, Zhang Y, Yin H (2019b) Genome editing with mRNA encoding ZFN, TALEN, and Cas9. Mol Ther 27(4):735–746

Zhang J, Zhang H, Botella JR, Zhu JK (2018b) Generation of new glutinous rice by CRISPR/Cas9-targeted mutagenesis of the Waxy gene in elite rice varieties. J Integr Plant Biol 60(5):369–375

Zhang K, Ge X, Shen P, Li W, Liu X, Cao Q, Zhu Y, Cao W, Tian Y (2019c) Predicting rice grain yield based on dynamic changes in vegetation indexes during early to mid-growth stages. Remote Sens 11(4):387

Zhang Y, Bai Y, Wu G, Zou S, Chen Y, Gao C, Tang D (2017) Simultaneous modification of three homoeologs of Ta EDR 1 by genome editing enhances powdery mildew resistance in wheat. Plant J 91(4):714–724

Zhang Y, Li J, Chen S, Ma X, Wei H, Chen C, Gao N, Zou Y, Kong D, Li T, Liu Z (2020) An APETALA2/ethylene responsive factor, OsEBP89 knockout enhances adaptation to direct-seeding on wet land and tolerance to drought stress in rice. Mol Gen Genomics 295(4):941–956

Zhang Y, Liang Z, Zong Y, Wang Y, Liu J, Chen K, Qiu JL, Gao C (2016) Efficient and transgene-free genome editing in wheat through transient expression of CRISPR/Cas9 DNA or RNA. Nat Commun 7(1):1–8

Zhang Z, Hua L, Gupta A, Tricoli D, Edwards KJ, Yang B, Li W (2019d) Development of an Agrobacterium-delivered CRISPR/Cas9 system for wheat genome editing. Plant Biotechnol J 17(8):1623–1635

Zhao K, Lu ZX, Park JW, Zhou Q, Xing Y (2013) GLiMMPS: robust statistical model for regulatory variation of alternative splicing using RNA-seq data. Genome Biol 14(7):1–15

Zhou X, Liao H, Chern M, Yin J, Chen Y, Wang J, Zhu X, Chen Z, Yuan C, Zhao W, Wang J (2018) Loss of function of a rice TPR-domain RNA-binding protein confers broad-spectrum disease resistance. Proc Natl Acad Sci 115(12):3174–3179

Chapter 6
Multiomics for Crop Improvement

Palak Chaturvedi, Iro Pierides, Shuang Zhang, Jana Schwarzerova, Arindam Ghatak, and Wolfram Weckwerth

Abstract The growing food demand in the world due to the increasing population and decreasing availability of agricultural land requires new crops that are more productive and resistant to harsher environmental conditions. Thus, rapid and effective exploration, identification, and validation of an important trait, gene, molecular mediator, and protein interaction are important for improving crop yield and quality in the near future. Integrating genomics, transcriptomics, proteomics, metabolomics and phenomics enables a deeper understanding of the mechanisms underlying the complex architecture of many phenotypic traits of agricultural relevance. Here, we cite several relevant examples that can appraise our understanding of the recent developments in omics technologies and how they drive our quest to breed climate-resilient crops. Large-scale genome resequencing, pangenomes, and genome-wide association studies aid in identifying and analysing species-level genome variations. RNA-sequencing-driven transcriptomics approach has provided unprecedented opportunities for performing crop abiotic and biotic stress response studies. Additionally, the high-resolution proteomics technologies necessitated a gradual shift from the general descriptive studies of plant protein abundances to large-scale analysis of protein-metabolite interactions. Especially, advent in metabolomics is currently receiving special attention, owing to the role metabolites

P. Chaturvedi (✉) · I. Pierides · S. Zhang · A. Ghatak
Molecular Systems Biology Lab (MOSYS), Department of Functional and Evolutionary Ecology, University of Vienna, Vienna, Austria
e-mail: palak.chaturvedi@univie.ac.at

J. Schwarzerova
Molecular Systems Biology Lab (MOSYS), Department of Functional and Evolutionary Ecology, University of Vienna, Vienna, Austria

Department of Biomedical Engineering, Faculty of Electrical Engineering and Communication, Brno University of Technology, Brno, Czech Republic

W. Weckwerth (✉)
Molecular Systems Biology Lab (MOSYS), Department of Functional and Evolutionary Ecology, University of Vienna, Vienna, Austria

Vienna Metabolomics Center (VIME), University of Vienna, Vienna, Austria
e-mail: wolfram.weckwerth@univie.ac.at

M. K. Pandey et al. (eds.), *Frontier Technologies for Crop Improvement*, Sustainable Agriculture and Food Security, https://doi.org/10.1007/978-981-99-4673-0_6

play as metabolic intermediates and close links to the phenotypic expression. Further, the high-throughput phenomics approach opened new research domains such as root system architecture analysis and plant root-associated microbes for improved crop health and climate resilience. Overall, integrating the PANOMICS approach to modern plant breeding and genetic engineering methods ensures the development of climate-smart crops with higher nutrition quality that can sustainably meet the current and future global food demands.

Keywords Climate change · Food and nutrition security · Crop improvement · Omics · Integrated omics · Multiomics · Breeding

6.1 Introduction

The global population has reached 7.8 billion and is expected to exceed the 10 billion by 2055 (https://countrymeters.info/cn/World). Such a rapid increase in the population causes a great challenge for the food supply. On the one hand, more food crops are needed to provide basic human calories. On the other hand, changing diet preferences toward higher average livestock and dairy products, especially in developing countries, imposes an additional burden on food consumption patterns (Ranganathan et al. 2018). Thus, a crop yield boost is needed to fill the gap between food production and demand. Meanwhile, food nutritional values are more interested in accommodating industrialised modern lives.

The unequivocal change in climate that includes variations in temperature, precipitation and changes in atmospheric CO_2 levels due to global warming has dramatically affected crop productivity (Lamaoui et al. 2018). Recently many studies have shown that a 1 °C rise in the global temperature would reduce the global yield of important crops such as wheat by 6.0%, rice by 3.2%, maize by 7.4% and soybean by 3.1% (Zhao et al. 2017). Thus there is an urgent need to develop climate-resilient crop varieties to reduce the threat of global food insecurity. Several innovative omics (multiomics) approaches have been developed in the past few decades and could be explored to study the regulation of stress response mechanisms in crops.

However, despite the notable progress in high-throughput next-generation sequencing, bioinformatics, functional genomics and availability of big data, their utilisation to identify novel molecular markers or stress regulators involved in biotic and abiotic stresses is still lagging in crop plants. Omics-based tools, such as genomics, transcriptomics, proteomics, metabolomics, and phenomics, can help us explore and elucidate molecular pathways and regulatory mechanisms driving crop development under stress conditions (Fig. 6.1). Recent advances in the multiomics approaches have facilitated the availability of enormous genomic and transcriptomic data for various crop species and have significantly improved the identification of unique and key traits in desired crops (Cortes and Lopez-Hernandez 2021; Muthamilarasan et al. 2019; Pathak et al. 2018; Weckwerth et al. 2020) that can facilitate crop improvement. It now permits us to routinely perform molecular and

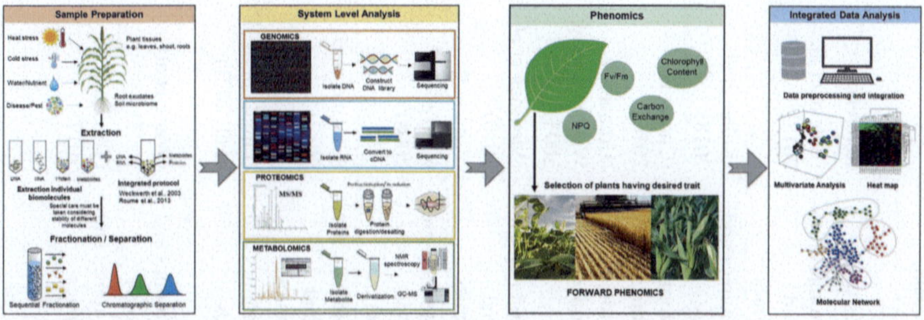

Fig. 6.1 Multiomics approach for crop improvement from sample preparation to system level analysis, phenomics and data integration

genetic analysis to underpin several phenotypic traits of agricultural importance (Scossa et al. 2021; Weckwerth et al. 2004). Furthermore, integrating modern crop improvement strategies such as speed breeding and gene editing technologies with omics approaches can now facilitate the rapid development of elite climate-smart cultivars with the desired traits of resilience and nutritive quality (Chaturvedi et al. 2022; Gao 2021; Kumar et al. 2021; Singh et al. 2021). Additionally, gaining insight through specific omics approaches will not be enough to address outstanding research questions. However, integrating omics technologies can effectively interpret the gene functions, pathways, and regulatory complex underlying the traits and support new breeding schemes to enable sustainable agriculture production (Table 6.1). This chapter will thus examine the recent progress of these omic approaches and their relevance in improving crop.

6.2 High-Throughput Genomic Sequencing, Pangenomics and Epigenomics for Crop Improvement

Genomics tools and progress in next-generation sequencing are vital aids in crop improvement through their use in 'genomics-assisted breeding' (Varshney et al. 2005). The structure elucidation and functional significance of genomic regions enable the link of the genome to the phenotype and thus provide plant breeders with the ability to select elite cultivars that are higher yielding and tolerant to various biotic and abiotic stresses. The first plant genome to be sequenced was *Arabidopsis thaliana*, with the famous but high-cost Sanger method (Sanger et al. 1977), achieving a high-quality, chromosome-level assembly (Arabidopsis Genome 2000). With the subsequent development of 'next-generation' sequencing (NGS) technologies (e.g. Illumina sequencing), sequencing costs were reduced, and the need to clone DNA fragments was eliminated. 'Third-generation' sequencing emerged soon after to address shortcomings such as the relatively short-read lengths that created a bottleneck on the assembly of plant genomes at the chromosome level.

Table 6.1 List of studies elucidating the multiomics approach to understand molecular regulation and stress response mechanism

Plant	Species	Plant material studied	Modification or stress	Method used	Multiomics approaches used	References
Tomato	*Solanum lycopersicum*	Glandular trichomes and leaves	-NA-	Real-time PCR, LC-MS and GC-MS	Transcriptomics proteomics and metabolomics	(Balcke et al. 2017)
Tomato	*Solanum lycopersicum*	Pollen	Heat	MACE sequencing and LC-MS/MS	Transcriptomics and proteomics	(Keller et al. 2018)
Potato	*Solanum tuberosum*	Tuber sprouts	Regulation of anthocyanin biosynthesis	RNA-seq and UPLC-Q-TOF-MS.	Transcriptomics and metabolomics	(Cho et al. 2016)
Potato	*Solanum tubersosum*	Developing tuber	-NA-	Microarrays and GC-MS	Transcriptomics and metabolomics	(Urbanczyk-Wochniak et al. 2003)
Common bean	*Phaseolus vulgaris*	Seeds	-NA-	Affymetrix microarrays, 2DGE-MALDI-TOF-MS, and UPLC-ESI-MS	Transcriptomics, proteomics and metabolomics	(Mensack et al. 2010)
Rice	*Oryza sativa*	Leaves	Ozone (O_3)	Hybridized microarrays and RT-PCR, 2DGE-LC-MS/MS, and CE-MS	Transcriptomics, proteomics, and metabolomics	(Cho et al. 2008)
Rice	*Oryza sativa*	Seeds – Endosperm and embryo	-NA-	Hybridized Affymetrix, LC-MS/MS, and GC-MS	Transcriptomics, proteomics, and metabolomics	(Galland et al. 2017)
Rice	*Oryza sativa*	Leaves, milky stage grains	Betaine aldehyde dehydrogenase 2 (BADH2) for rice fragrance	RNA-Seq and LC-MS/MS	Transcriptomics and proteomics	(Phitaktansakul et al. 2021)
Rice	*Oryza sativa*	Seedlings	Chemical stress (zaxinone)	RNAseq and qRT-PCR, GC-MS and LC-MS	Transcriptomics and metabolomics	(Wang et al. 2021)
Rice	*Oryza sativa*	Leaves and seedlings	Drought	Microarray and quantitative RT-PCR, GC-MS, LC-MS	Transcriptomics and metabolomics	(Todaka et al. 2017)

Rice	*Oryza sativa*	Leaves	Biotic stress (bacteria: Bacterial leaf blight)	Microarray, GC-TOF-MS and LC-TOF-MS	Transcriptomics and metabolomics	(Sana et al. 2010)
Wheat	*Triticum aestivum*	Grain development stages	-NA-	LC-MS/MS, and GC-MS	Proteomics and metabolomics	(Zhang et al. 2021)
Wheat	*Triticum aestivum*	Spikelets	Biotic stress (fungal: *Fusarium* head blight)	LC-hybrid MS, and LC-ESI-LTQ-Orbitrap	Proteomics and metabolomics	(Gunnaiah et al. 2012)
Wheat	*Triticum aestivum*	Young leaves	Drought	2DGE-LC-MS/MS and LC-MS/MS	Proteomics and metabolomics	(Michaletti et al. 2018)
Wheat	*Triticum turgidum*	Flag leaves	Drought	RNA-Seq and qRT-PCR, 2D-MALDI-TOF/TOF-MS	Transcriptomics and proteomics	(Peremarti et al. 2014)
Wheat	*Triticum aestivum*	Crown from seedlings	Cold	Illuminaseq and qPCR, LC-MS	Transcriptomics and metabolomics	(Zhao et al. 2019)
Maize	*Zea mays*	Kernels	Genetically modified (GM maize)	Hybridized microarrays, 2DGE-^{1}H-NMR, and GC-MS	Transcriptomics, proteomics, and metabolomics	(Barros et al. 2010)
Maize	*Zea mays*	Leaves and seedlings	NUE (nitrogen use efficiency)	Hybridized microarrays, 2DGE-LC-MS/MS, and GC-MS	Transcriptomics, proteomics, and metabolomics	(Amiour et al. 2012)
Maize	*Zea mays*	Endosperm and embryo	Transgenic maize lines with enhanced carotenoid synthesis	Affymetrix microarrays, 1D-gel-LC-MS/MS, and GC-MS	Transcriptomics, proteomics, and metabolomics	(Decourcelle et al. 2015)
Maize	*Zea mays*	Leaves	Unravelling lipid metabolism	RNA-Seq and UPLC-FT-MS	Transcriptomics and lipidomics	(de Abreu et al. 2018)
Maize	*Zea mays*	Leaves	Function of ZmUGTs in biotic stress	RNA-Seq and semi-quantitative RT-PCR, and UPLC-MS/MS	Transcriptomics and metabolomics	(Ge et al. 2021)
Chickpea	*Cicer arietinum*	Leaves	Salinity	RNAseq and RT-PCR, 2DGE-LC-MS/MS	Transcriptomics and proteomics	(Arefian et al. 2019)
Pigeon pea	*Cajanus cajan*	Pollen and anthers	Exogenous auxin	RNA-Seq, LC-MS/MS and GC-MS	Transcriptomics, proteomics, and metabolomics	(Pazhamala et al. 2020)

(continued)

Table 6.1 (continued)

Plant	Species	Plant material studied	Modification or stress	Method used	Multiomics approaches used	References
Berry	*Vitis vinifera*	Berry skin tissue	Drought	Microarray hybridization, LC-MS/MS, GC and LC-MS	Transcriptomics, proteomics, and metabolomics	(Ghan et al. 2015)
Soybean	*Glycine max*	Leaves	Fungal pathogen (Rhizoctonia foliar blight)	RNAseq and qRT-PCR, and ^{1}H-NMR	Transcriptomics and metabolomics	(Copley et al. 2017)
Sesame	*Sesamum indicum*	Leaves	Drought-tolerant and sensitive genotypes	RNAseq and qRT-PCR, and UPLC/MS, GC/MS	Transcriptomics and metabolomics	(You et al. 2019)
Pepper	*Capsicum annuum*	Fruit – Pericarp tissue	Fruit developmental stages	RNA-Seq, and gel-LC-MS/MS	Transcriptomics and proteomics	(Liu et al. 2019)
Rapeseed	*Brassica napus*	Seedlings	Cold	RNAseq and qRT-PCR, and UHPLC-LC-MS/MS	Transcriptomics and metabolomics	(Raza et al. 2021a)

Such long-read sequencing techniques (such as Oxford Nanopore technology, PacBio sequencing, MinION) enabled the assembly of several hundreds of plant genomes and the understanding of genome architecture and genomic variations in germplasm collections that might be linked to mechanisms for adaptation to climate change and plant-breeding-related adaptive traits. The association of genotype-to-phenotype starts with structural genomics, which encompasses the identification of sequence polymorphism and chromosomal organisation for the construction of physical, genetic maps (through linkage) to identify variations correlated to traits of interest for plant breeders. Functional genomics then aims to validate these genomic regions for their functional input to the trait by developing various biotechnological tools that isolate, clone, and overexpress or knock out the gene of interest in transgenic analyses (Kole et al. 2015). Examples include the CRISPR/Cas9 genome editing tool reported in many crops such as soybean, rice, maize, and sorghum (Jiang et al. 2013), as well as mutagenomic approaches that introduce genome-wide mutations that are screened for functional significance. The latter approach has been adapted to rice, tomato, barley, and soybean through the TILLING technology (Kashtwari et al. 2019; Kurowska et al. 2011), a useful tool for crop breeding.

Structural genomics is dependent on molecular markers that serve as tags of genes, for example, restriction fragment length polymorphisms (RFPL), random amplified polymorphic DNA (RAPD), amplified fragment length polymorphisms (AFPL) and single nucleotide polymorphisms (SNPs). SNPs are single DNA nucleotide variations dispersed all over the whole genome and occur more frequently than other markers, thus enabling shorter fragments of the genome to be correlated to a trait, which makes the identification of the tagged candidate gene easier and more accurate. SNP identification became feasible with the advent of NGS, and it is used in quantitative trait loci (QTL) mapping and genome-wide association studies (GWAS). Both are statistical methods; the first correlates the genomic variations to an observed trait within a population of closely related species enabling the mapping of QTLs that might contain the gene of interest. GWAS is a high-throughput approach that uses whole-genomic sequences in larger populations and finds more associations between genes and traits. GWAS was used in many genomic studies for crop improvement of yield in maize crops under heat and water stress (Millet et al. 2016), drought stress in sorghum (Lasky et al. 2015) as well as in rice (Liang et al. 2018). Breeders can then use these identified markers through marker-assisted breeding (MAS) to increase crop quality and yield (He et al. 2014). The multiparent advanced generation intercrosses (MAGIC) approach and nested association mapping (NAM) in model plants and crops (Yu et al. 2008) (Kover et al. 2009) are ideal for breeding improvement by exposing a large phenotypic diversity in crops under investigation.

6.2.1 Pangenomics

Pangenomics is another area of genomics that probes the architectural structure of a genome based on comparative genome studies in the wider gene pools of taxa under investigation. A pangenome is an integrated representation of the conserved gene space (the 'core genome') and the variable portions (the 'dispensable genome'), which are not widely conserved across many taxa. The assembly of plant pangenomes, facilitated by long-read sequencing, is crucial for crop improvement as rare beneficial alleles found in wild relatives that are part of part of the dispensable genome, can be reintroduced to modern elite crop varieties that lack these alleles. This lack might be due to the negative selection of rare beneficial alleles during crop domestication or genetic drift that randomly eliminates them. Core and dispensable genes differ in function, with the former enriched for functions related to essential processes such as primary metabolism. The latter is enriched in accessory functions related to disease resistance and response to abiotic stress (Bayer et al. 2019) and variants related to flowering time (Song et al. 2020). All functions are indispensable to crop improvement. A success story of reintroducing a rare beneficial allele through a pangenomic analysis was the assembly of the pangenome of over 700 tomato accessions, including wild species related to the domesticated tomato. Over 4000 novel protein-coding genes not predicted to be present in the reference tomato genome were identified using the map-to-pan approach, including a variant promoter allele *TomLoxC*, encoding a lipoxygenase involved in the synthesis of fruit volatiles (Gao et al. 2019). This allele was more frequent in the wild accessions and strongly counter selected during tomato domestication but reintroduced in some modern elite tomato varieties alongside the reference allele.

The identified markers can then be utilised by genomic selection (GS) tools to predict a specific crop's phenotypic traits of interest using parametric and non-parametric statistical models. GS is a useful method that provides an accurate evaluation of the effect of genomic variations identified through GWAS on the phenotype, eliminating the need to assess variants through costly and time-consuming breeding in the field that requires multiple generations to assess the effect of a variant on a trait. GS considers the combination and interaction of hundreds of genetic variants within a genome, thus providing a high-throughput method that considers additive and epistatic interactions that might explain the 'missing heritability' problem many GWAS studies face (Young 2019). GS parametric methods include Genomic Best Linear Unbiased Prediction (Habier et al. 2013), the Least Absolute Shrinkage and Selection Operator (Usai et al. 2009), Bayesian Ridge Regression (Gianola 2013) and BayesA, BayesB and BayesC (Habier et al. 2011). In a recent study (Abdollahi-Arpanahi et al. 2020), parametric GS methods were compared to the non-parametric ensemble and deep learning algorithms, such as random forests, gradient boosting, multilayer perceptron and convolutional neural networks. Gradient boosting, an ensemble method that converts weak learners to strong learners to reduce bias and variance, was identified as the superior machine learning method (ML) for predicting complex, multifactorial traits (Abdollahi-Arpanahi et al., 2020).

6.2.2 Epigenomics

Epigenetics refers to stable and heritable changes in gene expression, occurring without changes in the DNA sequence (Fujimoto et al. 2012). It is achieved via changes in chromatin conformation during plant development, attained by DNA methylation, post-translational histone modifications and non-coding RNA molecules that impact gene expression. Open chromatin conformation results in gene activation as it allows transcription factors to bind to the promoter regions of the genes, while closed chromatin conformation results in gene repression. Dynamic transitions from open to closed chromatin ensure proper cellular function at all stages of development and under different environmental conditions. This is especially important in adapting plants and crops to stress, which induces epigenetic changes resulting in differential gene expression plasticity. This kind of response is faster and more effective than changes in gene expression due to DNA mutations that require longer timescales to be transmitted to later generations.

In contrast, epigenetic changes inherited between generations within a specific environmental and developmental context can be reversible and thus confer much more flexibility and fast adaptability to direct external inputs and do not need to depend on random DNA mutations for survival. Thus, after stress, the epigenetic modifications can be 'memorized' by plant somatic cells and be utilized as marks inherited transgenerationally, enabling the same epigenetic modifications to occur when the progenies face similar stressful conditions. This can greatly impact the efficiency of breeding programs, a key factor in the plant's immediate response to stress and long-term adaptation (Fortes and Gallusci 2017).

DNA methylation involves adding a methyl group on the 5′ carbon of the cytosine base to form 5-methylcytosine, which mostly leads to gene silencing when binding to promoters but activates gene expression when it is present in the gene sequence. Two antagonizing classes of enzymes, cytosine-5 DNA methyltransferases such as *MET1* (Chan et al. 2005) and DNA demethylases such as *CMT2* (Stroud et al. 2014) govern the distribution and abundance of methylation in plant genomes. DNA methyltransferase mutants of *Arabidopsis thaliana* resulted in severe defects such as shorter stature, failure to enter the floral transition state, abnormal embryo development and reduced leaf size (Cokus et al. 2008), demonstrating the vital role of DNA methylation in all aspects of the plant's development. Histone modification is another epigenetic mark on the N-tails of nucleosomal histones. It includes methylation, acetylation, phosphorylation, ubiquitination, biotinylation and sumoylation (Engelhorn et al. 2014), leading to closed or open chromatin conformation. An example of how histone modifications control proper development is the regulation of the *FLOWERING LOCUS C (FLC)*, a repressor of flowering (Whittaker and Dean 2017). The transition from vegetative to floral state needs to happen under favourable temperature conditions and day lengths, requiring proper regulation. A long period of cold, known as vernalization, is required before flowering in the spring. Repression of floral transition in the fall, before vernalization, is achieved by activating chromatin marks (H3K4me3, H3K4me36 and H2Bub1) employed on *FLC* chromatin, resulting in high gene expression of *FLC* and consequent repression of floral transition. During vernalization in the winter, activating histone marks are reduced. In contrast, repressive marks such as H3K27me3 are enriched, resulting in suppression of the *FLC* gene expression and allowing transition to the flower state (Bastow et al. 2004).

Epigenetic control of adaptation to low temperatures has been investigated in chilling maize seedlings, with a decreased amount of DNA methyltransferase resulting in a 10% reduction of DNA methylation (Steward et al. 2002). Similar hypomethylation events were also observed in the tobacco genome under salt and cold stress, which might enhance alteration of chromatin structure, inducing transcriptional activation of stress-responsive genes. Under high chill conditions, dormancy break was linked to decreased total methylation leading to bud break and subsequent fruit set (Kumar et al. 2016). Under drought conditions, the H3K4me3 epigenetic mark of gene induction was modified in *Arabidopsis* by the histone methyltransferase *ATX1*, which activates *NCED3*, a gene in the ABA biosynthesis pathway, inducing the stress response pathway (van Dijk et al. 2010). In turn, ABA regulates gene expression through DNA methylation and histone acetylation status (Sah et al. 2016). For example, in response to drought, an increase in the histone acetylation level of *RD20, RD29A* and *RD29B* genes was observed, thought to be drought-induced. A plethora of stress-induced differentially methylated regions (DMRs) alter their level of methylation in plants exposed to different forms of biotic stresses as well. Some pathogen response genes contain many repeats in their promoter regions and are negatively regulated by DNA methylation. After pathogen invasion, these promoter loci decrease in DNA methylation to be transcriptionally activated (Le et al. 2014).

6.3 Transcriptomics: RNAseq to Regulatory Networks for Crop Improvement

The transcriptome represents all the expressed genes at a specific developmental stage. This is vital for the assessment of gene function as not all genes are expressed at any specific point in time and space, so even if there is a correlation between a genetic variant and a phenotypic trait, this might not be expressed under a certain condition and thus might be rendered irrelevant. Transcriptomics can, therefore, capture gene expression dynamics and link this to a function that depends on the downstream process of gene transcription and translation controlled by various regulatory processes such as transcription factors, environmental and epigenetic inputs, and alternative processes splicing. With the advent of robust techniques for

RNA profiling such as microarrays, NGS, RNAseq and SAGE, gene expression dynamics could be uncovered with spatial and time-specific resolution. This enables the transcriptome evaluation across various developmental stages that uncover the expression of different gene regulatory network modules representing the most upstream pathways in determining a trait of interest.

Comparative transcriptomics can reveal differential gene expression between abiotic stress and control conditions. Microarray analysis has revealed the differential expression of genes in soybean and barley during developmental and reproductive stages, respectively, under drought stress (Guo et al. 2009; Le et al. 2012). Various transcription factors have altered expression in response to abiotic stresses in *Arabidopsis*, soybean and rice crops (Xiong et al. 2002). Several TFs, such as *DREB2* and *DRE/CRT* in rice, were found to regulate abiotic stress responses in rice (Todaka et al. 2015). Sorghum exposed to drought, heat and osmotic stress, and hormone treatment expressed different genes uncovered by transcriptome studies (Dugas et al. 2011).

Furthermore, in situ RNA-seq, in which RNA is sequenced in living cells or tissues, can reveal expression patterns with a spatial resolution (Ke et al. 2013) to provide a molecular description of physiological processes in plants. RNA-seq analyses have unveiled tissue-specific expression in response to abiotic and biotic stress in millet and sweet potato crops (Lin et al. 2017; Qi et al. 2013). Comparative transcriptomics can also identify commonly expressed genes and cross-talk pathways between crops that reveal common gene regulatory modules that are core mechanisms to specific stress responses or developmental stages. Multiple cross-talk pathways have been revealed in cotton and potato under biotic and abiotic stresses and hormonal treatments (Massa et al. 2013). In addition, specialized NGS technologies facilitate scRNA (single-cell RNA) sequencing for refining gene expression with cell-specific resolution. This further characterizes the state of a cell that defines its identity and unveils the cell-specific states that each cell reaches under a condition that collectively with other cells make up for whole tissue function. With data analysis methods like tSNE, cell differentiation trajectories can be identified along developmental stages in different tissues. scRNA studies have described the development and differentiation of unique plant morphologies such as stomatal cells (Adrian et al. 2015) and female gametophytes (Schmid et al. 2015).

QTL mapping and identifying a specific gene involved in the trait of interest can be significantly improved using transcript profiling in a procedure referred to as 'expression genetics' (Varshney et al. 2005). Quantitative traits are now related to mRNA expression levels under different conditions, and these are subjected to QTL or GWAS analysis, making it possible to identify the 'ExpressedQTLs' (eQTLs). Thus eQTLs contain genes of interest that correspond to transcripts expressed under different biotic stresses or developmental stages. This mapping between genome and transcriptome is more accurate in identifying causative alleles. It is a mapping between two molecular networks, one downstream of the other; therefore, correlations are much more precise. Forty-five thousand one hundred thirty-nine differentially expressed genes, 288 miRNAs, 640 pathways and 435,828 putative marker variants were identified through RNAseq analysis in wheat root tissue under drought stress. QTL analysis was carried out, and drought-responsive QTLs, possessing 18 differentially regulated genes, were identified on chromosome 3B in wheat roots (Iquebal et al. 2019). In another study, upregulation of the auxin receptor (*AFB2*) and ABA-responsive transcription factors *MYB78, WRKY18* and *GBF3* was revealed with a similar method in wheat root tissue under drought stress (Dalal et al. 2018).

6.4 Proteomics: An Integral Part of Functional Omics Approach for Crop Improvement

Proteins are the functional molecules that describe the genes or provide a functional annotation. Proteomics is described a technique involved in the profiling of total expressed protein in an organism, cell or tissue and is divided into four parts: sequence, structural, functional, and expression proteomics. It can also be defined as an efficient and systematic high-throughput profiling or identification of proteins. Qualitative and quantitative measurements of the large number of proteins directly involved/influenced in cellular biochemistry can be attained via this technique (Chandramouli and Qian 2009; Ghatak et al. 2017; Zhang et al. 2013). Information about protein concentrations, post-translational modification (PTMs), protein-protein interaction, regulatory functions coded by genes with stress tolerance and relation can be obtained through proteomics (Chaturvedi et al. 2016; Weckwerth 2011). Approaches in proteomics have been transformed from descriptive to become highly useful for integration and data validation with other omics approaches. They also provide high-throughput information on stress tolerance mechanisms and biological process interpretation, applied in future crop breeding programs and climate-smart agriculture (Hu et al. 2015). Shotgun proteomics (gel-based and gel-free) nowadays is a well-practiced technique than the two-dimensional electrophoresis (2DE) (Chaturvedi et al. 2013, 2015, 2016; Paul et al. 2016; Pazhamala et al. 2020). This technique enables us to generate large protein datasets with quantifiable protein data-dependent differences between crop varieties and stress levels (Ghatak et al. 2017). Selected reaction-monitoring (SRM) strategies, multiple reaction monitoring (MRM), parallel monitoring reaction (PRM) and accurate inclusion mass screening (AIMS) allows to targeted quantitation of known protein of interest over a larger set of samples allowing double-haploid populations, near-isogenic lines, recombinant inbred lines, wild cultivars to be screened for the protein of interest and their abundance in various stresses (Borras and Sabido 2017; Gillet et al. 2016; Jacoby et al. 2013; Weckwerth et al. 2020; Wienkoop et al. 2008a; Wienkoop and Weckwerth 2006; Wienkoop et al. 2010). Ghatak and co-workers explained and summarized the effect of biotic and abiotic stress in all the members of the Solanaceae family (tomato, potato, tobacco etc) related to proteomics (Ghatak et al. 2017). Proteomics for abiotic stress in legumes has been extensively described in a recent review by Nelofer and co-workers (Jan et al. 2022). They have explicitly discussed every legume proteomics study and explained the rationale behind the studies undertaken by the different labs across the globe to understand the abiotic stresses that drastically impact legumes production in developing nations (Jan et al. 2022). For staple crops, the major question remains whether shotgun proteomics can be applied in the field studies under 'native' growth conditions. Hoehenwater and co-workers performed a large-scale proteomics screening of potatoes under field trials to help the breeders get a elite variety (Hoehenwarter et al. 2008). This study made it possible to implement a novel data processing and mining approach for high-throughput proteomics data using a machine learning approach. They also identified protein markers that predict specific traits of interest in potatoes, such as black spot disease and starch content sensitivity (Hoehenwarter et al. 2008). Furthermore, PROTMAX was used to identify protein polymorphisms from the mass spectrometric data measured for various potato genotypes, which was not predicted from the other databases available (Hoehenwarter et al. 2011b). Polyploidy level in the proteomics data is a major issue because of the multiplication of several protein isoforms from the whole genome sequencing and also their diversification in their potential function; by combining the shotgun proteomics and linear mathematical models, Hoehenwarter and co-workers resolve this protein isoform problem (Hoehenwarter et al. 2011a).

From an agricultural perspective, grain filling and development of grain is an essential biological processes in the crop life cycle, eventually contributing to the final seed yield and quality in all cereal crops and productivity. Proteomics and metabolomics studies in seed development (days after anthesis (DAA)) elucidate the molecular pathways and physiological transitions, which can contribute to the advancement of valuable and potentially agriculturally important strategies for improving yield, quality and stress tolerance in crop plants (Zhang et al. 2021). For plant breeders to analyse genes responsible for seed quality and generate predictive hypotheses, analyses of seed protein content and understanding the role of enzymes involved in the starch biosynthesis is important. Rice seed germination, which revealed the detailed mechanism of starch degradation in endosperm and starch biosynthesis in the embryo during seed development, was studied by Komatsu and Hossain using the proteomics approach (Komatsu and Hossain 2013). The combination of shotgun proteomics and cell biological studies recently unravelled the spatio-temporal expression and subcellular localization of hordoindoline across development in barley endosperm (Roustan et al. 2018; Shabrangy et al. 2018).

Similarly, in another study, barley seedlings were exposed to the magnetic field for the proteomics analysis, 2866 proteins were identified in total, and tissue-specific regulation showed carbohydrate metabolic process, cell redox homeostasis and oxidation-reduction process proteins were increased in abundance (Shabrangy et al. 2021). In another approach, shotgun proteomics on mature barley seeds was studied to fully characterise the barley seed proteome (Mahalingam, 2017). A key difference in hordoindoline (HINs) proteins was identified in this study between the two-rowed and six-rowed barley cultivars, which may contribute to the differences in seed hardness, suggesting that the different protein profiles may provide a useful tool for examining more complex traits and identifying novel protein marker for crop improvement (Mahalingam 2017). A full list of the studies in crops and legumes combining proteomics with other omics approaches is provided in Table 6.1.

Abiotic and biotic stresses are key limiting factors that impair crops' growth and yield. Reduced plant growth, pollen abortion, delayed seed germination and decreased crop yield are the major outcomes of the stress in the field. In this stressful environment and climate-changing scenario throughout the agricultural field across the globe, protein plays an important and crucial role in shaping new phenotypic changes via physiological trait adjustment where these proteins are directly involved (Chaturvedi et al. 2015; Jegadeesan et al. 2018; Komatsu and Hossain 2013). Subcellular proteins such as ion and water transporters, ROS scavengers, proteins related to transcriptional regulation, and signalling are frequently reported to be involved in stress tolerance (Nouri and Komatsu 2013). Ghatak and co-workers reported a drought stress response in the tissue-specific root, seed and leaf in pearl millet cereal. Two thousand two hundred eighty-one proteins were identified; 120 proteins for leaf, 25 for roots and 10 for seeds showed significant changes in comprehensive proteomics analysis (Ghatak et al. 2016). In a follow-up study by the same group, wheat (C3) and pearl millet (C4) cereals were compared under drought stress at the proteomics and physiological level. Pearl millet showed an exceptional drought-tolerant capacity compared to wheat (Ghatak et al. 2020). Similarly, drought-tolerant and sensitive varieties of wheat were exposed to drought stress to quantify and evaluate protein expression pattern and abscisic acid (ABA) effect on the root proteome (Alvarez et al. 2014), wheat-tolerant variety showed ABA-responsive and ABA-induced proteins in significantly higher number that plays an important role for wheat adaptation in the severe drought-prone area (Alvarez et al. 2014).

6.5 Metabolomics: Metabolic Readout of the Functional Gene for Crop Improvement

Metabolomics is one of the emerging fascinating omics tools, which has now been extensively applied for crop improvement. Metabolomics is crucial to study the abiotic stress tolerance, pathogen resistance, identifying robust ecotypes and developing metabolic-assisted breeding of crops because it helps to unravel the complexities of genotype-environment-phenotype complexities and provides information about phenotypic plasticity, which is not predicted from genome sequences alone (Ghatak et al. 2018; Weckwerth 2003, 2010, 2011; Weckwerth et al. 2020). Modern metabolomics platforms are exploited to explain complex biological pathways and hidden regulatory networks controlling the growth and development of crops. The metabolome consists of a complex set of low molecular weight metabolites in the biological systems responsible for trait development, crop yield, and nutritional quality control. The plant metabolome comprises of primary and secondary metabolites that provide extensive knowledge of the biochemical process during plant growth and development and response to stress conditions. The successful identification and detection of metabolites are possible through gas chromatography (GC-MS) and liquid chromatography (LC-MS) and non-destructive nuclear magnetic resonance spectroscopy (NMR) (Ghatak et al. 2018). On the flip side, GC-MS analysis needs chemical derivatization, which may cause exclusion of some metabolites from the analysis and may not produce sufficient information for the clear identification of the particular metabolite. However, combining multiple datasets from complementary analytical platforms offers a powerful strategy to analyse metabolomes. Primary metabolites consist of sugars, amino acids and lipids. During photosynthesis, they play an important role in the tricarboxylic acid and glycolysis cycle, affecting plant growth and development. Secondary metabolites like flavonoids, carotenoids, phytic acids etc., are not essential for plant survival. They are synthesized in response to stress conditions such as high temperatures, cold stress, drought, salinity and insect/pest attack. The plant metabolome also consists of specialized secondary metabolites such as phenolics, alkaloids and terpenoids that offer tolerance against biotic/abiotic stresses (Ghatak et al. 2018). Recently many specialized compounds have been identified as a unique biomarker that measures plant performance under stressful environments (Table 6.1).

Metabolomics analysis generates high-dimensional complex datasets that are difficult to analyse and interpret using univariate statistical analysis. Hence, multivariate data analysis and mathematical modelling approaches are widely used to obtain meaningful information (Ghatak et al. 2018; Schuhmacher et al. 2013; Weckwerth 2011; Weckwerth and Fiehn 2002; Weckwerth et al. 2020). Many multivariate statistical tools, mathematical modelling approaches, and structure elucidation of unknown metabolites are integrated into one toolbox known as COVAIN (Sun and Weckwerth 2012). This toolbox provides a novel approach to link causal relationships of biochemical networks with metabolite dynamics measured through the GC-MS or LC-MS platform. This approach goes beyond classical multivariate data analysis or correlation networks (Sun and Weckwerth 2012). It has been recently applied in several studies and plays an important role in understanding stress response mechanisms and developmental processes (Doerfler et al. 2013; Nagele et al. 2014; Nagele and Weckwerth 2013; Nukarinen et al. 2016; Wang et al. 2016; Zhang et al. 2021).

The application of metabolomics in plant breeding and identifying stress markers for crop improvement has been previously reviewed (Ghatak et al. 2018). Drought is one of the major limiting factors for agriculture productivity across the globe. In drought stress, plants adopt several physiological modifications, including leaf abscission, leaf area reduction, and greater nutrient uptake by plant roots (Chaves and Oliveira 2004). Plants synthesize many ubiquitous polyamines like spermidine and putrescine in response to drought stress as a defence mechanism (Bitrian et al. 2012). Metabolomic profiling of six wheat genotypes under drought stress identified several important metabolites such as gamma-aminobutyric acid (GABA), myo-inositol, threonine, proline, oxalic acid, malic acid, glucose and fructose (Marcek et al. 2019). Comparative metabolomics analysis of drought-tolerant and susceptible wheat cultivars indicated the high-level accumulation of lysine, arginine, methionine and proline in response to drought stress (Michaletti et al. 2018). Similarly, linamarin, proline, and tryptophan accumulation can be potentially used as a biomarker for screening drought-tolerant genotypes.

Plants need optimum temperature for their normal growth and development. Any fluctuation in temperature can cause severe damage to the developmental processes. High temperature disturbs the homeostasis and other physiological mechanisms of plants (Chaturvedi et al. 2021). Plants synthesize many secondary metabolites under heat stress, such as rhamnose, putrescine, myo-inositol etc. Untargeted metabolomics analysis of thermosensitive male sterile line of pigeon pea (*Cajanus cajan*) led to the identification of 27 metabolites in tetrad, microspore and mature pollen. In addition, fructose, glucose and sucrose metabolites were also associated with male sterility (Pazhamala et al. 2020). The high accumulation of sugars in the microspore stage indicates an impaired transport mechanism during microspore development (Pazhamala et al. 2020). Similarly, comparative metabolomics analysis of heat-tolerant and susceptible soybean cultivars identified 1,3-dihydroxyacetone, ribose and glycolate in the heat-tolerant varieties compared to susceptible ones. These tolerant varieties also produce low concentrations of many others such as chiro-inositol, pinitol, erythritol, arabitol etc. (Chebrolu et al. 2016).

A freezing temperature can seriously damage cellular membranes and plant cells. Many plant species develop freezing tolerance due to their exposure to non-freezing low temperatures, and this process is known as 'cold acclimatization'(Guy 1990). Wienkoop and co-workers investigated the dynamics of the metabolite-protein covariance network and the relation of starch metabolism during cold acclimation (Wienkoop et al. 2008b). In this study, raffinose accumulation belongs to general cold and heat temperature stress responses, and starch metabolism can be compensated by increased sucrose synthesis for cold adaptation processes (Wienkoop et al. 2008b).

Similarly, a follow-up study revealed the essential role of starch metabolism in response to cold stress in different *Arabidopsis* ecotypes, demonstrating that different ecotypes develop different biochemical strategies to cope with cold stress (Nagler et al. 2015). Doerfler and co-workers investigated the interface of primary and secondary metabolism as a response to cold and light stress. *Arabidopsis* accumulated a large flavonoid concentration, identified using the novel algorithm in non-targeted LC-MS metabolomics analysis. This study identified new potential flavonoid structures as a putative role of antioxidant response factors (Doerfler et al. 2013, 2014).

The application of metabolomics to understand belowground mitigation is important for crop improvement (Ghatak et al. 2021). Soil consists of organic/inorganic compounds and a wide range of microbial populations. The interrelationship/interconnection can be significantly elaborated using the metabolomics approach. Precise metabolomics analysis of the soil microbe populations in the soil metabolome and their connection with diverse metabolite analysis is important to understand and open new avenues in soil microbiome investigation. Metabolomics profiling of soil samples from wheat rhizosphere led to the identification of various bioactive metabolites such as glutarimide, consabatine, methlpyrrole, arachidonic acid, gibberellic acid etc. These identified unique metabolites play an important role in soil-plant signalling pathways and provide a defence mechanism against pathogens (Monreal and Schnitzer 2015). Recently, effect of drought stress on the root exudation process and composition was studied in Pearl millet (843-22B and ICTP 8203), an important cereal crop (Ghatak et al. 2021). This study determined comparative metabolic profiling (primary and secondary metabolite profiling) of the exudates from sensitive and tolerant genotypes of pearl millet after a period of drought stress. Primary metabolome characterization of root exudates under drought stress exhibited a stronger genotypic effect with stronger alteration in the concentration and composition of the exudates in the sensitive genotype (843-22B) compared to the tolerant genotype (ICTP8203). Drought treatment enhanced the concentration of organic acids (succinic acid, oxalic acid, lactic acid, fumaric acid, malonic acid and citric acid) in both the compared genotypes. Increased accumulation of these compounds in the exudate of drought-stressed plants can promote osmotic adjustment, which may help maintain root development to access water from deep layers of soil (Serraj and Sinclair 2002). Similarly, secondary metabolites such as adenosine were identified, which plays an important role in plant growth and development. A higher accumulation of adenosine under drought stress in both genotypes trigger protective mechanisms in plants under drought stress (e.g. defence mechanism against ROS) (Macková et al. 2013). A similar response was observed in holm oak under drought stress (Gargallo-Garriga et al. 2018). Furthermore, we also identified water-soluble vitamins such as riboflavin and pantothenic acid, showing enhanced accumulation under stress conditions in both genotypes. Additionally, the extracted root exudates were assessed for the biological nitrification inhibition (BNI) activity, and this study provides the first evidence of genotypic differences in the BNI capacity of pearl millet where BNI release is substantially influenced by drought stress, showing enhanced activity in 843-22B as compared to ICTP8203 (Ghatak et al. 2021).

6.6 Phenomics Facilitates Crop Improvement

Wilhelm Johannsen first proposed the concept of genotype and phenotype in 1909 in his textbook on heredity research, titled 'Elemente der exakten Ereblichkeitslehre (The Elements of an Exact Theory of Heredity)', and he developed the concept published in 1911 titled 'The Genotype Conception of Heredity', which refers to the visible physical traits presented by biological individuals or groups under specified situations (such as various environments and growth stages). The notion of phenomics was developed to complement genomics (Schork 1997). Plant phenome can be determined by the genome (G), environment (E), and management (M) interactions (Grosskinsky et al. 2018; Yang et al. 2021). Crop phenomics has developed significantly in recent years, allowing researchers to obtain multidimensional phenotypic data at numerous levels, including cell, organ, plant, and population levels (Bilder et al. 2009; Dhondt et al. 2013; Houle et al. 2010; Lobos et al. 2017; Pazhamala et al. 2021).

A series of high-throughput and high-precision phenotyping tools have been developed to address many problems such as multi-season repeated measurement of yield-related traits in multiple environments, errors in remote sensing in-field monitoring, and high-throughput analysis of key phenotypes (Furbank and Tester 2011). These techniques include environmental sensors, non-invasive imaging, reflectance spectroscopy, robotics, computer vision and high-content screening, etc. (Fiorani and Schurr 2013; Furbank and Tester 2011; Pieruschka and Schurr 2019), and have significantly assisted the plant phenotyping research. For instance, drones or unmanned aerial vehicles, as well as pocketPlant3D, are used to measure traits including leaf area index estimates, detect weeds and infections, and predict yield through various sensors (hyperspectral imaging and computed tomography imaging to targeted metabolic sensors) (Yang et al. 2020). Root systems of soil-grown plants and crops were imaged using electrical resistance tomography, electrical capacitance, X-ray computed tomography, and positron emission tomography without damaging plant samples (McGrail et al. 2020).

Plant phenomics can quickly and accurately provide quantitative information on important traits, including yield, quality, and stress resistance. The goal of systematic phenotyping research is to gather a variety of related phenotypic traits that provides big data and support for discovering the molecular mechanism and elucidating the gene functions of trait regulation (Cardona and Tomancak 2012; Grosskinsky et al. 2015). This big data support allows for rapid screening of high-quality genotypes and enhanced crop cultivars (Yang et al. 2020). The great potential of phenomics is also reflected in its integration with other omics research (such as genomics, epigenomics, transcriptomics, proteomics, metabolomics and ionomics, etc.) (Pazhamala et al. 2021; Raza et al. 2021b; Scossa et al. 2021; Weckwerth et al. 2020; Yang et al. 2021), providing quantitative analysis to the genetic laws of specific phenotypes, monitoring of crop phenotypes, and acquisition of dynamic traits at different developmental stages of crops.

Tardieu and Bennett proposed the concept of multiscale phenotyping research (Tardieu et al. 2017). They pointed out that converting the massive amounts of images and sensor data generated indoors and outdoors into useful biological information would be the next bottleneck. Namely, in phenotyping research, the importance of data interpretation cannot be further emphasized.

According to different application carrier platforms, phenotyping acquisition technologies can be roughly divided into handheld, man-carried, vehicle-mounted, field real-time monitoring, large-scale indoor and outdoor automation platforms, aviation airborne, and satellite imaging platforms of different levels. The Phenospex field phenotyping platform equipped with a three-dimensional laser imager (PlantEyeTM) can dynamically compare the developmental traits of near-isogenic lines in wheat (Kirchgessner et al. 2016). Leaf-GP, open-source software based on crop image sequences captured by mobile devices (such as smartphones) can automatically analyse various growth phenotyping indicators (Zhou et al. 2017). SeedGerm system based on machine learning algorithms can screen large-scale seed germination and seedling phenotypes (Colmer et al. 2020).

It is expected to address the bottleneck of data analysis in phenotyping by using open-source software libraries such as OpenCV (Bradski and Kaehler 2008), SciKit-Image (van der Walt et al. 2014), Scikit-Learn (Pedregosa et al. 2012), and TensorFlow (Abadi et al. 2016). Teams with professional backgrounds in mathematics, algorithm design, software and hardware development, and biological sciences are needed to develop such analytical techniques (Goff et al. 2011; McCouch et al. 2013).

Phenomics is a crucial research field for future crop improvement, and phenotypic analysis can provide big data-based decision support for breeding, cultivation, and agricultural practices by identifying essential traits. Thus, phenotyping research should have a clear purpose, based on quick screening of essential traits, to provide decision-making data for targeted solutions to scientific problems. A clear purpose should also guide the selection of phenotyping equipment and analytic methods: to improve the solution of specific biological issues.

6.7 Systems Biology and Bioinformatics Approach for Crop Improvement

We are in the middle of a golden age of genetics in which we are discovering a plethora of polymorphisms associated with phenotypic variance. The recent flood of different types of GWAS and other data has greatly enabled a leap forward concerning quantitative genetics that attempt to use statistical techniques to associate collections of genes with measurable phenotypic variation. These correlations provide insight into the genotype-phenotype map as a 'black box' process by associating the inputs (genotypes) with the outputs (phenotypes) and have so far been proven valuable in identifying genetic drivers to predict phenotypes for different crop cultivars in plant breeding. However, statistical correlations provide little insight into the mechanisms through which genetic changes are manifested as phenotypic change. The pure statistical nature of GWAS limits its ability to provide more accurate predictions for phenotypic change, evident in the 'missing' heritability problem that many GWAS studies face that can partly be explained by non-additive, epistatic interactions between genes, as well as the small size effects of causal polymorphisms and their low-frequency occurrence in the genome that makes them difficult to detect (Marjoram et al. 2014). To account for the limitations of the one-to-one mapping of genotype to a phenotype that most GWAS studies offer, the topology and effect of the gene of interest within a gene regulatory network must be considered following from the understanding that a gene is not directly linked to a phenotype but has a direct effect only on intermediate processes such as networks of genes, proteins or metabolites that in turn affect higher-order traits. The most impressive development in this paradigm has occurred in plant breeding. There is a deep tradition of modelling phenotypes with process-based approaches, termed the genotype-phenotype problem (Cooper et al. 2002). One recent trend is to estimate phenotypic outcomes by modelling the underlying gene and gene interactions at the expression level. In these cases, genes can be modelled using Boolean logic, oscillators or differential equation-based expression level models of important plant genes (Locke et al. 2005). The logical next step is to integrate GWAS with network models where parameters important for network function are perturbed to infer which ones affect a particular phenotype. Then the polymorphisms affecting this parameter are searched through a GWAS-like approach focused on specific nodes or pathways of the network (Marjoram et al. 2014). Thus with the integration of different omics layers in network models, systems biology aims to bridge the genotype-phenotype (GP) map by considering complex, non-linear interactions between components, shifting the paradigm into the post-GWAS era. Various 'omics' integration approaches are outlined below, followed by mathematical modelling approaches of omics networks.

6.8 Data Integration

Early studies of the GP map had probed biological mechanisms through reductionist approaches where each molecular layer is situated within a hierarchy, with each layer comprising a basic biological entity that, when taken together, form a new entity at a higher level (Nichol et al. 2019). Reductionism fails in bridging between levels of the biological hierarchy, failing to account for feedback mechanisms that act across different scales. Concerning the GP mapping, it is known that gene expression is regulated by environmental factors and a bi-directional coupling among different levels. Molecular networks are thus heterarchical and not simply hierarchical (Cumming 2016). Therefore, it is imperative that we consider the interaction between different omics data that feedforward between each other to functionally determine the final phenotypic trait and facilitate the dissection of complex plant regulatory networks (Urano et al. 2010). There are three main ways integration can be conceptualized (Jamil et al. 2020): the statistical method that is correlation-based and finds associations between components of all omics networks using Pearson's or Spearman's coefficients, clustering (such as hierarchical, k-means or random forests) and multivariate analysis (principal component analysis (PCA), partial least squares discriminant analysis (PLS-DA)); the pathway-based method that maps associations on available pathway databases such as KEGG, AraCyc or SolCyc; and finally the mathematical-based approach that takes into account the dynamics within and between each network. The model-based approach follows four main steps: identification of system components, their regulation and topology within a network, formulation of mathematical equations such as ODEs and finally parameter selection and optimization, outlined in more detail below.

Within the purely statistical framework of multiomics integration, several studies found a weak correlation between transcripts and their cognate proteins, for example, in the methyl jasmonate hormone stress treatment on *Persicaria minor* herbal plants, with poor overall transcriptome-proteome correlation ($r = 0.34$) (Aizat et al.

2018) and in the ripening process of the tomato where transcripts and proteins related to the ethylene pathway were weakly correlated (Mata et al. 2018), implying the presence of post-translational and post-transcriptional mechanisms for the majority of stress and ripening pathways. The association of the proteome and metabolome was evaluated in a recent study on *Gingko Biloba* during the leaf maturity process (Guo et al. 2020), where correlation analysis was performed between all differentially expressed proteins and metabolites. K-means clustering groups data points such as distinct, clear groupings emerge to different expression patterns. This clustering approach has been used to integrate metabolomics and proteomics from developing cacao seeds (Wang et al. 2017). Multivariate analysis can handle more complex datasets and enables the discovery and prediction of variance and covariance trends within datasets. Multiple co-inertia analysis (MCIA) was used to integrate the metabolome and proteome of a near-isogenic maize line and its transgenic counterpart, identifying important metabolic differences between the two, such as sugar metabolism and polyamine biosynthesis (Mesnage et al. 2016). The multivariate technique GFLASSO was also applied in maize to integrate the transcriptome and metabolome in deciphering its lipid biosynthesis (de Abreu et al. 2018).

Software tools such as MapMan and PathVisio have been used within the pathway-based integration approach to study and integrate multiomics datasets from plants. For example, integration between transcriptomics, proteomics and metabolomics was performed using PathVisio to study *Arabidopsis* signalling in mutant plants (Bjornson et al. 2017). PathVisio, Paintomics and InCroMap produce a p-value for common pathways by enumerating the number of differentially expressed components within each omics layer before combining them (Cavill et al. 2016). More software tools are expected to be developed for specific plant species with different metabolic pathways. Co-expression analysis relies on statistical correlations between different omics datasets, which are then transformed into a weighted network and can be visualized using various tools such as Weighted Gene Co-expression Network Analysis (WGCNA) in R or the Cytoscape tool.

The use of machine learning (ML) is an advanced statistical approach that aids in integrating highly diverse omics platforms (Picard et al. 2021). ML methods use various integration approaches, the most straightforward one being early integration. After being independently pre-processed, all omics datasets are concatenated to a single matrix, which acts as an input to an ML model. This results in a more complex, noisy and high-dimensional matrix that needs to undergo feature selection or dimensionality reduction to address its increasing complexity. Feature selection includes wrapper methods that repeatedly apply an ML model to different features, and those that improve the prediction accuracy are kept and used as inputs to further models. One of the most powerful ML models is artificial neural networks (AAN) that are composed of many neurons organized into layers. They can be used directly on concatenated omics or separately on each omic layer. The hidden layers of an ANN are considered feature extraction layers, while the final output layer can produce a prediction (e.g. high yield). A most recent approach is to include intermittent molecular layers, such as the proteome or metabolome, in the hidden layer of neural networks to simulate biological information flow from the genome to the phenome, in efforts to increase prediction accuracies. Although they are highly adaptive, one of the most challenging issues with ANNs is their black-box nature, which does not directly infer how genes and other molecules are implicated in the biological process within the GP map. Thus the incorporation of transparent models in prediction tools, for example inspired by adaptive landscape models such as Fisher's Geometric Model (Orr 1998), is more likely to lead to a useful dissection of mechanism in the GP map.

Integration of omics layers using ML methods has recently been used to distinguish genes responsible for specialized metabolism important for plant-environment interactions (Moore et al. 2019) and precision breeding for certain traits of interest (Weckwerth et al. 2020). In addition, deep learning methods to iteratively correct genome-scale models will undeniably become a future direction in plant multiomics research (Silva et al. 2019). While multiomics integration leads to better results as it provides more information and can reveal pathways from different biological layers, the question arises on whether more data is better, which might not always be a good idea. Worse performances could arise if a model is unsuitable for a particular goal or certain multiomics datasets. Thus, the aim should not be only more data but also more biology in the way we can extract meaning using appropriate conceptual abstractions that lead to formulations of accurate mathematical models.

6.9 Systems Modelling

Several systemic modelling approaches include ordinary differential equations (ODEs), Boolean networks and linear programming-based models (Ji et al. 2017). ODE-based modelling is widely used for continuous dynamic modelling in complex biological systems, representing interactions among various molecules which reflect time-varying effects of biological processes (Wittmann et al. 2009). An example of successful use of the ODE model was a study on lignin biosynthesis in poplar (Wang et al. 2018). Transcript-protein equations were developed to model the effect of gene silencing. A mass-balance kinetic ODE was generated to predict the effect of gene perturbation on lignin content and wood properties, useful for the breeding program of this tree species. The challenge of such models is the need to estimate and have knowledge of many unknown parameters, which prevents ODE models from being used in larger networks such as integrated multiomics networks. This can be circumvented by using data-driven approaches that do not require explicit models and parameters but are derived from the data. For ODE-based approximations, the inverse calculation of the Jacobian matrix is deduced from a Lyapunov function that takes as input the covariance matrix (which could include one or more omics layers) and a noise matrix that takes into account input fluctuations (Weckwerth 2019). The equation assumes that the system is in a homeostatic steady state, allowing the Jacobian to be derived. The Jacobian matrix consists of partial derivatives between variables and functions of variables that delineate the rate of change of one variable concerning another within a homeostatic state. It can also capture input-output relationships where external or intrinsic inputs induce a change in one variable that can be observed in the Jacobian as an upregulated entry on another variable. This is especially useful for metabolomic data where the direct influence of the environment that acts as an input can be determined in changes within the Jacobian entries, as metabolomic networks are the most downstream molecular networks that have a closer association to external perturbations from the environment. In plant breeding, this is useful as the effect of biotic and abiotic stresses can be captured through changes in the Jacobian matrix of metabolomics and other omics layers integrated later on. In addition, one of the major utilities of the Jacobian is its central role in stability analysis that depends directly on the Jacobian's eigenvalues (Nägele T 2013). This gives clues to stable strategies that a network attains as well as transitory states from time-series data. Such information is useful for plant studies that aim to capture developmental and transitory changes to various biotic or abiotic stresses, as well as the study of molecular and physiological plasticity.

Finally, linear programming approaches have been extensively used in plant breeding through Flux Balance Analysis (FBA) for genome-scale metabolic network reconstructions. In simple terms, linear programming optimises an outcome based on some set of constraints using a linear mathematical model. According to the concept of FBA (Orth et al. 2010), a metabolic network reconstruction consists of a list of stoichiometrically balanced biochemical reactions, each controlled by an enzyme. A mathematical model is constructed by forming a stoichiometric matrix S, in which each row represents a metabolite, and each column represents a reaction. The model assumes a steady state, such that $S*V = 0$, where V is the flux of each reaction that we aim to find. By defining an objective function, such as max biomass rate, the linear programming with a set of constraints identifies the solution vector V in a subspace. This allows the reconstruction of a metabolic network in the scale of an entire genome and unveils the reaction fluxes that need to be maximized to achieve a certain goal, such as an increase in biomass. Anthocyanin biosynthesis in grapes, *Vitis vinifera*, has been successfully modelled using FBA using metabolomic and proteomic data in a growth experiment (Soubeyrand et al. 2018). The result suggests that metabolic flux is strongly induced upon nitrogen deprivation. Recently, metabolic fluxes have been predicted by genomic selection (GS) models employed in estimating growth for a given genotype in an *Arabidopsis thaliana* population (Tong et al. 2020). The study proposed an approach for network-based GS (termed netGS) that uses metabolic models like FBA to improve the prediction accuracy of classical GS for growth within and across environments.

The combination of integration-based methods for multiomics network reconstruction with mathematical models enhances GWAS approaches' power to accurately predict polymorphisms associated with agriculturally important traits as well as advance and challenge current abstract understandings of the genotype-phenotype map that effectively creeps in every area of biological research.

6.10 Conclusion and Future Perspectives

We are in an era with enormous high-throughput omics data. Integration of omics approaches will enable rapid identification of many genes for a relevant trait. This will fundamentally change our current research paradigm from single gene analysis to pathway or network analysis. The enormous knowledge of improvement gained from omics data, combined with the new gene-editing technologies, can create future-resilient crops with enhanced stress tolerance. Similarly, the engineering of metabolic pathways could open new areas to develop climate-resilient ready-to-grow plants. Recently, speed breeding has emerged as the most powerful tool to enhance plant growth and produce a specific environment. Thus, the combination of omics, genome editing, and speed breeding can ensure success in surprising ways for sustainable agriculture. However, such envisioned integrated approaches also face some challenges that researchers must be aware of before designing a research proposal or a project. Below, we outline some points of consideration for omics-driven studies.

1. One inherent limitation is the multifactorial impact of various internal and external parameters, which are rather difficult to control in the field experiment due to the spatiotemporal variation of a multitude of biotic and abiotic factors, the individual status of each plant, and the intense interaction of the individual factors (G × E × P) (genotype x environment x phenotype). This issue can be overcome by performing studies under controlled laboratory conditions. However, only specific studies might be possible to a certain extent.
2. Although different approaches seem promising to elucidate crop response mechanisms in more detail, this will generate an enormous amount of data, which poses challenges for precise terminology, storage, handling, and evaluation. The huge file sizes derived from non-invasive imaging approaches ultimately impose a greater data management challenge.
3. Another challenging aspect is the need to correlate the diverse datasets derived from various independent approaches to achieve a holistic view.
4. The association of these big individual datasets is a complex task to (i) obtain meaningful correct information and (ii) not lose valuable information. Therefore, it is crucial to develop and implement advanced mathematical and statistical methods and models for the required uni- and multivariate analysis. This will need considerations of its own that might currently be limited and thus not sufficient to extract all this information according to the needs of such suggested holistic approaches.

References

Abadi M et al. (2016) Tensor flow: large-scale machine learning on heterogeneous distributed systems

Abdollahi-Arpanahi R, Gianola D, Penagaricano F (2020) Deep learning versus parametric and ensemble methods for genomic prediction of complex phenotypes. Genet Sel Evol 52:12. https://doi.org/10.1186/s12711-020-00531-z

Adrian J et al (2015) Transcriptome dynamics of the stomatal lineage: birth, amplification, and termination of a self-renewing population dev. Cell 33:107–118. https://doi.org/10.1016/j.devcel.2015.01.025

Aizat WM, Ibrahim S, Rahnamaie-Tajadod R, Loke KK, Goh HH, Noor NM (2018) Proteomics (SWATH-MS) informed by transcriptomics approach of tropical herb Persicaria minor leaves upon methyl jasmonate elicitation. PeerJ 6:e5525. https://doi.org/10.7717/peerj.5525

Alvarez S, Roy Choudhury S, Pandey S (2014) Comparative quantitative proteomics analysis of the ABA response of roots of drought-sensitive and drought-tolerant wheat varieties identifies proteomic signatures of drought adaptability. J Proteome Res 13:1688–1701. https://doi.org/10.1021/pr401165b

Amiour N et al (2012) The use of metabolomics integrated with transcriptomic and proteomic studies for identifying key steps involved in the control of nitrogen metabolism in crops such as maize. J Exp Bot 63:5017–5033. https://doi.org/10.1093/jxb/ers186

Arabidopsis Genome I (2000) Analysis of the genome sequence of the flowering plant Arabidopsis thaliana. Nature 408:796–815. https://doi.org/10.1038/35048692

Arefian M, Vessal S, Malekzadeh-Shafaroudi S, Siddique KHM, Bagheri A (2019) Comparative proteomics and gene expression analyses revealed responsive proteins and mechanisms for salt tolerance in chickpea genotypes. BMC Plant Biol 19:300. https://doi.org/10.1186/s12870-019-1793-z

Balcke GU et al (2017) Multi-omics of tomato glandular Trichomes reveals distinct features of central carbon metabolism supporting high productivity of specialized metabolites. Plant Cell 29:960–983. https://doi.org/10.1105/tpc.17.00060

Barros E, Lezar S, Anttonen MJ, van Dijk JP, Rohlig RM, Kok EJ, Engel KH (2010) Comparison of two GM maize varieties with a near-isogenic non-GM variety using transcriptomics, proteomics and metabolomics. Plant Biotechnol J 8:436–451. https://doi.org/10.1111/j.1467-7652.2009.00487.x

Bastow R, Mylne JS, Lister C, Lippman Z, Martienssen RA, Dean C (2004) Vernalization requires epigenetic silencing of FLC by histone methylation. Nature 427:164–167. https://doi.org/10.1038/nature02269

Bayer PE, Golicz AA, Tirnaz S, Chan CK, Edwards D, Batley J (2019) Variation in abundance of predicted resistance genes in the Brassica oleracea pangenome. Plant Biotechnol J 17:789–800. https://doi.org/10.1111/pbi.13015

Bilder RM et al (2009) Phenomics: the systematic study of phenotypes on a genome-wide scale. Neuroscience 164:30–42. https://doi.org/10.1016/j.neuroscience.2009.01.027

Bitrian M, Zarza X, Altabella T, Tiburcio AF, Alcazar R (2012) Polyamines under abiotic stress: metabolic crossroads and hormonal crosstalks in plants. Metabolites 2:516–528. https://doi.org/10.3390/metabo2030516

Bjornson M et al (2017) Integrated omics analyses of retrograde signaling mutant delineate interrelated stress-response strata. Plant J 91:70–84. https://doi.org/10.1111/tpj.13547

Borras E, Sabido E (2017) What is targeted proteomics? A concise revision of targeted acquisition and targeted data analysis in mass spectrometry. Proteomics 17. https://doi.org/10.1002/pmic.201700180

Bradski G, Kaehler A (2008) Learning OpenCV–computer vision with the OpenCV library: software that sees

Cardona A, Tomancak P (2012) Current challenges in open-source bioimage informatics. Nat Methods 9:661–665. https://doi.org/10.1038/nmeth.2082

Cavill R, Jennen D, Kleinjans J, Briede JJ (2016) Transcriptomic and metabolomic data integration. Brief Bioinform 17:891–901. https://doi.org/10.1093/bib/bbv090

Chan SW, Henderson IR, Jacobsen SE (2005) Gardening the genome: DNA methylation in Arabidopsis thaliana. Nat Rev Genet 6:351–360. https://doi.org/10.1038/nrg1601

Chandramouli K, Qian PY (2009) Proteomics: challenges, techniques and possibilities to overcome biological sample complexity. Hum Genomics Proteomics 2009:239204. https://doi.org/10.4061/2009/239204

Chaturvedi P et al (2015) Heat-treatment-responsive proteins in different developmental stages of tomato pollen detected by targeted mass accuracy precursor alignment (tMAPA). J Proteome Res 14:4463–4471. https://doi.org/10.1021/pr501240n

Chaturvedi P, Ghatak A, Weckwerth W (2016) Pollen proteomics: from stress physiology to developmental priming plant. Reproduction 29:119–132. https://doi.org/10.1007/s00497-016-0283-9

Chaturvedi P, Govindaraj M, Govindan V, Weckwerth W (2022) Editorial: sorghum and pearl millet as climate resilient crops for food and nutrition security. Front Plant Sci 13:851970. https://doi.org/10.3389/fpls.2022.851970

Chaturvedi P, Ischebeck T, Egelhofer V, Lichtscheidl I, Weckwerth W (2013) Cell-specific analysis of the tomato pollen proteome from pollen mother cell to mature pollen provides evidence for developmental priming. J Proteome Res 12:4892–4903. https://doi.org/10.1021/pr400197p

Chaturvedi P, Wiese AJ, Ghatak A, Zaveska Drabkova L, Weckwerth W, Honys D (2021) Heat stress response mechanisms in pollen development. New Phytol 231:571–585. https://doi.org/10.1111/nph.17380

Chaves MM, Oliveira MM (2004) Mechanisms underlying plant resilience to water deficits: prospects for water-saving agriculture. J Exp Bot 55:2365–2384. https://doi.org/10.1093/jxb/erh269

Chebrolu KK, Fritschi FB, Ye SQ, Krishnan HB, Smith JR, Gillman JD (2016) Impact of heat stress during seed development on soybean seed metabolome. Metabolomics 12:28. https://doi.org/10.1007/s11306-015-0941-1

Cho K et al (2016) Network analysis of the metabolome and transcriptome reveals novel regulation of potato pigmentation. J Exp Bot 67:1519–1533. https://doi.org/10.1093/jxb/erv549

Cho K et al (2008) Integrated transcriptomics, proteomics, and metabolomics analyses to survey ozone responses in the leaves of rice seedling. J Proteome Res 7:2980–2998. https://doi.org/10.1021/pr800128q

Cokus SJ et al (2008) Shotgun bisulphite sequencing of the Arabidopsis genome reveals DNA methylation patterning. Nature 452:215–219. https://doi.org/10.1038/nature06745

Colmer J et al (2020) SeedGerm: a cost-effective phenotyping platform for automated seed imaging and machine-learning based phenotypic analysis of crop seed germination. New Phytol 228:778–793. https://doi.org/10.1111/nph.16736

Cooper M, Chapman SC, Podlich DW, Hammer GL (2002) The GP problem: quantifying gene-to-phenotype relationships. In Silico Biol 2:151–164

Copley TR, Aliferis KA, Kliebenstein DJ, Jabaji SH (2017) An integrated RNAseq-(1)H NMR metabolomics approach to understand soybean primary metabolism regulation in response to Rhizoctonia foliar blight disease. BMC Plant Biol 17:84. https://doi.org/10.1186/s12870-017-1020-8

Cortes AJ, Lopez-Hernandez F (2021) Harnessing crop wild diversity for climate change adaptation. Genes (Basel) 12:783. https://doi.org/10.3390/genes12050783

Cumming GS (2016) Heterarchies: reconciling networks and hierarchies. Trends Ecol Evol 31:622–632. https://doi.org/10.1016/j.tree.2016.04.009

Dalal M, Sahu S, Tiwari S, Rao AR, Gaikwad K (2018) Transcriptome analysis reveals interplay between hormones, ROS metabolism and cell wall biosynthesis for drought-induced root growth in wheat. Plant Physiol Biochem 130:482–492. https://doi.org/10.1016/j.plaphy.2018.07.035

de Abreu ELF, Li K, Wen W, Yan J, Nikoloski Z, Willmitzer L, Brotman Y (2018) Unraveling lipid metabolism in maize with time-resolved multi-omics data. Plant J 93:1102–1115. https://doi.org/10.1111/tpj.13833

Decourcelle M et al (2015) Combined transcript, proteome, and metabolite analysis of transgenic maize seeds engineered for enhanced carotenoid synthesis reveals pleotropic effects in core metabolism. J Exp Bot 66:3141–3150. https://doi.org/10.1093/jxb/erv120

Dhondt S, Wuyts N, Inze D (2013) Cell to whole-plant phenotyping: the best is yet to come. Trends Plant Sci 18:428–439. https://doi.org/10.1016/j.tplants.2013.04.008

Doerfler H et al (2013) Granger causality in integrated GC-MS and LC-MS metabolomics data reveals the interface of primary and secondary metabolism. Metabolomics 9:564–574. https://doi.org/10.1007/s11306-012-0470-0

Doerfler H, Sun XL, Wang L, Engelmeier D, Lyon D, Weckwerth W (2014) mzGroupAnalyzer-predicting pathways and novel chemical structures from untargeted high-throughput metabolomics data. Plos One 9:e96188. https://doi.org/10.1371/journal.pone.0096188

Dugas DV, Monaco MK, Olsen A, Klein RR, Kumari S, Ware D, Klein PE (2011) Functional annotation of the transcriptome of Sorghum bicolor in response to osmotic stress and abscisic acid. BMC Genomics 12:514. https://doi.org/10.1186/1471-2164-12-514

Engelhorn J, Blanvillain R, Carles CC (2014) Gene activation and cell fate control in plants: a chromatin perspective. Cell Mol Life Sci 71:3119–3137. https://doi.org/10.1007/s00018-014-1609-0

Fiorani F, Schurr U (2013) Future scenarios for plant phenotyping. Annu Rev Plant Biol 64:267–291. https://doi.org/10.1146/annurev-arplant-050312-120137

Fortes AM, Gallusci P (2017) Plant stress responses and phenotypic plasticity in the Epigenomics era: perspectives on the grapevine scenario, a model for perennial crop plants front. Plant Sci 8:82. https://doi.org/10.3389/fpls.2017.00082

Fujimoto R, Sasaki T, Ishikawa R, Osabe K, Kawanabe T, Dennis ES (2012) Molecular mechanisms of epigenetic variation in plants. Int J Mol Sci 13:9900–9922. https://doi.org/10.3390/ijms13089900

Furbank RT, Tester M (2011) Phenomics–technologies to relieve the phenotyping bottleneck. Trends Plant Sci 16:635–644. https://doi.org/10.1016/j.tplants.2011.09.005

Galland M et al (2017) An integrated "multi-omics" comparison of embryo and endosperm tissue-specific features and their impact on Rice seed quality. Front Plant Sci 8:1984. https://doi.org/10.3389/fpls.2017.01984

Gao C (2021) Genome engineering for crop improvement and future agriculture. Cell 184:1621–1635. https://doi.org/10.1016/j.cell.2021.01.005

Gao L et al (2019) The tomato pan-genome uncovers new genes and a rare allele regulating fruit flavor. Nat Genet 51:1044–1051. https://doi.org/10.1038/s41588-019-0410-2

Gargallo-Garriga A, Preece C, Sardans J, Oravec M, Urban O, Peñuelas J (2018) Root exudate metabolomes change under drought and show limited capacity for recovery. Sci Rep 8 (1):12696. https://doi.org/10.1038/s41598-018-30150-0

Ge C, Wang YG, Lu S, Zhao XY, Hou BK, Balint-Kurti PJ, Wang GF (2021) Multi-omics analyses reveal the regulatory network and the function of ZmUGTs in maize defense response front. Plant Sci 12:738261. https://doi.org/10.3389/fpls.2021.738261

Ghan R et al (2015) Five omic technologies are concordant in differentiating the biochemical characteristics of the berries of five grapevine (Vitis vinifera L.) cultivars. BMC Genomics 16: 946. https://doi.org/10.1186/s12864-015-2115-y

Ghatak A et al (2020) Physiological and proteomic signatures reveal mechanisms of superior drought resilience in pearl Millet compared to wheat front. Plant Sci 11:600278. https://doi.org/10.3389/fpls.2020.600278

Ghatak A et al (2016) Comprehensive tissue-specific proteome analysis of drought stress responses in Pennisetum glaucum (L.) R. Br. (Pearl millet). J Proteome 143:122–135. https://doi.org/10.1016/j.jprot.2016.02.032

Ghatak A et al (2017) Proteomics survey of Solanaceae family: current status and challenges ahead. J Proteome 169:41–57. https://doi.org/10.1016/j.jprot.2017.05.016

Ghatak A, Chaturvedi P, Weckwerth W (2018) Metabolomics in plant stress physiology. Adv Biochem Eng Biot 164:187–236. https://doi.org/10.1007/10_2017_55

Ghatak A et al (2021) Root exudation of contrasting drought-stressed pearl millet genotypes conveys varying biological nitrification inhibition (BNI) activity. Biol Fert Soils 58:291. https://doi.org/10.1007/s00374-021-01578-w

Gianola D (2013) Priors in whole-genome regression: the bayesian alphabet returns. Genetics 194: 573–596. https://doi.org/10.1534/genetics.113.151753

Gillet LC, Leitner A, Aebersold R (2016) Mass spectrometry applied to bottom-up proteomics: entering the high-throughput era for hypothesis testing. Annu Rev Anal Chem (Palo Alto Calif) 9:449–472. https://doi.org/10.1146/annurev-anchem-071015-041535

Goff SA et al (2011) The iPlant collaborative: cyberinfrastructure for plant biology front. Plant Sci 2:34. https://doi.org/10.3389/fpls.2011.00034

Grosskinsky DK, Svensgaard J, Christensen S, Roitsch T (2015) Plant phenomics and the need for physiological phenotyping across scales to narrow the genotype-to-phenotype knowledge gap. J Exp Bot 66:5429–5440. https://doi.org/10.1093/jxb/erv345

Grosskinsky DK, Syaifullah SJ, Roitsch T (2018) Integration of multi-omics techniques and physiological phenotyping within a holistic phenomics approach to study senescence in model and crop plants. J Exp Bot 69:825–844. https://doi.org/10.1093/jxb/erx333

Gunnaiah R, Kushalappa AC, Duggavathi R, Fox S, Somers DJ (2012) Integrated metabolo-proteomic approach to decipher the mechanisms by which wheat QTL (Fhb1) contributes to resistance against fusarium graminearum. PLoS One 7:e40695. https://doi.org/10.1371/journal.pone.0040695

Guo J, Wu Y, Wang G, Wang T, Cao F (2020) Integrated analysis of the transcriptome and metabolome in young and mature leaves of Ginkgo biloba L. Ind Crop Prod 143:111906. https://doi.org/10.1016/j.indcrop.2019.111906

Guo P et al (2009) Differentially expressed genes between drought-tolerant and drought-sensitive barley genotypes in response to drought stress during the reproductive stage. J Exp Bot 60: 3531–3544. https://doi.org/10.1093/jxb/erp194

Guy CL (1990) Cold-acclimation and freezing stress tolerance–role of protein-metabolism. Annu Rev Plant Phys 41:187–223. https://doi.org/10.1146/annurev.pp.41.060190.001155

Habier D, Fernando RL, Garrick DJ (2013) Genomic BLUP decoded: a look into the black box of genomic prediction. Genetics 194:597–607. https://doi.org/10.1534/genetics.113.152207

Habier D, Fernando RL, Kizilkaya K, Garrick DJ (2011) Extension of the bayesian alphabet for genomic selection. BMC Bioinformatics 12:186. https://doi.org/10.1186/1471-2105-12-186

He J, Zhao X, Laroche A, Lu ZX, Liu H, Li Z (2014) Genotyping-by-sequencing (GBS), an ultimate marker-assisted selection (MAS) tool to accelerate plant breeding. Front Plant Sci 5: 484. https://doi.org/10.3389/fpls.2014.00484

Hoehenwarter W, Chen Y, Recuenco-Munoz L, Wienkoop S, Weckwerth W (2011a) Functional analysis of proteins and protein species using shotgun proteomics and linear mathematics. Amino Acids 41:329–341. https://doi.org/10.1007/s00726-010-0669-1

Hoehenwarter W et al (2011b) MAPA distinguishes genotype-specific variability of highly similar regulatory protein isoforms in potato tuber. J Proteome Res 10:2979–2991. https://doi.org/10.1021/pr101109a

Hoehenwarter W et al (2008) A rapid approach for phenotype-screening and database independent detection of cSNP/protein polymorphism using mass accuracy precursor alignment. Proteomics 8:4214–4225. https://doi.org/10.1002/pmic.200701047

Houle D, Govindaraju DR, Omholt S (2010) Phenomics: the next challenge. Nat Rev Genet 11: 855–866. https://doi.org/10.1038/nrg2897

Hu J, Rampitsch C, Bykova NV (2015) Advances in plant proteomics toward improvement of crop productivity and stress resistancex front. Plant Sci 6:209. https://doi.org/10.3389/fpls.2015.00209

Huang XY, Chao DY, Gao JP, Zhu MZ, Shi M, Lin HX (2009) A previously unknown zinc finger protein, DST, regulates drought and salt tolerance in rice via stomatal aperture control. Genes Dev 23:1805–1817. https://doi.org/10.1101/gad.1812409

Iquebal MA et al (2019) RNAseq analysis reveals drought-responsive molecular pathways with candidate genes and putative molecular markers in root tissue of wheat. Sci Rep 9:13917. https://doi.org/10.1038/s41598-019-49915-2

Jacoby RP, Millar AH, Taylor NL (2013) Application of selected reaction monitoring mass spectrometry to field-grown crop plants to allow dissection of the molecular mechanisms of abiotic stress tolerance front. Plant Sci 4:20. https://doi.org/10.3389/fpls.2013.00020

Jamil IN, Remali J, Azizan KA, Nor Muhammad NA, Arita M, Goh HH, Aizat WM (2020) Systematic multi-omics integration (MOI) approach in plant systems biology front. Plant Sci 11:944. https://doi.org/10.3389/fpls.2020.00944

Jan N et al (2022) Proteomics for abiotic stresses in legumes: present status and future directions. Crit Rev Biotechnol 43:1–20. https://doi.org/10.1080/07388551.2021.2025033

Jegadeesan S et al (2018) Proteomics of heat-stress and ethylene-mediated thermotolerance mechanisms in tomato pollen grains. Front Plant Sci 9:9. https://doi.org/10.3389/fpls.2018.01558

Ji Z, Yan K, Li W, Hu H, Zhu X (2017) Mathematical and computational modeling in complex biological systems. Biomed Res Int 2017:5958321. https://doi.org/10.1155/2017/5958321

Jiang W, Zhou H, Bi H, Fromm M, Yang B, Weeks DP (2013) Demonstration of CRISPR/Cas9/sgRNA-mediated targeted gene modification in Arabidopsis, tobacco, sorghum and rice. Nucleic Acids Res 41:e188. https://doi.org/10.1093/nar/gkt780

Kashtwari M, Wani AA, Rather RN (2019) TILLING: an alternative path for crop improvement. J Crop Improv 33:83–109. https://doi.org/10.1080/15427528.2018.1544954

Ke R, Mignardi M, Pacureanu A, Svedlund J, Botling J, Wahlby C, Nilsson M (2013) In situ sequencing for RNA analysis in preserved tissue and cells. Nat Methods 10:857–860. https://doi.org/10.1038/nmeth.2563

Keller M, Consortium S-I, Simm S (2018) The coupling of transcriptome and proteome adaptation during development and heat stress response of tomato pollen. BMC Genomics 19:447. https://doi.org/10.1186/s12864-018-4824-5

Kirchgessner N, Liebisch F, Yu K, Pfeifer J, Friedli M, Hund A, Walter A (2016) The ETH field phenotyping platform FIP: a cable-suspended multi-sensor system. Funct Plant Biol 44:154–168. https://doi.org/10.1071/FP16165

Kole C et al (2015) Application of genomics-assisted breeding for generation of climate resilient crops: progress and prospects front. Plant Sci 6:563. https://doi.org/10.3389/fpls.2015.00563

Komatsu S, Hossain Z (2013) Organ-specific proteome analysis for identification of abiotic stress response mechanism in crop front. Plant Sci 4:71. https://doi.org/10.3389/fpls.2013.00071

Kover PX et al (2009) A Multiparent advanced generation inter-cross to fine-map quantitative traits in Arabidopsis thaliana. Plos Genet 5:e1000551. https://doi.org/10.1371/journal.pgen.1000551

Kumar A et al (2021) Integrating omics and gene editing tools for rapid improvement of traditional food plants for diversified and sustainable food security. Int J Mol Sci 22:22. https://doi.org/10.3390/ijms22158093

Kumar G, Rattan UK, Singh AK (2016) Chilling-mediated DNA methylation changes during dormancy and its release reveal the importance of epigenetic regulation during winter dormancy in apple (malus x domestica Borkh.). PLoS One 11:e0149934. https://doi.org/10.1371/journal.pone.0149934

Kurowska M, Daszkowska-Golec A, Gruszka D, Marzec M, Szurman M, Szarejko I, Maluszynski M (2011) TILLING - a shortcut in functional genomics. J Appl Genet 52:371. https://doi.org/10.1007/s13353-011-0061-1

Lamaoui M, Jemo M, Datla R, Bekkaoui F (2018) Heat and drought stresses in crops and approaches for their mitigation. Front Chem 6:26. https://doi.org/10.3389/fchem.2018.00026

Lasky JR et al (2015) Genome-environment associations in sorghum landraces predict adaptive traits. Sci Adv 1:e1400218. https://doi.org/10.1126/sciadv.1400218

Le DT et al (2012) Differential gene expression in soybean leaf tissues at late developmental stages under drought stress revealed by genome-wide transcriptome analysis. PLoS One 7:e49522. https://doi.org/10.1371/journal.pone.0049522

Le TN et al (2014) DNA demethylases target promoter transposable elements to positively regulate stress responsive genes in Arabidopsis. Genome Biol 15:458. https://doi.org/10.1186/s13059-014-0458-3

Liang J et al (2018) Constitutive expression of REL1 confers the rice response to drought stress and abscisic acid. Rice (N Y) 11:59. https://doi.org/10.1186/s12284-018-0251-0

Lin Y et al (2017) Transcriptome profiling and digital gene expression analysis of sweet potato for the identification of putative genes involved in the defense response against fusarium oxysporum f. sp. batatas. PLoS One 12:e0187838. https://doi.org/10.1371/journal.pone.0187838

Liu Z et al (2019) Integrative transcriptome and proteome analysis identifies major metabolic pathways involved in pepper fruit development. J Proteome Res 18:982–994. https://doi.org/10.1021/acs.jproteome.8b00673

Lobos GA, Camargo AV, Del Pozo A, Araus JL, Ortiz R, Doonan JH (2017) Editorial: plant phenotyping and Phenomics for plant breeding front. Plant Sci 8:2181. https://doi.org/10.3389/fpls.2017.02181

Locke JC, Millar AJ, Turner MS (2005) Modelling genetic networks with noisy and varied experimental data: the circadian clock in Arabidopsis thaliana. J Theor Biol 234:383–393. https://doi.org/10.1016/j.jtbi.2004.11.038

Macková H, Hronková M, Dobrá J, Turečková V, Novák O, Lubovská Z, Motyka V, Haisel D, Hájek T, Prášil IT, Gaudinová A, Štorchová H, Ge E, Werner T, Schmülling T, Vanková R (2013) Enhanced drought and heat stress tolerance of tobacco plants with ectopically enhanced cytokinin oxidase/dehydrogenase gene expression. J Exp Bot 64(10):2805–2815. https://doi.org/10.1093/jxb/ert131

Mahalingam R (2017) Shotgun proteomics of the barley seed proteome. BMC Genomics 18:44. https://doi.org/10.1186/s12864-016-3408-5

Marcek T, Hamow KA, Vegh B, Janda T, Darko E (2019) Metabolic response to drought in six winter wheat genotypes. PLoS One 14:e0212411. https://doi.org/10.1371/journal.pone.0212411

Marjoram P, Zubair A, Nuzhdin SV (2014) Post-GWAS: where next? More samples, more SNPs or more biology? Heredity (Edinb) 112:79–88. https://doi.org/10.1038/hdy.2013.52

Massa AN, Childs KL, Buell CR (2013) Abiotic and biotic stress responses in Solanum tuberosum group Phureja DM1-3 516 R44 as measured through whole transcriptome sequencing. Plant Genome 6(3). https://doi.org/10.3835/plantgenome2013.05.0014

Mata CI et al (2018) Ethylene receptors, CTRs and EIN2 target protein identification and quantification through parallel reaction monitoring during tomato fruit ripening. Front Plant Sci 9:1626. https://doi.org/10.3389/fpls.2018.01626

McCouch S et al (2013) Agriculture: feeding the future. Nature 499:23–24. https://doi.org/10.1038/499023a

McGrail R, Sanford D, McNear D (2020) Trait-based root phenotyping as a necessary tool for crop selection and improvement. Agronomy 10:1328. https://doi.org/10.3390/agronomy10091328

Mensack MM, Fitzgerald VK, Ryan EP, Lewis MR, Thompson HJ, Brick MA (2010) Evaluation of diversity among common beans (Phaseolus vulgaris L.) from two centers of domestication using 'omics' technologies. BMC Genomics 11:686. https://doi.org/10.1186/1471-2164-11-686

Mesnage R et al (2016) An integrated multi-omics analysis of the NK603 roundup-tolerant GM maize reveals metabolism disturbances caused by the transformation process. Sci Rep 6:37855. https://doi.org/10.1038/srep37855

Michaletti A, Naghavi MR, Toorchi M, Zolla L, Rinalducci S (2018) Metabolomics and proteomics reveal drought-stress responses of leaf tissues from spring-wheat. Sci Rep 8:5710. https://doi.org/10.1038/s41598-018-24012-y

Millet EJ et al (2016) Genome-wide analysis of yield in Europe: allelic effects vary with drought and heat scenarios. Plant Physiol 172:749–764. https://doi.org/10.1104/pp.16.00621

Monreal CM, Schnitzer MI (2015) Labile organic matter in soil solution: II. Separation and identification of metabolites from plant-microbial communication in soil solutions of wheat rhizospheres. Sssa Spec Publ 62:173–193. https://doi.org/10.2136/sssaspecpub62.2014.0074

Moore BM et al (2019) Robust predictions of specialized metabolism genes through machine learning. Proc Natl Acad Sci U S A 116:2344–2353. https://doi.org/10.1073/pnas.1817074116

Muthamilarasan M, Singh NK, Prasad M (2019) Multi-omics approaches for strategic improvement of stress tolerance in underutilized crop species: a climate change perspective. Adv Genet 103: 1–38. https://doi.org/10.1016/bs.adgen.2019.01.001

Nagele T, Mair A, Sun XL, Fragner L, Teige M, Weckwerth W (2014) Solving the differential biochemical Jacobian from metabolomics covariance data. Plos One 9:e92299. https://doi.org/10.1371/journal.pone.0092299

Nagele T, Weckwerth W (2013) A workflow for mathematical modeling of subcellular metabolic pathways in leaf metabolism of Arabidopsis thaliana. Front Plant Sci 4:541. https://doi.org/10.3389/fpls.2013.00541

Nägele TWW (2013) Eigenvalues of jacobian matrices report on steps of metabolic reprogramming in a complex plant-environment interaction. Appl Math Ser B 4:44–49. https://doi.org/10.4236/am.2013.48A007

Nagler M, Nukarinen E, Weckwerth W, Nagele T (2015) Integrative molecular profiling indicates a central role of transitory starch breakdown in establishing a stable C/N homeostasis during cold acclimation in two natural accessions of Arabidopsis thaliana Bmc. Plant Biol 15:284. https://doi.org/10.1186/s12870-015-0668-1

Nichol D, Robertson-Tessi M, Anderson ARA, Jeavons P (2019) Model genotype-phenotype mappings and the algorithmic structure of evolution. J R Soc Interface 16:20190332. https://doi.org/10.1098/rsif.2019.0332

Nouri MZ, Komatsu S (2013) Subcellular protein overexpression to develop abiotic stress tolerant plants front. Plant Sci 4:2. https://doi.org/10.3389/fpls.2013.00002

Nukarinen E et al (2016) Quantitative phosphoproteomics reveals the role of the AMPK plant ortholog SnRK1 as a metabolic master regulator under energy deprivation. Sci Rep-UK 6: 31697. https://doi.org/10.1038/srep31697

Orr HA (1998) The population genetics of adaptation: the distribution of factors fixed during adaptive evolution. Evolution 52:935–949. https://doi.org/10.1111/j.1558-5646.1998.tb01823.x

Orth JD, Thiele I, Palsson BO (2010) What is flux balance analysis? Nat Biotechnol 28:245–248. https://doi.org/10.1038/nbt.1614

Pathak RK, Baunthiyal M, Pandey D, Kumar A (2018) Augmentation of crop productivity through interventions of omics technologies in India: challenges and opportunities 3. Biotech 8:454. https://doi.org/10.1007/s13205-018-1473-y

Paul P et al (2016) The membrane proteome of male gametophyte in Solanum lycopersicum. J Proteome 131:48–60. https://doi.org/10.1016/j.jprot.2015.10.009

Pazhamala LT et al (2020) Multiomics approach unravels fertility transition in a pigeonpea line for a two-line hybrid system the plant. Genome 13:e20028. https://doi.org/10.1002/tpg2.20028

Pazhamala LT, Kudapa H, Weckwerth W, Millar AH, Varshney RK (2021) Systems biology for crop improvement. Plant Genome-Us 14:e20098. https://doi.org/10.1002/tpg2.20098

Pedregosa F et al (2012) Scikit-learn: machine learning in python, J Mach Learn Res:12

Peremarti A, Mare C, Aprile A, Roncaglia E, Cattivelli L, Villegas D, Royo C (2014) Transcriptomic and proteomic analyses of a pale-green durum wheat mutant shows variations in photosystem components and metabolic deficiencies under drought stress. BMC Genomics 15:125. https://doi.org/10.1186/1471-2164-15-125

Phitaktansakul R et al (2021) Multi-omics analysis reveals the genetic basis of rice fragrance mediated by betaine aldehyde dehydrogenase 2. J Adv Res 42:303. https://doi.org/10.1016/j.jare.2021.12.004

Picard M, Scott-Boyer MP, Bodein A, Perin O, Droit A (2021) Integration strategies of multi-omics data for machine learning analysis Comput Struct. Biotechnol J 19:3735–3746. https://doi.org/10.1016/j.csbj.2021.06.030

Pieruschka R, Schurr U (2019) Plant phenotyping: past, present, and future. Plant Phenomics 2019: 7507131. https://doi.org/10.34133/2019/7507131

Qi X, Xie S, Liu Y, Yi F, Yu J (2013) Genome-wide annotation of genes and noncoding RNAs of foxtail millet in response to simulated drought stress by deep sequencing. Plant Mol Biol 83: 459–473. https://doi.org/10.1007/s11103-013-0104-6

Ranganathan J, Waite R, Searchinger T, Hanson C (2018) How to sustainably feed 10 billion people by 2050, in 21 Charts

Raza A et al (2021a) Integrated analysis of metabolome and transcriptome reveals insights for cold tolerance in rapeseed (Brassica napus L.). Front Plant Sci 12:721681. https://doi.org/10.3389/fpls.2021.721681

Raza A, Tabassum J, Kudapa H, Varshney RK (2021b) Can omics deliver temperature resilient ready-to-grow crops? Crit Rev Biotechnol 41:1209–1232. https://doi.org/10.1080/07388551.2021.1898332

Roustan V et al (2018) Microscopic and proteomic analysis of dissected developing barley endosperm layers reveals the starchy endosperm as prominent storage tissue for ER-derived Hordeins alongside the accumulation of barley protein disulfide isomerase (HvPDIL1-1). Front Plant Sci 9:1248. https://doi.org/10.3389/fpls.2018.01248

Sah SK, Reddy KR, Li J (2016) Abscisic acid and abiotic stress tolerance in crop plants. Front Plant Sci 7:571. https://doi.org/10.3389/fpls.2016.00571

Sana TR, Fischer S, Wohlgemuth G, Katrekar A, Jung KH, Ronald PC, Fiehn O (2010) Metabolomic and transcriptomic analysis of the rice response to the bacterial blight pathogen Xanthomonas oryzae pv. Oryzae. Metabolomics 6:451–465. https://doi.org/10.1007/s11306-010-0218-7

Sanger F, Nicklen S, Coulson AR (1977) DNA sequencing with chain-terminating inhibitors. Proc Natl Acad Sci U S A 74:5463–5467. https://doi.org/10.1073/pnas.74.12.5463

Schmid MW, Schmidt A, Grossniklaus U (2015) The female gametophyte: an emerging model for cell type-specific systems biology in plant development. Front Plant Sci 6:907. https://doi.org/10.3389/fpls.2015.00907

Schork NJ (1997) Genetics of complex disease: approaches, problems, and solutions. Am J Respir Crit Care Med 156:S103–S109. https://doi.org/10.1164/ajrccm.156.4.12-tac-5

Schuhmacher R, Krska R, Weckwerth W, Goodacre R (2013) Metabolomics and metabolite profiling. Anal Bioanal Chem 405:5003–5004. https://doi.org/10.1007/s00216-013-6939-5

Scossa F, Alseekh S, Fernie AR (2021) Integrating multi-omics data for crop improvement. J Plant Physiol 257:153352. https://doi.org/10.1016/j.jplph.2020.153352

Serraj R, Sinclair TR (2002) Osmolyte Accumulation: Can It Really Help Increase Crop Yield under Drought Conditions? Plant Cell Environ 25:333–341. https://doi.org/10.1046/j.1365-3040.2002.00754.x

Shabrangy A, Ghatak A, Zhang S, Priller A, Chaturvedi P, Weckwerth W (2021) Magnetic field induced changes in the shoot and root proteome of barley (Hordeum vulgare L.). Front Plant Sci 12:622795. https://doi.org/10.3389/fpls.2021.622795

Shabrangy A et al (2018) Using RT-qPCR, proteomics, and microscopy to unravel the Spatio-temporal expression and subcellular localization of Hordoindolines across development in barley endosperm. Front Plant Sci 9:775. https://doi.org/10.3389/fpls.2018.00775

Silva JCF, Teixeira RM, Silva FF, Brommonschenkel SH, Fontes EPB (2019) Machine learning approaches and their current application in plant molecular biology: a systematic review. Plant Sci 284:37–47. https://doi.org/10.1016/j.plantsci.2019.03.020

Singh RK, Muthamilarasan M, Prasad M (2021) Biotechnological approaches to dissect climate-resilient traits in millets and their application in crop improvement. J Biotechnol 327:64–73. https://doi.org/10.1016/j.jbiotec.2021.01.002

Song JM et al (2020) Eight high-quality genomes reveal pan-genome architecture and ecotype differentiation of Brassica napus. Nat Plants 6:34–45. https://doi.org/10.1038/s41477-019-0577-7

Soubeyrand E et al (2018) Constraint-based modeling highlights cell energy, redox status and alpha-ketoglutarate availability as metabolic drivers for anthocyanin accumulation in grape cells under nitrogen limitation. Front Plant Sci 9:421. https://doi.org/10.3389/fpls.2018.00421

Steward N, Ito M, Yamaguchi Y, Koizumi N, Sano H (2002) Periodic DNA methylation in maize nucleosomes and demethylation by environmental stress. J Biol Chem 277:37741–37746. https://doi.org/10.1074/jbc.M204050200

Stroud H et al (2014) Non-CG methylation patterns shape the epigenetic landscape in Arabidopsis. Nat Struct Mol Biol 21:64–72. https://doi.org/10.1038/nsmb.2735

Sun XL, Weckwerth W (2012) COVAIN: a toolbox for uni- and multivariate statistics, time-series and correlation network analysis and inverse estimation of the differential Jacobian from metabolomics covariance data. Metabolomics 8:S81–S93. https://doi.org/10.1007/s11306-012-0399-3

Tardieu F, Cabrera-Bosquet L, Pridmore T, Bennett M (2017) Plant Phenomics, from sensors to knowledge. Curr Biol 27:R770–R783. https://doi.org/10.1016/j.cub.2017.05.055

Todaka D, Shinozaki K, Yamaguchi-Shinozaki K (2015) Recent advances in the dissection of drought-stress regulatory networks and strategies for development of drought-tolerant transgenic rice plants. Front Plant Sci 6:84. https://doi.org/10.3389/fpls.2015.00084

Todaka D et al (2017) Temporal and spatial changes in gene expression, metabolite accumulation and phytohormone content in rice seedlings grown under drought stress conditions. Plant J 90: 61–78. https://doi.org/10.1111/tpj.13468

Tong H, Kuken A, Nikoloski Z (2020) Integrating molecular markers into metabolic models improves genomic selection for Arabidopsis growth. Nat Comm 11:2410. https://doi.org/10.1038/s41467-020-16279-5

Urano K, Kurihara Y, Seki M, Shinozaki K (2010) 'Omics' analyses of regulatory networks in plant abiotic stress responses. Curr Plant Biol 13:132. https://doi.org/10.1016/j.pbi.2009.12.006

Urbanczyk-Wochniak E, Luedemann A, Kopka J, Selbig J, Roessner-Tunali U, Willmitzer L, Fernie AR (2003) Parallel analysis of transcript and metabolic profiles: a new approach in systems biology. EMBO Rep 4:989–993. https://doi.org/10.1038/sj.embor.embor944

Usai MG, Goddard ME, Hayes BJ (2009) LASSO with cross-validation for genomic selection. Genet Res (Camb) 91:427–436. https://doi.org/10.1017/S0016672309990334

van der Walt S et al (2014) scikit-image: image processing in python. PeerJ 2:e453. https://doi.org/10.7717/peerj.453

van Dijk K et al (2010) Dynamic changes in genome-wide histone H3 lysine 4 methylation patterns in response to dehydration stress in Arabidopsis thaliana. BMC Plant Biol 10:238. https://doi.org/10.1186/1471-2229-10-238

Varshney RK, Graner A, Sorrells ME (2005) Genomics-assisted breeding for crop improvement. Trends Plant Sci 10:621–630. https://doi.org/10.1016/j.tplants.2005.10.004

Wang JP et al (2018) Improving wood properties for wood utilization through multi-omics integration in lignin biosynthesis. Nat Comm 9:1579. https://doi.org/10.1038/s41467-018-03863-z

Wang JY et al (2021) Multi-omics approaches explain the growth-promoting effect of the apocarotenoid growth regulator zaxinone in rice. Commun Biol 4:1222. https://doi.org/10.1038/s42003-021-02740-8

Wang L et al (2016) System level analysis of cacao seed ripening reveals a sequential interplay of primary and secondary metabolism leading to polyphenol accumulation and preparation of stress resistance. Plant J 87:318–332. https://doi.org/10.1111/tpj.13201

Wang L, Sun X, Weiszmann J, Weckwerth W (2017) System-level and granger network analysis of integrated proteomic and Metabolomic dynamics identifies key points of grape berry development at the Interface of primary and secondary metabolism. Front Plant Sci 8:1066. https://doi.org/10.3389/fpls.2017.01066

Weckwerth W (2003) Metabolomics in systems biology. Annu Rev Plant Biol 54:669–689. https://doi.org/10.1146/annurev.arplant.54.031902.135014
Weckwerth W (2010) Metabolomics: an integral technique in systems biology. Bioanalysis 2:829–836. https://doi.org/10.4155/bio.09.192
Weckwerth W (2011) Green systems biology–from single genomes, proteomes and metabolomes to ecosystems research and biotechnology. J Proteome 75:284–305. https://doi.org/10.1016/j.jprot.2011.07.010
Weckwerth W (2019) Toward a unification of system-theoretical principles in biology and ecology—the stochastic lyapunov matrix equation and its inverse application. Front Appl Math Stat 5. https://doi.org/10.3389/fams.2019.00029
Weckwerth W, Fiehn O (2002) Can we discover novel pathways using metabolomic analysis? Curr Opin Biotechnol 13:156–160. https://doi.org/10.1016/s0958-1669(02)00299-9
Weckwerth W, Ghatak A, Bellaire A, Chaturvedi P, Varshney RK (2020) PANOMICS meets germplasm. Plant Biotechnol J 18:1507–1525. https://doi.org/10.1111/pbi.13372
Weckwerth W, Wenzel K, Fiehn O (2004) Process for the integrated extraction, identification and quantification of metabolites, proteins and RNA to reveal their co-regulation in biochemical networks. Proteomics 4:78–83. https://doi.org/10.1002/pmic.200200500
Whittaker C, Dean C (2017) The FLC Locus: a platform for discoveries in epigenetics and adaptation. Annu Rev Cell Dev Biol 33:555–575. https://doi.org/10.1146/annurev-cellbio-100616-060546
Wienkoop S, Larrainzar E, Glinski M, Gonzalez EM, Arrese-Igor C, Weckwerth W (2008a) Absolute quantification of Medicago truncatula sucrose synthase isoforms and N-metabolism enzymes in symbiotic root nodules and the detection of novel nodule phosphoproteins by mass spectrometry. J Exp Bot 59:3307–3315. https://doi.org/10.1093/jxb/ern182
Wienkoop S, Morgenthal K, Wolschin F, Scholz M, Selbig J, Weckwerth W (2008b) Integration of metabolomic and proteomic phenotypes. Mol Cell Proteomics 7:1725–1736. https://doi.org/10.1074/mcp.M700273-MCP200
Wienkoop S, Weckwerth W (2006) Relative and absolute quantitative shotgun proteomics: targeting low-abundance proteins in Arabidopsis thaliana. J Exp Bot 57:1529–1535. https://doi.org/10.1093/jxb/erj157
Wienkoop S et al (2010) Targeted proteomics for Chlamydomonas reinhardtii combined with rapid subcellular protein fractionation, metabolomics and metabolic flux analyses. Mol Biosyst 6:1018. https://doi.org/10.1039/b920913a
Wittmann DM, Krumsiek J, Saez-Rodriguez J, Lauffenburger DA, Klamt S, Theis FJ (2009) Transforming Boolean models to continuous models: methodology and application to T-cell receptor signaling. BMC Syst Biol 3:98. https://doi.org/10.1186/1752-0509-3-98
Xiong L, Schumaker KS, Zhu JK (2002) Cell signaling during cold, drought, and salt stress. Plant Cell 14:S165. https://doi.org/10.1105/tpc.000596
Yang W et al (2020) Crop Phenomics and high-throughput phenotyping: past decades, current challenges, and future perspectives. Mol Plant 13:187–214. https://doi.org/10.1016/j.molp.2020.01.008
Yang Y et al (2021) Applications of multi-omics technologies for crop improvement. Front Plant Sci 12:563953. https://doi.org/10.3389/fpls.2021.563953
You J et al (2019) Transcriptomic and metabolomic profiling of drought-tolerant and susceptible sesame genotypes in response to drought stress. BMC Plant Biol 19:267. https://doi.org/10.1186/s12870-019-1880-1
Young AI (2019) Solving the missing heritability problem. Plos Genet 15:e1008222. https://doi.org/10.1371/journal.pgen.1008222
Yu J, Holland JB, McMullen MD, Buckler ES (2008) Genetic design and statistical power of nested association mapping in maize. Genetics 178:539–551. https://doi.org/10.1534/genetics.107.074245

Zhang S et al (2021) Spatial distribution of proteins and metabolites in developing wheat grain and their differential regulatory response during the grain filling process. Plant J 107:669–687. https://doi.org/10.1111/tpj.15410

Zhang YY, Fonslow BR, Shan B, Baek MC, Yates JR (2013) Protein analysis by shotgun/bottom-up proteomics. Chem Rev 113:2343–2394. https://doi.org/10.1021/cr3003533

Zhao C et al (2017) Temperature increase reduces global yields of major crops in four independent estimates P. Natl Acad Sci USA 114:9326–9331. https://doi.org/10.1073/pnas.1701762114

Zhao Y, Zhou M, Xu K, Li J, Li S, Zhang S, Yang X (2019) Integrated transcriptomics and metabolomics analyses provide insights into cold stress response in wheat. The Crop J 7:857–866. https://doi.org/10.1016/j.cj.2019.09.002

Zhou J et al (2017) Leaf-GP: an open and automated software application for measuring growth phenotypes for arabidopsis and wheat. Plant Methods 13:117. https://doi.org/10.1186/s13007-017-0266-3

Chapter 7
Sequence-Based Breeding for Plant Improvement

Pallavi Sinha, Mallana Gowdra Mallikarjuna, Vinay Nandigam, Sonali Habade, Krishna Tesman Sundaram, Prasanna Rajesh, Uma Maheshwar Singh, and Vikas Kumar Singh

Abstract This book chapter discusses the significant improvement in breeding programs' utilizing modern genomic tools and technologies. Next-generation sequencing technologies have enabled the availability of genome sequence assemblies, high-density genetic maps, various marker genotyping platforms, and the identification of markers associated with numerous agronomic traits in several economically important crops. While marker-assisted backcrossing and selection approaches have been employed to develop improved lines, there is a need for continuous population improvement after every breeding cycle to hasten the genetic gains in crops. This book chapter proposes a sequencing-based breeding approach involving parental selection, enhancement of genetic diversity of breeding programs, forward breeding for early generation selection, and genomic selection using sequencing/genotyping technologies. Additionally, we discussed the speed breeding technology, which allows for 4–6 generations per year and can accelerate genetic gain much faster. The book chapter is described in two sections: the first focuses on sequencing-based trait mapping, and the second is geared toward sequencing-based

---

P. Sinha · K. T. Sundaram
IRRI, South Asia Hub, ICRISAT Campus, Patancheru, India

M. G. Mallikarjuna
Division of Genetics, ICAR-Indian Agricultural Research Institute, New Delhi, India

V. Nandigam
IRRI, South Asia Hub, ICRISAT Campus, Patancheru, India

Agricultural College, Acharya NG Ranga Agricultural University, Bapatla, India

S. Habade · U. M. Singh
IRRI, South Asia Regional Center (ISARC), Varanasi, India

P. Rajesh
Agricultural College, Acharya NG Ranga Agricultural University, Bapatla, India

V. K. Singh (✉)
IRRI, South Asia Hub, ICRISAT Campus, Patancheru, India

IRRI, South Asia Regional Center (ISARC), Varanasi, India
e-mail: v.k.singh@irri.org

M. K. Pandey et al. (eds.), *Frontier Technologies for Crop Improvement*, Sustainable Agriculture and Food Security, https://doi.org/10.1007/978-981-99-4673-0_7

breeding for crop improvement. The potential of sequence-based breeding applications, including trait discovery, is immense and could revolutionize crop improvement programs.

7.1 Introduction

Over the past decade, the efficiency of breeding programs for important crops has significantly improved due to the utilization of modern genomic tools and technologies. The advent of next-generation sequencing technologies has allowed for the availability of genome sequence assemblies, the re-sequencing of hundreds of lines, high-density genetic maps, various marker genotyping platforms, and the identification of markers associated with numerous agronomic traits in different crops. While marker-assisted backcrossing and selection approaches have been employed to develop superior lines in several cases, there is an increasing need for continuous population improvement after every breeding cycle to hasten genetic gain in breeding programs. A sequence-based breeding approach involving independent or combined parental selection, enhancement of genetic diversity of breeding programs, forward breeding for early generation selection, and genomic selection using sequencing/genotyping technologies is necessary. In addition, the adoption of speed breeding technology, which allows for the generation of 4–6 generations per year, can aid in accelerating genetic gain. The potential of sequence-based breeding applications, including trait discovery, is immense and could revolutionize crop improvement programs. This book chapter describes prominent approaches in two sections: the first section focuses on sequencing-based trait mapping, and the second section is geared towards sequencing-based breeding for crop improvement.

7.2 Sequencing-Based Trait Mapping

The mapping of agronomic traits is a crucial step in crop improvement, as it can enhance the genetic gain per selection cycle. Trait mapping in plants using molecular markers was initiated with the identification of quantitative trait loci (QTLs) for yield and its contributing factors in maize and was further used in the tomato to map QTLs for fruit size and other traits. Since then, several molecular tools and statistical algorithms have been devised and widely used to map traits of importance in both model plants and crops. However, many of these techniques have several drawbacks, including being costly in terms of time and resources and requiring a large number of polymorphic markers and large segregating populations. To overcome the issue of the nonavailability of polymorphic markers, bulked segregant analysis (BSA) was proposed by Michelmore et al. (1991). BSA analysis creates conceptual near-isogenic lines by pooling DNA samples from inbred lines showing contrasting phenotypes. Markers exhibiting polymorphism between bulked pools of contrasting traits, as well as parental lines, are considered putatively linked to the trait of interest. However, BSA has some limitations, such as the nonavailability of sufficient DNA markers and low-throughput genotyping.

Advancements in genomics have led to the development of next-generation sequencing (NGS) technologies, which have revolutionized crop improvement programs and provided the foundation for genomics-assisted breeding (GAB). NGS technologies have allowed for the modification and improvement of traditionally difficult and time-consuming bulked segregant analysis (BSA) into rapid and whole-genome sequencing-based high-resolution trait mapping. With cheaper sequencing costs and high-throughput NGS tools, whole genome sequencing has become routine. The availability of draft sequences of a species accelerates the resequencing of multiple individuals of that species, allowing for the rapid identification of genomic variations and mapping and isolation of genes for causative mutations/target traits. There are several sequencing-based trait mapping tools, including SHOREmap (Schneeberger et al. 2009), NGM (Austin et al. 2011), bulked segregant RNA-seq (BSR-seq) (Liu et al. 2012), MutMap and its variants (Abe et al. 2012; Fekih et al. 2013; Takagi et al. 2013a), QTL-Seq (Takagi et al. 2013b), Seq-BSA (Singh et al. 2015), and Indel-Seq (Singh et al. 2017), which have been developed and are capable of efficiently mapping causal mutations or traits in various organisms, including plants.

Trait mapping by sequencing combines both classical genetics and NGS platforms to map the traits on genomic regions. The whole-genome resequencing data of population/lines and sequencing of recombinant pools/nonrecombinant mutant genomes allow direct linking of phenotype(s) to causal mutations in the genome. Sequencing-based trait mapping can be broadly grouped into (I) trait mapping through pooled sequencing-based approach and II) trait mapping through the sequencing of complete populations.

7.2.1 Trait Mapping through Pooled Sequencing-Based Approach

The various DNA-based approaches, which were devised to map the genes, conceptually trace back to the BSA of Michelmore et al. (1991). In BSA, two extreme pools are created, and markers are screened for polymorphism between the parent 1, trait 1 pool, and parent 2, trait 2 pool. It was observed that such polymorphic markers possess a very low probability to be unlinked to the trait of interest.

The schematic representation of trait mapping approaches through pooled sequencing are mentioned in Fig. 7.1. Few of the prominent approach's details are mentioned here.

MutMap is a very powerful approach when we have to map the trait of interest in an EMS-induced mutant; this is the most simple and robust approach (Abe et al. 2012). In MutMap, DNA from multiple F2 progenies possessing mutant phenotype

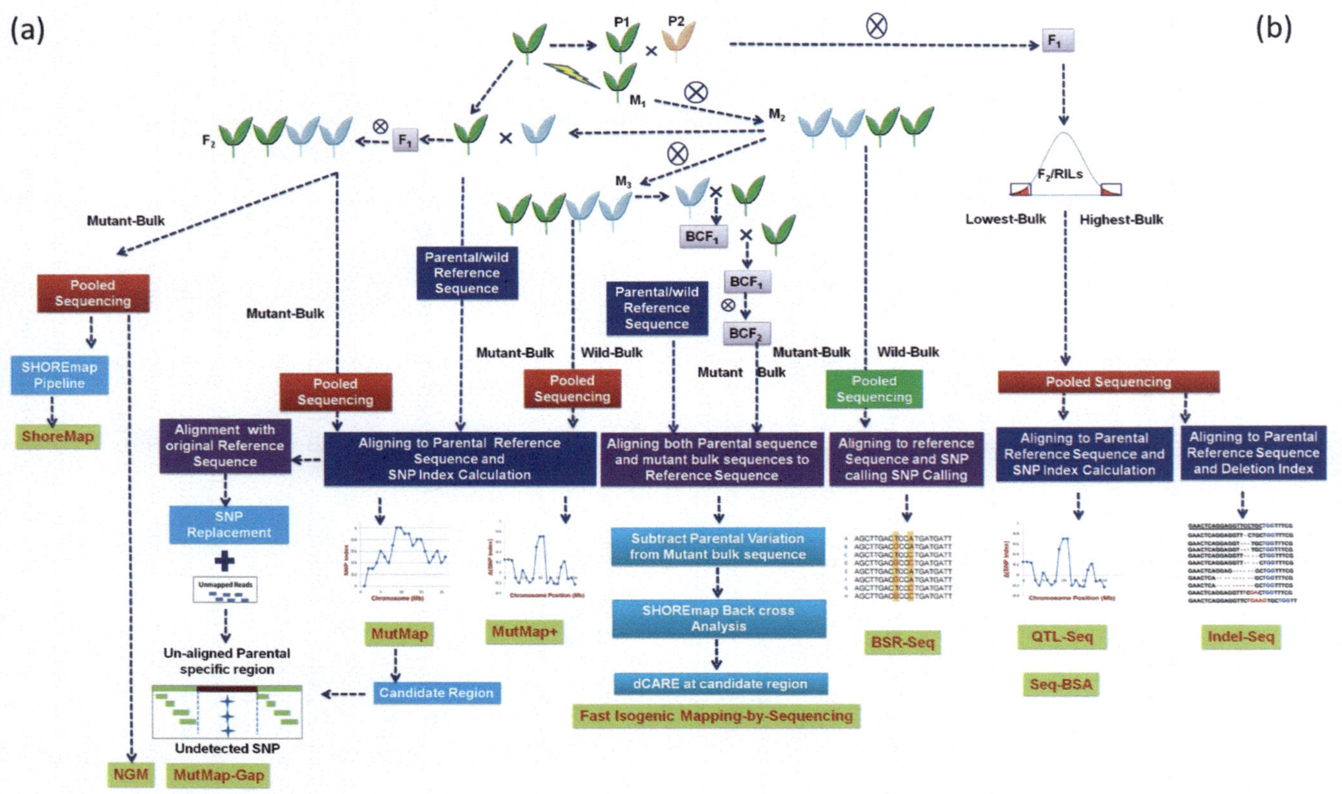

Fig. 7.1 A comprehensive view of trait mapping through next-generation sequencing, utilizing the extreme pools-based approach to identify genes/genomic regions associated with the targeted traits in crops. The red boxes indicated the pooling of DNA samples; the green box indicated the pooling of RNA samples. The dotted line separates the workflow into two halves, (**a**) mutant-based pooling and (**b**) natural variant-based pooling

is bulk sequenced and aligned with the reference sequence of the parental line to point to the genomic location responsible for the phenotype. However, MutMap approach is not as effective when mutant plants possess lethality and sterility and are found not suitable for crossing, and sometimes if mutations are in those regions which do not present in the reference genome. MutMap+ (Fekih et al. (2013) and MutMap-Gap (Takagi et al. 2013a) were proposed to address these limitations of MutMap.

QTL-Seq combines the BSA and NGS approaches for rapid identification of the genomic regions and causal SNPs for a target trait that differs between two parents of the biparental population (Takagi et al. 2013b). The pooled DNA collected from contrasting plants of F2 or RILs are sequenced through the NGS platform and aligned with the reference genome sequence to identify the QTLs and causal SNPs for the target trait. QTL-Seq has been successfully employed in the localization of the genomic regions in many crops (see Varshney et al. 2019).

Seq-BSA is an NGS-based simple and robust approach for the identification of candidate SNPs in the targeted genomic regions. This approach works on the calculation of genome-wide SNP-index of both the extreme bulks using high trait parent as reference parent assembly using QTL-seq pipeline (Takagi et al. 2013a). Identified SNPs that were monomorphic for high trait parent and high trait bulk will show an SNP index of “0” due to the presence of a similar genomic region of a particular locus. However, identification of the SNP index value of “1” in low trait bulk, with the same genomic positions, might be the putative SNPs linked to the target traits. This approach has been successfully utilized for the identification of putative SNPs associated with fusarium wilt and sterility mosaic disease in pigeon pea (Singh et al. 2015).

Indel-Seq has emerged as a promising method for trait mapping, focusing on insertions and deletions. Currently, trait mapping approaches typically rely on identifying single-nucleotide polymorphisms (SNPs) to pinpoint candidate genomic regions or genes. However, these approaches largely ignore the presence of insertions and deletions in the candidate genomic region, which could be important for trait mapping. Indel-Seq is a practical alternative, as many cloned genes in rice and other crops contain Indels in reported candidate genes. This method has already proven successful in identifying candidate genes for fusarium wilt and sterility mosaic disease in pigeon pea, as reported by Singh et al. (2017).

BSR-Seq is a novel approach compared to the others, as in this approach, in the case of DNA, we can pool the RNA-Seq data to identify regions of the genome associated with the trait of interest (Liu et al. 2012). Moreover, analyzing the RNA-Seq data can provide information on the effects of the mutant on global gene expression patterns at no additional cost. BSR-Seq, the modification of BSA that utilizes RNA-Seq data, can effectively validate its results using gene cloning experiments. Overall, BSR-Seq is an efficient and cost-effective approach for mapping genes responsible for mutant phenotypes, especially in populations lacking known polymorphic markers.

7.2.2 *Trait Mapping through Sequencing of Complete Populations*

Trait discovery program is associated with diverse panels like the 3 K Subset, association mapping panel, nested association mapping (NAM), multiparent advanced generation intercross (MAGIC), and some biparental crosses, which segregate for multiple traits required to be genotyped fully through any of the proposed approaches (Fig. 7.2).

Genotyping-by-sequencing (Elshire et al. 2011) has evolved as a promising approach for rapid identification of a large number of genome-wide SNPs for diversity assessment, trait mapping, genome-wide association studies, and genomic selection in several crops (see (Varshney et al. 2019). GBS-based sequencing approach are also being utilized in a significant number of segregating mapping population in pigeon pea and groundnut. Whole genome sequencing (WGS) is a powerful approach for a range of studies, including trait mapping, genome-wide association studies, and genomic selection. However, sequencing entire mapping populations or diverse germplasm sets through the GBS approach can lead to missing regions of the genome, which is a limitation. Whole genome resequencing (WGRS) overcomes this challenge but can be expensive.

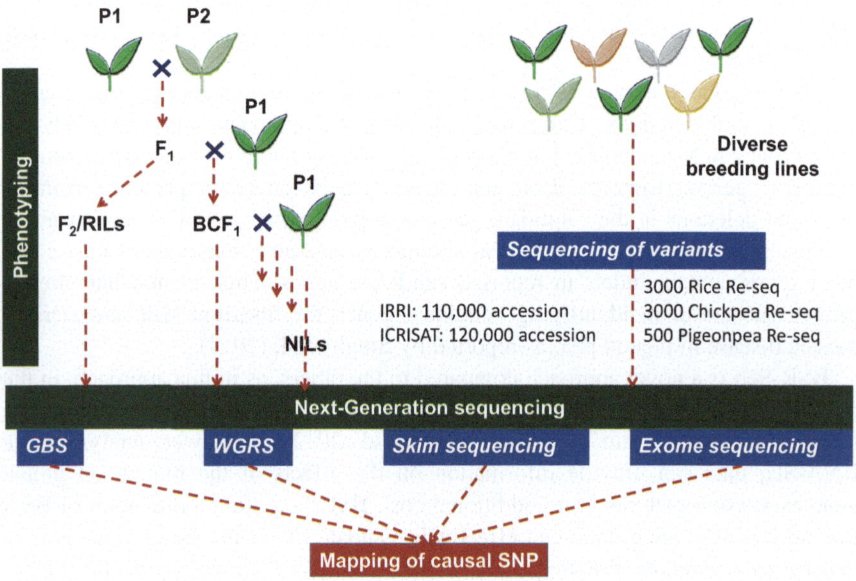

Fig. 7.2 Next-generation sequencing-based (NGS) methods to detect genetic variations linked with specific traits of interest. The choice of sequencing approach depends on various factors, such as the accessibility of the reference genome, financial considerations, and the desired depth of genome coverage

To save costs, WGRS can be done at a lower depth, referred to as skim sequencing. WGRS for targeting direct candidate genes from the segregating mapping population and/or diverse germplasm set in a plant species with significant genome sizes is cumbersome and time-consuming. To save costs, WGRS can be done at a lower depth, referred to as skim sequencing. However, targeting direct candidate genes from a segregating mapping population or diverse germplasm set in plant species with large genomes can be time-consuming and costly in crops with larger genomes like maize, barley, and wheat. Exome sequencing offers a solution by allowing the sequencing of all protein-coding genes in a genome, making it easier and faster to identify non-synonymous SNP substitutions in candidate genomic regions or through the identification of putative nsSNPs through principal component analysis between different sets of genotypes. This approach has been successfully used in crops with large genomes, such as maize, barley, and wheat, to identify nsSNPs of target candidate genes, and an example of complex traits like drought in maize is reported by Xu et al. (2014).

The generation of high-quality data is crucial in association studies and other genotype-to-phenotype investigations. Advances in next-generation sequencing have made it possible to generate genotypic data for many lines with high genome coverage quickly. However, generating high-quality phenotypic data on a large scale has become a bottleneck. Machine learning (ML) and deep learning (DL) techniques have emerged as efficient and sophisticated tools to overcome this limitation. ML-based models provide a suite of methods that can address the challenges in genome-wide association studies (GWAS) by connecting genomic information to phenotypic expression. These models can incorporate additive, nonadditive, and environmental interactions to understand the genetic architecture of a complex trait. An AI-based framework can predict a larger population's phenotypes by utilizing a smaller subset's manual phenotypic data. For example, a framework based on AI can be used to identify patterns in the genotypic data of rice germplasm and predict the value of targeted traits while considering the transcriptome dynamics of the plant system. AI framework can be designed to incorporate epistatic interactions and environmental impact, making it applicable across various crops. The QTLs/MTAs identified from the predicted phenotypic values can be compared with the results from the original phenotypic values to validate the prediction power of the framework (Sundaram et al. unpublished). The combined phenotypic data of original and predicted values can serve as the whole diversity panel for further association analysis to identify QTLs in the future.

7.3 Sequencing-Based Breeding

Marker-assisted selection (MAS) using SSR markers has shown promise in developing crop varieties with up to 1–3 gene combinations. However, stacking a large number of genes in one genetic background through backcrossing or assembling all genes through forward breeding utilizing low-cost genotyping systems remains a significant challenge. For example, targeting 10 genes would require generating 1024 distinct types of F1 gametes (2n, where n is the number of target genes), 59,049 different genotypes in the F2 population (3n), and the smallest perfect F2 population would be 1,048,576 (4n). Achieving these numbers using low-throughput marker systems like SSRs would be impractical before the flowering stage. Although generating such a large number of lines in any breeding program is possible, the higher cost associated with genotyping often results in a reduction in the number of lines targeted, which ultimately affects selection efficiency.

Recently, a new approach was reported, in which superior haplotypes of targeted traits can be identified and utilized in the breeding programs (Abbai et al. 2019). Superior haplotypes of already clones genes with high economic importance as well as superior haplotypes for the newly identified trait-marker associations can be introgressed and pyramided in elite breeding lines through sequenced-based breeding strategies (Sinha et al. 2020; Selvaraj et al. 2021; Varshney et al. 2021). The establishment of a trait identification pipeline to identify superior haplotypes for different traits of economic importance in different crops and thereby to establish trait-specific breeding panels on the basis of a haplotype blueprint can assist in the effective selection of parental lines for use in breeding programs.

Currently, two high-throughput approaches are being utilized in the breeding programs for the selection of the breeding lines with superior alleles or superior haplotypes: the first is the selection of lines through fixed arrays, and the second one is the selection of lines through sequencing (targeted sequencing or WGRS).

7.3.1 Selection of Lines through Fixed Arrays

Identification and development of SNP assays is a viable method for performing high-throughput genotyping that is reliable, has a quick turnaround time, is easy to retrieve information from, and is cost-effective in marker-assisted breeding programs, including genomic selection. Depending on the genotyping requirements, various SNP assays can be used in breeding programs. For example, flexible arrays with 1–10 SNPs can be used for foreground selection, while arrays with a large number of markers (>50 K) can be used for trait mapping, background selection, and diversity analysis. SNP genotyping arrays can be especially beneficial if they are designed to target high-value functional alleles for traits of interest, particularly if a breeding program lacks the informatics support needed to interpret NGS information in a timely manner. Although the density of SNPs on an array is typically lower than the SNPs assayed by NGS, the selection of array-based SNPs can be optimized for breeding applications. In several crops, SNP genotyping arrays constructed from NGS datasets have been developed and used to enhance breeding efficiency (Varshney et al. 2019).

7.3.2 Selection of Lines through Sequencing

Sequencing-based breeding or selection based on sequencing can be effectively applied to crops with relatively small platinum or gold genomes, such as rice (387 Mb), chickpea (738 Mb), and pigeon pea (833 Mb). This approach allows for simultaneous marker discovery, marker validation, and genotyping. In backcross breeding, sequencing of recombinants is crucial to quickly identify individuals carrying the critical recombination breakpoints that break the linkage drag. However, it can be challenging to identify such recombinants using few SSR or SNP markers when landraces are used as donors in breeding programs.

In trait-based introgression programs, sequencing should be a common approach, as it can be extremely helpful in identifying recombinants to break linkage drag and liberate new forms of genetic variation for breeding use. For example, a sequencing approach was used to identify recombinants that broke linkage between a favorable allele for rice blast disease resistance and a deleterious gene affecting grain quality (Fukuoka et al. 2009), as well as between a favorable allele for drought tolerance in rice and an unfavorable allele for tall plant stature (Vikram et al. 2015). Recombinant sequencing has also been used to identify recombinants with a minimum number of donor parent SNPs. For instance, while developing salt-tolerant lines through backcross breeding, Takagi et al. (2015) subjected the final developed lines (BC_1F_3) to WGRS and selected lines with more than 200 SNPs from the respective wild type. Low-cost sequencing approaches, such as skim sequencing or genotyping-by-sequencing, are currently being utilized in many applications of genomics-assisted breeding (GAB). It is anticipated that in the future, these approaches will be mainly used in breeding applications. Consequently, sequencing-based breeding may replace fixed SNP array-based systems in relatively small genome-sized crops, while fixed SNP arrays will continue to be useful, particularly when they assay gene-specific or genome-specific markers that facilitate accurate mapping (Varshney et al. 2019).

Acknowledgments The authors express sincere thanks to the Department of Biotechnology (DBT), Government of India for financial support under the project of "Development of superior haplotype-based near-isogenic lines (Haplo-NILs) for enhanced genetic gain in rice" grant (BT/PR32853/AGill/103/1159/2019). IRRI is a member of the CGIAR Consortium.

References

Abbai R, Singh VK, Nachimuthu VV, Sinha P, Selvaraj R, Vipparla AK, Singh AK et al (2019) Haplotype analysis of key genes governing grain yield and quality traits across 3K RG panel reveals scope for the development of tailor-made rice with enhanced genetic gains. Plant Biotechnol J 17(8):1612–1622

Abe A, Kosugi S, Yoshida K, Natsume S, Takagi H, Kanzaki H, Matsumura H, Yoshida K et al (2012) Genome sequencing reveals agronomically important loci in rice using MutMap. Nat Biotechnol 30:174–179

Austin RS, Vidaurre D, Stamatiou G, Breit R, Provart NJ, Bonetta D, Zhang J et al (2011) Next-generation mapping of Arabidopsis genes. Plant J 67(4):715–725

Elshire RJ, Glaubitz JC, Sun Q, Poland JA, Kawamoto K, Buckler ES, Mitchell SE (2011) A robust, simple genotyping-by-sequencing (GBS) approach for high diversity species. PLoS One 6(5): e19379

Fekih R, Takagi H, Tamiru M, Abe A, Natsume S (2013) MutMap+: genetic mapping and mutant identification without crossing in rice. PLoS One 10:e68529

Fukuoka S, Saka N, Koga H, Ono K, Shimizu T, Ebana K, Hayashi N et al (2009) Loss of function of a proline-containing protein confers durable disease resistance in rice. Science 325:998–1001

Liu S, Yeh C-T, Tang HM, Nettleton D, Schnable PS (2012) Gene mapping via bulked segregant RNA-Seq (BSR-Seq). PLoS One 7:36406

Michelmore RW, Paran I, Kesseli RV (1991) Identification of markers linked to disease-resistance genes by bulked segregant analysis: a rapid method to detect markers in specific genomic regions by using segregating populations. Proc Natl Acad Sci U S A 88:9828–9832

Schneeberger K, Ossowski S, Lanz C, Juul T, Petersen A, Nielsen K, Jørgensen WD, Andersen S (2009) SHOREmap: simultaneous mapping and mutation identification by deep sequencing. Nat Methods 6:550–551

Selvaraj R, Singh AK, Singh VK, Abbai R, Habde SV, Singh UM, Kumar A (2021) Superior haplotypes towards development of low glycemic index rice with preferred grain and cooking quality. Sci Rep 11(1):10082

Singh VK, Khan AW, Saxena RK, Kumar V, Kale SM, Sinha P, Chitikineni A et al (2015) Next generation sequencing for identification of candidate genes for fusarium wilt and sterility mosaic disease in pigeonpea (*Cajanus cajan*). Plant Biotechnol J 14:1183. https://doi.org/10.1111/pbi.12470

Singh VK, Khan AW, Saxena RK, Sinha P, Kale SM, Parupalli S, Kumar V et al (2017) Indel-seq: a fast-forward genetics approach for identification of trait-associated putative candidate genomic regions and its application in pigeonpea (*Cajanus cajan*). Plant Biotechnol J 15(7):906–914

Sinha P, Singh VK, Saxena RK, Khan AW, Abbai R, Chitikineni A, Desai A et al (2020) Superior haplotypes for haplotype-based breeding for drought tolerance in pigeonpea (*Cajanus cajan* L.). Plant Biotechnol J 18(12):2482–2490

Takagi H, Abe A, Yoshida K, Kosugi S, Natsume S, Mitsuoka C, Uemura A et al (2013a) QTL-seq: rapid mapping of quantitative trait loci in rice by whole genome resequencing of DNA from two bulked populations. Plant J 74:174–183

Takagi H, Tamiru M, Abe A, Yoshida K, Uemura A, Yaegashi H, Obara T et al (2015) MutMap accelerates breeding of a salt-tolerant rice cultivar. Nat Biotechnol 33(5):445–449

Takagi H, Uemura A, Yaegashi H, Tamiru M, Abe A, Mitsuoka C, Utsushi H et al (2013b) MutMap-gap: whole-genome resequencing of mutant F_2 progeny bulk combined with de novo assembly of gap regions identifies the rice blast resistance gene *Pii*. New Phytol 200: 276–283

Varshney RK, Pandey MK, Bohra A, Singh VK, Thudi M, Saxena RK (2019) Toward the sequence-based breeding in legumes in the post-genome sequencing era. Theor Appl Genet 132(3):797–816

Varshney RK, Roorkiwal M, Sun S, Bajaj P, Chitikineni A, Thudi M, Singh NP, Du X et al (2021) A chickpea genetic variation map based on the sequencing of 3,366 genomes. Nature 599(7886):622–627

Vikram P, Swamy BP, Dixit S, Singh R, Singh BP, Miro B, Kohli A et al (2015) Drought susceptibility of modern rice varieties: an effect of linkage of drought tolerance with undesirable traits. Sci Rep 5:14799

Xu J, Yuan Y, Xu Y, Zhang G, Guo X, Wu F, Wang Q et al (2014) Identification of candidate genes for drought tolerance by whole-genome resequencing in maize. BMC Plant Biol 14(1):83

Chapter 8
Forward Breeding for Efficient Selection

Rajaguru Bohar, Susanne Dreisigacker, Hannele Lindqvist-Kreuze, Moctar Kante, Manish K. Pandey, Vinay Sharma, Sunil Chaudhari, and Rajeev K. Varshney

Abstract Global food security is the *numero uno* priority in the current global situation, threatened by a number of challenges catalyzed by accelerated climate change and population growth. Crop improvement coupled with the modern plant breeding approaches, such as genomic-assisted breeding, is a proven solution to meet the food security. One of the key mandates in the modern plant breeding program is to combine the power of genomic selection into the breeding pipeline employing a low-cost genotyping solution. Several SNP marker-based platforms are now available depending on the objectives and field of application; despite the

---

R. Bohar (✉)
CGIAR Excellence in Breeding (EiB), International Maize and Wheat Improvement Center (CIMMYT), c/o ICRISAT, Hyderabad, India
e-mail: R.Bohar@cgiar.org

S. Dreisigacker
International Maize and Wheat Improvement Center (CIMMYT), Veracruz, Mexico
e-mail: S.Dreisigacker@cgiar.org

H. Lindqvist-Kreuze · M. Kante
International Potato Center (CIP), Lima, Peru
e-mail: h.lindqvist-kreuze@cgiar.org; m.kante@cgiar.org

M. K. Pandey
Center of Excellence in Genomics and Systems Biology (CEGSB), International Crops Research Institute for the Semi-Arid Tropics (ICRISAT), Hyderabad, India
e-mail: manish.pandey@icrisat.org

V. Sharma
International Crops Research Institute for the Semi-Arid Tropics (ICRISAT), Hyderabad, Telangana, India
e-mail: vinay.sharma@icrisat.org

S. Chaudhari
World Vegetable Center, South Asia, c/o ICRISAT, Hyderabad, India
e-mail: sunil.chaudhari@cgiar.org

R. K. Varshney
Murdoch's Centre for Crop and Food Innovation, State Agricultural Biotechnology Centre, Food Futures Institute, Murdoch University, Murdoch, WA, Australia
e-mail: rajeev.varshney@murdoch.edu.au

M. K. Pandey et al. (eds.), *Frontier Technologies for Crop Improvement*, Sustainable Agriculture and Food Security, https://doi.org/10.1007/978-981-99-4673-0_8

availability of different platforms, the public sectors face challenges in terms of funding and accessibility to the latest technology when compared to private sectors. Shared genotyping platform coupled with open breeding informatics involving different stakeholders with active support from donors will address several constraints faced by the public breeding program. Here, we summarize the available forward breeding genomic resources in the space of low-mid-density genotyping platform with special emphasis on shared services for four crop groups:

1. Wheat (cereal)
2. Potato (roots, tubers, and bananas (RTB crops))
3. Groundnut (grain legumes)
4. Vigna species (legumes)

Keywords Forward breeding · Genomics-assisted breeding · Marker-assisted selection · Single-nucleotide polymorphism · Genomic selection · Wheat · Potato · Groundnut · Vigna · Cowpea · Mung bean · Black gram

8.1 Introduction

Global food security is the *numero uno* priority in the current global situation, threatened by a number of challenges catalyzed by accelerated climate change and with the expected global population growth to bypass 9.75 billion figures, where more than 2.5 billion people reside in Africa by 2050 (FAOSTAT (2022)). Crop improvement coupled with the modern plant breeding approaches is one of the proven solutions to meet the food security along with number of other factors including, but not limited to exploiting the natural variations. Next-generation breeding, such as genomic-assisted breeding (GAB), employing molecular markers has a significant impact and is considered as a practical way forward to accelerate the crop improvement with specific focus to improved crop yield (Razzaq, Ali, et al. 2021). Importance of GAB is evident from the commitment made by the Consultative Group on International Agricultural Research (CGIAR) with its 2030 Research and Innovation Strategy (Action Area 3) to support the breeding optimization pipelines and implementation of GAB approaches (CGIAR 2021). Methods of plant breeding was positively changed by the implementation of molecular marker technology especially after the boom of the genomic sequencing era that enhanced the pace of crop improvement. From the era of restriction fragment length polymorphisms (RFLPs) to simple sequence repeats (SSRs) to single-nucleotide polymorphisms (SNPs), the marker technology rapidly advanced, and currently, SNPs is considered as the most advanced and commonly used marker systems in plant breeding application (Bohar et al. 2020). One of the key mandates in the modern plant breeding program is to combine the power of genomic selection (GS) into the breeding pipeline. Low-cost genotyping solution, which is affordable and meaningful, makes the practical application of GS integration. Many genotyping platforms are available, among which SNP genotyping method utilizing the high-throughput multiplexed approach is a predominantly used genotyping method, such as genotyping-by-sequencing (GBS). Though GBS is a preferred choice, it lacks practical suitability, such as requirement of imputation to fill the missing data, and in turn requires a complex bioinformatics to suit the multiyear data interpretation. Diversity Arrays Technology's DArTag (Diversity Arrays Technology, Bruce, Australia), Integrated DNA Technologies' rhAmpSeq (Integrated DNA Technologies, Coralville, IA, USA), and Illumina's AmpliSeq (Illumina, San Diego, CA, USA,) are some of the commercially available pooled and multiplexed sequencing technology targeting the specific SNPs. Though the available platforms demand a high up-front cost for designing, they have a comparative advantage with GBS in terms of repeatability, heterozygotes identification, and requirement of less bioinformatics support (Sneller et al. 2021).

Several SNPs' marker-based platforms are now available, depending on the objectives and field of application such as the following: high-density SNP genotyping platforms (HDSG) for discovery studies and linkage mapping; medium-density SNP genotyping platforms (MDSG) for GS and background studies; and low-density SNP genotyping platforms (LDSG), such as Kompetitive Allele Specific PCR (KASP), for forward breeding application through marker-assisted selection (MAS), marker-assisted backcrossing (MABC), and quality control (QC) analysis (Roorkiwal et al. 2020). Despite the availability of different platforms, the public sectors face challenges in terms of funding and accessibility to latest technology when compared to private sectors. To tackle such situation, Xu et al. suggested that there should be coordinated efforts by the scientific community, particularly in developing countries backed by big donors, such as Bill and Melinda Gates Foundation (BMGF). These efforts, especially in the field of modernization of plant breeding through establishment of public–private partnerships, will improve the international crop improvement system (Xu et al. 2017; Cobb et al. 2019). Shared genotyping platform coupled with open breeding informatics involving different stake holders with active support from donors, such as BMGF, will address several constraints faced by the public breeding program. CGIAR's strategic approach in this direction includes the development of world-class shared genotyping service through High-Throughput Genotyping Project (HTPG) (http://cegsb.icrisat.org/high-throughput-genotyping-project-htpg/), the Genomics Open-Source Breeding Informatics Initiative (GOBii) (http://gobiiproject.org/), and the Excellence in Breeding (EiB) (https://excellenceinbreeding.org/) platform led by the CGIAR institutes with the able funding of the BMGF (Bohar et al. 2020). HTPG facilitated low-cost and high-throughput genotyping for CGIAR and National Agricultural Research Systems (NARS) led by ICRISAT (2016–2020) and further transitioned to genotyping/sequencing tools and services module of EiB platform through service provider, Intertek AgriTech (http://www.intertek.com/agriculture/agritech/) (Bohar et al. 2020).

EiB has been coordinating and supporting the use of genotyping by NARS and has launched shared genotyping services, such as EiB-LDSG and EiB-MDSG, for the benefit of national, CGIAR, and other breeding programs. Currently, the EiB shared services is functional with all CGIAR centers and their partner programs in rice, wheat, maize, several millets, legumes, and several other crops, which are readily deployable in the breeding programs with continuous addition of new crops and marker resources. These services are targeted at CGIAR and NARS breeding institutions, aggregating demand across institutions to offer high-quality, low-cost genotyping with faster turnaround time of 10–15 business working days. Implementation of shared services is one among the six requests from the funder consortium Crops to End Hunger (CtEH), whose overall objective is to modernize CGIAR breeding programs and networks ensuring those programs deliver the highest possible rate of genetic gains in farmers' field (Hunt et al. 2021). The EiB-LDSG service (formerly HTPG) based on KASP platform, which is cost-effective up to 200 markers, is suited for applications including specific trait screening (foreground selection), QC, and MAS. The markers available for use in EiB-LDSG can be accessed at https://excellenceinbreeding.org/module3/kasp, which is continuously updated and improved (EiB-LDSG 2022). The EiB-MDSG service is a DArTAg genotyping method with a density of up to 4000 markers, primarily suited to GS applications, but can also be used for diversity studies, DNA fingerprinting, and MABC for background recovery analysis (EiB-MDSG 2022).

Success stories and publications incorporating the HTPG, EiB-LDSG, and/or EiB-MDSG are reported in several crops, such as groundnut (Parmar et al. 2021), potato (Kante et al. 2021; Sood et al. 2022), cassava (Le Thuy et al. 2021), sorghum (Mwamahonje et al. 2021), and banana (Garcia Oliveira et al. 2021). Here, we summarize the available forward breeding genomic resources in the space of low-mid-density genotyping platform with special emphasis on the resources available with EiB genotyping shared services for four crop groups: (1) wheat (cereal), (2) potato (roots, tubers, and bananas (RTB crops)), (3) groundnut (grain legumes), and (4) *Vigna* species (legumes).

8.2 Genomic Resources and Forward Breeding in Wheat

Wheat is a key food staple that provides around 20% of protein and calories consumed worldwide. The demand for wheat is projected to continue to grow over the coming decades, particularly in the developing world to feed an increasing population and, with wheat being a preferred food, continuing to account for a substantial share of human energy needs in 2050 (Wageningen 2016). Current annual genetic gains for grain yield of about 1% are being realized in the CIMMYT Global Wheat Program (GWP) for a number of target populations of environments (Crespo-Herrera et al. 2017, 2018; Honsdorf et al. 2018; Gerard et al. 2020). Thus, higher-yielding, more productive varieties continue to be released in developing countries, resulting in enhanced productivity (Lantican et al. 2016). In addition to grain yield improvement together with yield stability, key breeding objectives of the program (similar to others) are resistance/tolerance to biotic and abiotic stresses, end use, and nutritional quality characteristics. In order to keep up with the pressing future demands of wheat production and to adapt to changing environmental factors, wheat breeders overall and at CIMMYT are constantly turning to new and emerging technologies and breeding strategies. For example, advanced genetics and genomics tools are progressively deployed and related operating processes optimized (Fig. 8.1) (Dreisigacker et al. 2016, 2021).

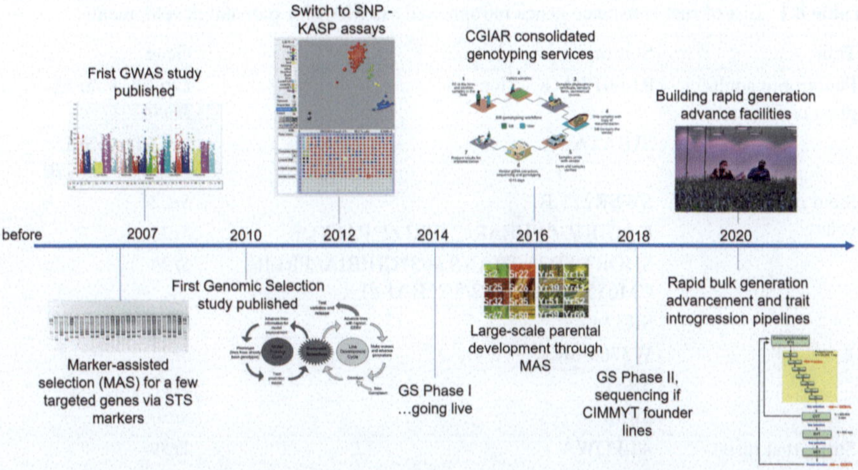

Fig. 8.1 Implementation of genetic and genomic tools in the CIMMYT Global Wheat Program

Marker-assisted forward and backcross breeding are approaches that can be successfully deployed in crops, mainly when (i) a target trait is rather difficult to manage in the field because it is expensive or time-consuming to measure, has low penetrance, or has complex inheritance; (ii) trait selection depends on specific environments or host developmental stages; (iii) recessive alleles during backcrossing need to be maintained or for speeding up backcross breeding in general; and (iv) pyramiding multiple monogenic traits or several QTL for single traits is looked for (Miedaner and Korzun 2012). However, often the number of markers per trait with enough information content about their relevance and usefulness to a breeding program is low. To increase response to selection using marker-assisted forward and backcross breeding in wheat, markers related to genes for disease resistance, end use, and nutritional quality are mainly used, because they show reasonable effect size. In the CIMMYT GWP, the targeted development of rust-resistant wheat germplasm is probably the most important example for which markers are adopted. The aim is to develop elite breeding lines that carry a combination of non-race-specific adult-plant resistance genes and race-specific genes, to avoid applying extremely high selection pressure on the pathogen that might endanger the avirulence of individual genes in developing countries. Rust research in the GWP has mapped and officially designated several rust genes (reviewed by Lan and Basnet 2016). Multiple pleiotropic non-race-specific genes (including *Lr34/Yr18/Sr57/Pm38*, *Lr46/Yr29/Sr58/Pm39*, and *Sr2/Yr30/Lr27/Pm*) are present in the CIMMYT wheat germplasm pool and build the basis of resistance against the three rusts (Singh et al. 2014). A larger number of race-specific stem and yellow rust resistance genes not present in CIMMYT germplasm have recently been introgressed into a set of elite genetic backgrounds via MABC to develop new parental lines (Table 8.1). In addition to rust, resistance to fusarium head blight (FHB) and Septoria tritici blotch are targets for forward breeding. For example, recombinant inbred lines that have the resistant *Fhb1* and *Sr2* alleles in coupled

Table 8.1 List of rust resistance genes introgressed via MAS for parental development

Trait	Source	Gene
Pleiotropic adult-plant resistance	RL6077/AOC-YR	*Lr67/Yr46/Sr55/ Pm46*
	SUJATA	*Lr67/Yr46/Sr55/ Pm46, YrSuj-7BL*
Stem rust resistance genes	SWSR22T.B.	*Sr22*
	KACHU/3/WHEAR//2*PRL/2*PASTOR	*Sr25*
	SHORT SR26 TRANS./4/3*CHIBIA//PRLII/ CM65531/3/MISR 2/5/2*BAJ #1	*Sr26*
	SR32	*Sr32*
	W3763-SR35	*Sr35*
	SR47	*Sr47*
	SR50	*Sr50*
Stripe rust resistance genes	ALPOWA	*Yr39*
	CHUAN NONG 19	*Yr41*
	BLANCA GRANDE 515	*Yr5, Yr15*
	SUMMIT 515	*Yr5, Yr15*
	YR51#5515–1	*Yr51*
	KOELZ W 11192:AE	*Yr52*
	YR57#5474–6	*Yr57*
	IRAGI	*Yr59*
	LALBMONO1*4/PVN	*Yr60*

phase linkage in the background of the cultivar HARTOG were crossed with CIMMYT bread wheat lines and selected with molecular markers for both genes in addition to the use of pseudo-black chaff (PBC) as a phenotypic marker for the selection of *Sr2* (He et al. 2020). The *Fhb1*-resistant allele has previously been absent in CIMMYT germplasm as the gene is usually tightly linked in repulsion phase (the case where each homologous chromosome has one dominant and one recessive allele from the two genes) on chromosome 3BS and CIMMYT wheat breeding focused much time and energy on stem rust resistance. Durum wheat (*Triticum turgidum* subsp. *durum*) is also an important crop worldwide, while its production runs secondary to that of bread wheat. One aspect to consider is the relatively restricted food functionalities of durum wheat, primarily attributed to its kernel texture and gluten strength limitations (Morris et al. 2019). During the last few years, soft kernel durum was crossed with CIMMYT elite durum lines to produce soft kernel progeny with a high degree of genetic variance for milling and baking quality. Selection for the novel soft kernel types was routinely supported with associated KASP markers.

Breeding in Mexico routinely utilizes two crop seasons per year that cuts the breeding time by about half but also allows selection for a range of traits at contrasting field sites that have distinct daylength and temperature regimes. MAS is deployed in both crop-seasons. In 2020, a field screenhouse and rapid-generation advance (RGA) greenhouse facility at the CIMMYT Toluca research station were constructed. Donor parents derived from the previous parental development pipelines are now used and crossed with selected elite lines for rapid introgression, pyramiding, and trait augmentation through RGA. The scaling-up marker-assisted forward and backcross breeding in the public sector has long been hampered by high genotyping costs and insufficient data management support. And still, the cost of genotyping for several high-throughput SNP platforms is inversely proportional to sample quantity, which in the case of smaller public sector organizations that have low individual demand partly impedes the routine deployment of molecular markers. In 2016, the HTPG project supported by the BMGF developed a shared industrial-scale service of low-density SNP genotyping serving the CGIAR and partner breeding programs. A low-cost, fast turnaround service (EiB-LDSG) was established by EIB platform. EiB-LDSG is routinely used in the CIMMYT GWP. Information on the wheat markers that are routinely applied is also available at https://excellenceinbreeding.org/module3/kasp.

8.3 Genomic Resources and Forward Breeding in Potato

To come up with a satisfactory variety, potato breeders must concentrate their efforts on selecting for a multitude of traits besides yield. Depending on the breeding program, goals as many as 12 traits have been proposed (Bonierbale et al. 2019). Although stacking of multiple genes is complicated in highly heterozygous tetraploid potato, it is possible to make progress through a dedicated MAS procedure (Bradshaw 2017; Stefańczyk et al. 2020; Rakosy-Tican et al. 2020). Furthermore, MAS facilitates the traits-based selection without heavy phenotypic evaluation at early stages, as compared to conventional breeding methods. In International Potato Center (CIP), the main breeding targets were for decades centered on disease resistance and climate resilience to achieve varieties that are productive under stressful conditions. The main diseases of potato in the breeding target areas of CIP are late blight (LB) caused by the oomycete *Phytophthora infestans* (Mont.) de Bary, potato virus Y (PVY), potato virus X (PVX), and potato leaf roll virus (PLRV). Molecular markers have been identified to trace the presence of many genes that provide control to these diseases (Nie et al. 2016; Fulladolsa et al. 2015; Ottoman et al. 2009; Whitworth et al. 2009; Tiwari et al. 2013), and significant time and cost savings can be achieved by using them instead of traditional phenotyping. We had calculated that the use of MAS for *Rladg* gene that provides resistance to PLRV (Mihovilovich et al. 2014; Velásquez et al. 2007) can save up to 88% of the costs as compared to phenotypic assays (RTB. 2019). This is particularly valuable for traits such as PLRV resistance that is extremely difficult to measure because the virus is phloem limited and thus infection assays require the use of vector insects or grafting to indicator plants (Mihovilovich et al. 2014). The *Rladg* assay that was developed at CIP is a gel-based SCAR marker (Mihovilovich et al. 2014) that is well suited for the screening of a relatively small number of samples. However, this screening method is not ideal for screening in the early breeding stages when thousands of clones would be evaluated. Furthermore, a breeding program is usually interested in screening for more than one single trait, and therefore, screening for multiple markers in a single assay is more efficient. KASP offers an excellent cost-efficient option for breeding programs to develop custom sets of low-density markers for purposes, such as QC or MAS (Semagn et al. 2014; Caruana et al. 2021). Therefore, at CIP we started to convert some of the most important trait markers for our breeding program into KASP markers. The first trait markers converted and validated are two markers for PVY resistance and two for LB resistance (Kante et al. 2021). The PVY markers were designed for the *Ryadg* gene that is the main source of PVY resistance in CIP breeding populations (de Herrera et al. 2018). Based on high assay power and low error rate, the new markers work very efficiently in CIP breeding program (Kante et al. 2021). The LB markers were discovered in GWAS studies using CIP breeding germplasm and located in or near the QTL in chromosome 9 that has been shown to contain *R8* resistance gene (Lindqvist-Kreuze et al. 2014, 2021; Jiang et al. 2018). These markers have a variable performance, depending on the germplasm they are applied to, but have consistently good performance in the populations targeting LB as a main trait (Kante et al. 2021; M. Gastelo, pers. Comm). These markers are available at https://excellenceinbreeding.org/module3/kasp through EiB-LDSG. Other markers well worth pursuing as a KASP assay for disease resistance traits in CIP germplasm include at least gene *Rx* for PVX resistance and *Rladg* for PLRV resistance.

Another useful application for KASP markers is identity verification with the help of a set of QC markers. QC marker sets have been reported for maize, rice, and sweet potato (Gemenet et al. 2020a; Ndjiondjop et al. 2018; Semagn et al. 2012). Mislabeled genotypes at any stage of the breeding process are problematic and should be avoided as they waste time and resources and negatively affect genetic gains (Gemenet et al. 2020a). We selected a set of SNP markers from the SolCAP Infinium array that had discriminatory power to differentiate CIP breeding germplasm. Using the tetraploid calls, it was possible to discriminate full and half sibs with as few as 20 SNP markers (Kante et al. 2021). Several cases of mislabeling were discovered when this set of markers was applied to test the identity of clones across different stages (Kante et al. 2021). Although the KASP assay for a relatively low number of markers is cost-efficient, there is a need to carefully define the best approach for routine use of these markers in the breeding program. At the very least, the CIP breeding program will strive to verify the identity of the progenitors in the crossing block and the advanced clones that are shared with partners for variety evaluation.

The rate of progress in genetic gains by recurrent selection in potato is largely limited by the number of vegetative generations needed to complete all phenotyping (Bradshaw 2017). GS and genomic estimations of breeding values are a powerful tool that can shorten the breeding cycle of potato and lead to increased genetic gains for several traits (Ortiz 2020). Most studies published in potato GS up to date have utilized hybridization-based Illumina SNP SolCAP array (Stich and Van Inghelandt 2018; Enciso-Rodriguez et al. 2018; Endelman et al. 2018) or GBS (Sverrisdóttir et al. 2017; Caruana et al. 2019; Byrne et al. 2020). SolCAP array is a relatively expensive assay keeping in mind that not all markers are applicable across different breeding populations (Slater et al. 2014). GBS can be more affordable particularly if sequencing depth is kept at a moderate level, but it is computationally intensive (Gemenet et al. 2020b). However, if one wishes to utilize the allele dosage information in the GS model, it is advisable to consider sequencing depth of 60–80x so that the different heterozygous states can be called reliably in autotetraploid potato (Uitdewilligen et al. 2013).

Targeted sequencing with selected SNP across the genome is therefore an attractive option to reduce the number of markers to a more manageable level allowing for a sufficient sequencing depth (Slater et al. 2014). DArTag is a targeted Diversity Arrays Technology (DArT, http://www.diversityarrays.com), genotyping method where a single oligonucleotide is used to capture selected SNPs or indels. Further advantage of this system is that the sequencing of the fragments allows for identification of other polymorphisms beyond the originally targeted SNP/indel and thus a capture of additional haplotypes. Tetraploid calls can be obtained from the raw read counts, even for samples with low read depth (<15). As part of the EiB-MDSG efforts, we developed such a reduced, but representative set of markers using previously available data points from the SolCAP SNP (Hamilton et al. 2011; Felcher et al. 2012) and SolSTW SNP (Uitdewilligen et al. 2013; Vos et al. 2015) genotyping efforts with CIP and Jeff Endelman's Lab at the University of Wisconsin potato germplasm. The marker selection and primer design yielded a set of 2503 markers, with a number of markers per chromosome ranging between 177 and 272. In a standard DArTag assay, these markers have a mean median read depth of 164, which allows for successful dosage calling in tetraploid potato. Like the KASP marker set, the DArTag marker set is in public domain and available for all interested breeding programs through EiB-MDSG. The mid-density DArTag marker set will allow a cost-effective genotyping of the breeding material for genomic prediction. Further, the inclusion of two trait markers for *Ryadg* and *Rysto* into the panel will allow a characterization of the breeding material without necessarily increasing the genotyping costs. Inclusion of more trait markers is and will allow a one-time genotyping of clones through the breeding cycle, for trait characterization and genomic prediction. CIP clones from intermediate trials DArTag genotyping for GS can be done from tuber and/or leaf samples, assuring therefore a rapid availability of molecular data of the prediction population.

8.4 Genomic Resources and Forward Breeding in Groundnut

Groundnut or peanut (*Arachis hypogaea* L.) is an important food and oil crop, cultivated across the tropics and subtropics regions. Globally, it is cultivated on an estimated 29.6 million hectares of area, producing 48.86 million tons (FAOSTAT 2022; http://www.fao.org/faostat/en/#data/QC). Moreover, half of groundnut crop produce is used to extract oil, with the remaining eaten raw or processed. Groundnut is often referred to as the "poor man's almond" owing to its high nutritional content, like oil, protein, fiber, polyphenols, antioxidants, vitamins, and minerals. Gibbons began breeding work in groundnut around 1976. Groundnut breeding's major goal is to develop cultivars having high-yielding, early maturity, adaptability to specific conditions and production systems, resilience or endurance to environmental challenges, disease and insect resistance, and improved nutritional quality. Efforts to enhance groundnut breeding via genomics have been very successful, over the previous decades. Availability of genomic resources, such as extensive whole-genome assemblies and transcriptome assemblies, high-density saturated genetic maps, and, most interestingly, linked markers with traits of interest, has eased routine breeding programs (Pandey et al. 2020). Modern-day advanced breeding strategies, such as GS, speed breeding, or RGA, are projected to boost groundnut genetic gains (Parmar et al. 2021). Here, we commence by highlighting recent advances in groundnut genomics, with an emphasis on next-generation technologies.

8.4.1 Genomic Resources in Modern Era

Recent groundnut genomic advancements have enabled the emergence of a diverse array of advanced genomic methods and technologies for crop improvement. The groundnut researchers have succeeded in terms of leveraging genetic resources, notably during the past decade (Fig. 8.2). A wide range of genomic resources, including reference genomes assemblies, genome-wide genetic markers, expression atlas, genotyping platforms (low, mid, and high density), QC panels, and diagnostic markers, have facilitated the translational genomics in groundnut research.

8.4.2 Reference Genomes Assemblies

A reference genome provides a standardized, important basis for genomic and genetic research. Utilizing advanced sequencing technologies, a well-assembled and annotated genome has been developed, which facilitates the understanding of complex genome structure, aids gene discovery, and helps in the exploiting genetic resources for faster crop improvement (Pandey et al. 2020). In 2016, IPGI through Peanut Genome Consortium (PGC) completed the sequencing of two diploid ancestors (*A. duranensis* V14167 and *A. ipaensis* K30076) of cultivated groundnut (Bertioli et al. 2016) in order to solve the complexity associated with assembling chromosomal pseudomolecules. Meanwhile, the Diploid Progenitor Peanut A-Genome Sequencing Consortium (DPPAGSC) sequenced both diploid progenitor species, *A. duranensis* (Chen et al. 2016) and *A. ipaensis* (Lu et al. 2018). Furthermore, with the advent of next-generation sequencing (NGS) technology, it aided research in further improving, refining, and developing high-quality reference

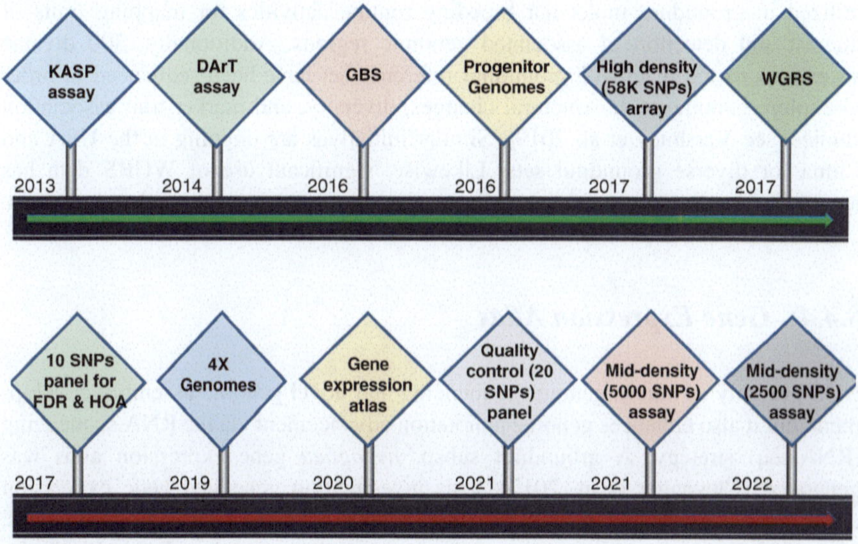

Fig. 8.2 A road map of significant developments in the field of groundnut genomics

genome assemblies for the cultivated tetraploid groundnut. In 2018, the genome assembly for the allotetraploid wild groundnut *A. monticola* was made publicly available (Yin et al. 2018). A major breakthrough among groundnut researchers took place with the availability of two reference genomes for subsp. *fastigiata* (Chen et al. 2019; Zhuang et al. 2019) and one for subsp. *hypogaea* (Bertioli et al. 2019). These reference genome assemblies revealed new knowledge on groundnut genes that have been impacted by domestication and breeding, highlighting breeding possibilities for future.

8.4.3 Whole-Genome Resequencing and Genome-Wide Markers

Adoption of NGS approaches enabled the discovery and deployment of a diverse array of molecular markers in groundnut. Both diploid genomes (*A. duranensis* and *A. ipaensis*) were mined for markers, resulting in the discovery of 135,529 and 199,957 SSRs, respectively (Zhao et al. 2017). Additional comparative diploid genome study between *A. duranensis* and *A. ipansis* detected 515,223 InDels, i.e., 269,973 insertions and 245,250 deletions (Vishwakarma et al. 2017). Furthermore, high-coverage WGRS data for 41 distinct genotypes have been generated, resulting in the finding of 98,375 SNPs in the A-subgenome and 65,407 SNPs in the B-subgenome (Pandey et al. 2017a; Clevenger et al. 2017). Notably, 58,233 high-quality informative SNPs were employed to develop high-density genotyping platform, "Axiom_*Arachis*" Array (Pandey et al. 2017a). Nowadays, it is commonly utilized in groundnut molecular breeding routine activities for mapping traits of interest and detection of associated genomic regions. Additionally, 300 diverse accessions of the ICRISAT groundnut reference set have been sequenced in order to explore genome-wide structural changes, diversity, and marker trait association studies (see Varshney et al. 2019). Similar initiatives are ongoing in the USA and China for diverse groundnut sets. Likewise, significant use of WGRS data has resulted in the discovery of unique alleles, signature sequences, and related markers.

8.4.4 Gene Expression Atlas

NGS not only advances genome sequencing and novel genome assembly development, but it also enhances genome annotation advancement via the RNA sequencing (RNA-seq) strategy. A groundnut subsp. *hypogaea* gene expression atlas was generated (Clevenger et al. 2017). This investigation generated gene expression data from 22 groundnut tissues and reported 8816 putative homeologous genes, over 9000 alternative splicing events, and over 6000 non-coding RNA. Similarly, the second groundnut transcriptome map has been generated using RNA-seq of 39 samples from various tissues and environments. This map represents 91.73% of all annotated genes (Zhuang et al. 2019). Subsequently, a gene atlas for subsp. *fastigiata* has been employed by utilizing the same subspecies tetraploid genome (Sinha et al. 2020). Furthermore, genome annotation expansion facilitates the integration of proteome and metabolome atlases. These findings will help us to better understand genomes that have more RNA genes than protein-coding ones. Crop trait development and expression will be better understood if the genome, metabolome, and epigenome are all studied in greater depth.

8.4.5 Rapid and Cost-Effective Genotyping Assays

The current crop genomics scenario has been transformed by NGS technologies, which give a multitude of sequence information with significant increases in coverage, speed, and expense. Such advances have permitted the design of low- to high-throughput genotyping platforms. SNPs are regarded the ideal marker due to genome-wide uniform distribution and cost-effectiveness. For the present, efficient and cost-effective genotyping platforms are being developed in groundnut genomics, which will be used according to the study's objective. For example: (a) High-density genotyping platform ("Axiom_*Arachis*" 58 K SNP arrays) has been developed and utilized in several studies, marker trait analysis, dense genetic map construction, and QTL analysis. (b) A mid-density assay (2500–5000 SNPs) has been successfully deployed in genetic mapping; QTL analysis (Bomireddy et al. 2022) is suggested for performing GS, genetic diversity assessment, and background selection; using Agri-seq 5 K assay, genotyping of about 3000 lines has been done so far in order to optimize genomic prediction in groundnut breeding. Similar DArT-seq assay with 2500 SNPs has been developed and ready for deployment through EiB-MDSG. And (c) low-density assay (10–100 SNPs) can be effectively utilized for QC assessment and purity testing of founder parents, hybridity confirmation, and early generation selection through EiB-LDSG. Panel of highly informative 20 SNPs has successfully tested for quality check among founder parents of IC-Asia, IC-WCA, IC-ESA, ICAR-DGR, and UAS-Dharwad breeding (Unpublished). So far, QC panel has effectively used in hybridity testing of more than 700 F_1 plants (IC-Asia, ICAR-DGR, and UAS-Dharwad) and for determination of homozygosity among advanced breeding lines.

All the abovementioned genotyping assays are successfully deployed in routine breeding activities. In future, such genotyping platforms will become more widely used in breeding programs around the world and pave a new era of genomics in molecular breeding.

8.4.6 Sequencing-Based Trait Mapping

New inventions in sequencing and high-throughput genotyping have accelerated trait discovery studies significantly while also enabling candidate gene identification at a fast pace. The substantial decrease in the cost of sequencing and the increased advent of high reference genomes in groundnut have facilitated sequence-based trait mapping more feasible and time-effective, over the past 2–3 years. In fact, right now, sequencing-based mapping is now accomplished out by either sequencing the entire populations or pooling samples exhibiting extreme phenotypes of the desired trait (Varshney et al. 2019). Inside this aspect, the low-to high-density sequencing methodologies are being employed to develop high-density genetic maps and improve trait mapping precision in groundnut. Recently, the GBS method is employed to develop high-density maps, to perform QTL mapping and detection of candidate genes (Pandey et al. 2017c; Han et al. 2018; Dodia et al. 2019). By using SLAF-seq technology, two high-density maps (with 2266 and 2808 SNP loci) have been constructed, for QTL mapping of quality and growth habit traits (Hu et al. 2018; Li et al. 2019), respectively. Similar approach has been successfully deployed to develop dense map (with 7184 SNPs), which facilitates to narrow down the co-localized regions associated with seed size traits, resulting in the detection of trait-linked candidate genes. Additionally, same genetic map has been used for testa color trait mapping (Zhuang et al. 2019). Association mapping of 158 groundnut diverse accessions with 17,338 polymorphic SNPs has allowed to identify marker trait association for agronomic traits (Zhang et al. 2017). To obtain large number of polymorphic SNPs, the WGRS approach has been deployed in RIL population (Tifrunner GT-C20), which enabled the generation of high-density genetic maps (8869 to 11,106 SNP loci), fine mapping of genomic region, and candidate gene discovery for ELS, LLS, and TSWV resistance (Agarwal et al. 2018, 2019). However, both WGRS and GBS platforms have been extensively deployed for mapping traits and identification of QTLs in groundnut, but each has its own shortcomings. Recent methods for trait mapping (QTL-seq, Indel-seq, Seq-BSA, MutMap, and BSR-seq) generally rely on high- and low-bulk sample pooled sequencing, which exhibits extreme phenotypes related to the trait of interest under investigation. Among all, the "QTL-Seq" approach has been effective in identifying genomic regions and candidate genes in groundnut (Pandey et al. 2017b; Clevenger et al. 2018; Luo et al. 2019; Zhuang et al. 2019; Kumar et al. 2020; Zhao et al. 2020). Such research efforts have been facilitating the development of diagnostic markers and will help in pyramiding of QTLs associated with traits of interest in groundnut breeding.

8.4.7 Genomics-Assisted Breeding to Accelerate Groundnut Breeding

The groundnut rcscarchcr has taken several initiatives over the past decade to provide efficient genetic resources for crop improvement. Groundnut achieved the status of crop with the plentiful of genomic resources, which enable identification of genes and associated markers used in GAB. Based on the available genomic resources, diagnostic markers for high-oleic acid and resistance to nematode, rust, and LLS have been effectively generated in groundnut. The three GAB procedures, notably MAS or MABC, marker-assisted recurrent selection (MARS), and GS, have proven the most effective in groundnut. The development of a nematode-resistant variety, NemaTAM, was the first successful use of MAS (Simpson et al. 2003). Later, by pyramiding the nematode resistance and high-oleic trait, employing molecular markers linked with these traits, a high-oleic "Tifguard High O/L cultivar" has been developed (Chu et al. 2011). Following that, favorable alleles (*ahFAD2A and ahFAD2B*) for oleic acid content from the donor parent SunOleic 95R were transferred to ICGV 06110, ICGV 06142, and ICGV 06420 using MAS (Pasupuleti et al. 2016). Similarly, utilizing the MABC approach, high-oleate lines have been developed in ICGV 05141 (Bera et al. 2018), GPBD 4 (Ndeve et al. 2019), and ICGV 06100 (Bera et al. 2018). In another study, high-oleate lines have been developed through MABC utilizing favorable alleles from parents (KN176, DF12, and KX016) (Huang et al. 2019). Likewise, work has been done in developing cultivars resistant to foliar diseases (rust and LLS) together on high-oleic background (Janila et al. 2016; Bera et al. 2018; Shasidhar et al. 2020). There have certain attempts to combine high-oleic acid with foliar disease resistance in GPBD 4, G 2–52, TMV 2, and JL 24, in addition to ICGV 06189 variety (Bhat et al. 2022). Among the molecular breeding lines developed through marker-assisted backcrossing for high oleic acid (Janila et al. 2016), two Virginia bunch high oleic varieties namely Girnar 4 (ICGV 15083) and Girnar 5 (ICGV 15090) and two Spanish bunch high oleic varieties namely GG 39 (ICGV 16697) and GG 40 (ICGV 16688) are released for cultivation in India. More recently, two elite varieties, GPBD 4 and G-252, have been improved for oleic acid content using MABC (Jadhav et al. 2021). Now, development of groundnut backcross lines has been made more precise and efficient using the available genetic resources for LLS and rust resistance (Pandey et al. 2020). The first effort has been made by utilizing SSR markers to enhance resistance for rust in three susceptible varieties, ICGV 91114, JL 24, and TAG 24 (Varshney et al. 2014). Using two to three backcrosses and selfing, 200 backcross lines were selected. From these, 81 lines were examined in the field and found to be more resistant to rust. As contrast to susceptible parents, these lines produced far more pod yield (56–96%) in affected environments (Pasupuleti et al. 2016). With GPBD 4 as the donor, various attempts have been made to develop LLS- and rust-resistant backcross lines in JL 24 (Yeri and Bhat 2016) and TMV 2 (Kolekar et al. 2017; Ramakrishnan et al. 2020). Two foliar disease resistance varieties namely Improved JL 24 (DBG 3) and Super TMV 2 (DBG 4) using donor source GPBD 4 (Bhat et al. 2022) were developed using marker-assisted selection and released for cultivation in Karnataka state of India. The most of lines are being tested in multiple locations or on large-scale farms in order to facilitate varietal development (Bhat et al. 2022).

8.5 Genomic Resources and Forward Breeding in *Vigna* Species

The genus *Vigna* with around nine domesticated crops, viz., mung bean (*Vigna radiata* (L.) R. Wilczek *var. radiata*), black gram, (*V. mungo* (L.) Hepper), cowpea (*V. unguiculata* (L.) Walp), moth bean (*V. aconitifolia* (Jacq.) Marechal), azuki bean (*V. angularis* (Willd.) Ohwi & H. Ohashi), rice bean (*V. umbellata* (Thunb.) Ohwi & H. Ohashi), creole bean (*V. reflexo-pilosa*), tuber cowpea (*V. vexillata* (L.) A. Rich), and bambara groundnut (*V. subterranean* (L.) Verdc.), is one of the important genera contributing toward food and nutritional security across the world (Takahashi and Tomooka 2020). Except for mung bean, cowpea, and black gram, the other species have limited progress on development of genomic resources and use of forward breeding approaches for crop improvement. The forward breeding approaches have been successfully proven efficient in enhancing selection accuracies and intensity for complex traits in many crops, thus helping breeders in selecting individuals with the most desirable traits of the product profile. The initial efforts used isoenzymes, random amplified polymorphic DNA (RAPD), and RFLP markers and later shifted to microsatellite markers derived from various *Vigna* species for the construction of genetic maps and to find QTLs associated with the traits of interests. The recent advancement in whole-genome sequence has strongly improved the access to molecular markers for these crops. Several thousands of SNP markers have been produced by GBS and WGRS, which are being used to assess the diversity at the genomic level and for developing new genomic resources including diagnostic markers and tools for deployment in breeding programs. The genomic resources available in cowpea (Muñoz-Amatriaín et al. 2017; Boukar et al. 2019), mung bean (Kim et al. 2015), and black gram (Bisht and Singh 2013) are discussed in detail in earlier studies. The important genomic resources that are useful for the genetic improvement of important *Vigna* crops through forward breeding approaches are summarized below.

8.5.1 Cowpea

Cowpea (*Vigna unguiculata*), 2n = 22 with a genome size of 620 MB, is one of the largely grown *Vigna* species in the dry agroecologies of the tropics in Latin America, Africa, and South Asia. The cultivated cowpea belonging to the cultivar group *Unguiculata* is divided into five cultivar groups, namely, *unguiculata*, *sesquipedalis* (yard-long-bean), *textilis*, *biflora*, and *melanophthalmus* (Padulosi and Ng 1997). Initial efforts for use of molecular markers for the cowpea improvement used RFLP (Fatokun et al. 1993), RAPD (Kaga et al. 1996; Simon et al. 2007), inter-simple sequence repeat (ISSR) (Ajibade et al. 2000), amplified fragment length polymorphisms (AFLPs) (Fang et al. 2007), and SSR (Gupta and Gopalakrishna 2010; Ogunkanmi et al. 2008) to characterize accessions belonging to the four subgenera of the genus *Vigna* and to study genetic diversity within cultivated as well as wild relatives of cowpea accessions. The world collection of landraces and African ancestral wild cowpea were characterized at molecular level with >1200 SNP markers. The study identified two major gene pools in cultivated cowpea in Africa, each with landraces mostly distributed in Western Africa (GP1) and Eastern Africa (GP2) (Huynh et al. 2013). GBS was used to discover more SNPs in cowpea that could be used to study genetic diversity, population structure, and phylogenetic relationships (Xiong et al. 2016).

Four QTLs on Vu01 with 24 to 95% PV associated with root-galling and egg masses per root system were reported most effective against resistance to root-knot nematode caused by *M. javanica* (Ndeve et al. 2019). The trait-specific SRR markers associated with different traits, such as seed size, pod fiber thickness, seed weight (Andargie et al. 2011), pod length (Kongjaimun et al. 2012), days to flower (Andargie et al. 2013), pod number per plant (Xuet al. 2013), and pod tenderness (Kongjaimun et al. 2012), were identified in different studies. Similarly, useful SNP markers were reported for cowpea bacterial blight (Agbicodo et al. 2010), foliar thrips (Lucas et al. 2012), leaf senescence (Xu et al. 2013), heat tolerance (Lucas et al. 2013a), seed size (Lucas et al. 2013b), aphid infestation (Huynh et al. 2015), and fusarium wilt (Pottorff et al. 2012) that could be deployed in the breeding programs. The microsatellite marker SSR1 was successfully used to transfer striga resistance gene from the breeding line IT93K-693-2 into three farmers' preferred varieties, viz., IT90K-372-1-2, KVx30–309-6G, and TN5–78 through MABC (Salifou et al. 2016). The resistance to three striga races SG1, SG3, and SG5 from IT97K-499-35 into an elite farmer preferred cowpea cultivar Borno Brown was successfully introgressed using three markers SSR-1, 61RM-2 and C42–26

(Omoigui et al. 2017). Around 28 introgression lines selected in the BC1$F_{2:4}$ generation with large seed size, brown seed coat color carrying marker alleles were evaluated in the field for resistance to striga resistance. The SSR-1 was identified as best for screening genotypes for striga resistance. A rare haplotype associated with large seeds at the Css-1 locus was successfully stacked from an African buff seed-type cultivar IT82E-18 (18.5 g/100 seeds) into a blackeye seed-type cultivar CB27 (22 g/100 seed) (Lucas et al. 2015). The foreground and background selections using genome-wide SNPs identified introgression lines with very large seed size (28–35 g/100 seeds) and desirable seed quality traits. For bacterial blight, one major QTL on linkage group (LG) Vu09 (qtlblb-1) accounting for 30.58% phenotypic variation (PV) and two QTLs, i.e., qtlblb-2 and qtlblb-3 on LG Vu04 with 10.77% and 10.63%, PV, respectively, were reported (Dinesh et al. 2016). The major QTL on Vu09 was successfully introgressed from cultivar V-16 into the bacterial leaf blight susceptible variety C-152 through marker-assisted backcrossing (MABC) (Dinesh et al. 2016).

A set of informative SNP panel, i.e., Cowpea iSelect Consortium Array with 51,128 SNPs, was developed to facilitate researchers with useful genomic resources (Muñoz-Amatriaín et al. 2017). The array was further used to develop a mid-density marker platform for cowpea with 2602 SNP markers distributed evenly throughout 11 chromosomes. The SNPs for the mid-density panel were selected based on iSelect data from 2714 diverse cultivated cowpea accessions with more weightage on 184 most commonly used accessions in African breeding programs while selecting these SNPs. This mid-density array is quite suitable for marker-assisted breeding, genomic-based predictions, QTL studies, molecular diversity analyses, and germplasm management applications. KASP fluorescence-based methodology offers rapid and cost-effective genotyping useful for target trait screening, QC, and MAS in the breeding programs across several crops. The KASP assay-based SNP markers were developed and being used in cowpea for screening against resistance to aphid infestation (Huynh et al. 2015) and bacterial blight (Agbicodo et al. 2010) (Table 8.2). The SNP-based foreground and background selections with KASP genotyping platform were successfully used to combine drought tolerance along with nematode and striga resistance into Moussa local, a cowpea variety from Burkina Faso, using MABC (Batieno et al. 2016). Six promising families were identified based on MAS and preliminary field testing for yield under well-watered and water-stress, and striga resistance field trials demonstrated the high efficiency of using SNP markers for foreground and background selections to combine target traits (Batieno et al. 2016). Around 17 KASP-based SNP markers were used to determine parental diversity and to confirm hybridity of cowpea crosses (Ongom et al. 2021). These QC markers differentiated 222 cowpea parental genotypes with mean efficiency of 37.9% and a range of 3.4–82.8%, revealing unique fingerprints of the parents. These markers demonstrated an effective application of KASP-based SNP assay in fingerprinting, confirmation of hybridity, and early detection of true F_1 plants (Ongom et al. 2021).

Table 8.2 Diagnostic trait-specific and QC SNP markers available for KASP genotyping platform useful for forward breeding in cowpea and mung bean

Target trait	SNP ID	Chr.	Position	SNP/ Indel	Favorable allele	Alternate allele	Reference
Cowpea							
Aphid resistance	snpVU0031	2	25,345,278–25,345,401	A/G	A	G	Huynh et al. (2015)
	snpVU0032	2	25,479,793-25,480,105	A/G	A	G	
	snpVU0024	5	3,748,293–3,748,425	A/T	A	T	
	snpVU0025	5	4,562,162–4,562,294	C/G	C	G	
Bacterial blight	snpVU0041	3	992,603–993,420	C/T	C	T	Agbicodo et al. (2010)
Quality control and hybridity test	snpVU0007	1	36,773,526–36,773,649	T/C	–	–	Ongom et al. (2021)
	snpVU0011	2	22,941,996–22,942,128	T/C	–	–	
	snpVU0018	4	16,415,787–16,415,919	A/G	–	–	
	snpVU0019	4	24,230,438–24,230,570	T/G	–	–	
	snpVU0001	5	399,824–399,956	C/G	–	–	
	snpVU0002	5	43,326,556–43,327,417	A/G	–	–	
	snpVU0009	6	30,511,313–30,511,445	A/C	–	–	
	snpVU0010	6	34,246,871–34,247,003	T/G	–	–	
	snpVU0003	7	4,914,544–491,665	T/C	–	–	
	snpVU0004	7	39,680,298–39,680,430	T/C	–	–	
	snpVU0008	8	34,271,840–34,271,972	A/G	–	–	
	snpVU0012	9	29,111,205–29,111,337	A/C	–	–	
	snpVU0013	9	37,010,557–37,010,817	A/T	–	–	
	snpVU0016	10	37,900,312–37,900,440	A/G	–	–	
	snpVU0017	10	967,432–967,564	C/G	–	–	
	snpVU0014	11	34,083,600–34,083,732	A/G	–	–	

(continued)

Table 8.2 (continued)

Target trait	SNP ID	Chr.	Position	SNP/ Indel	Favorable allele	Alternate allele	Reference
	snpVU0015	11	12,936,036–12,936,168	T/C	–	–	
Mung bean[a]							
Bruchid resistance	snpVR00001	5	5,178,332	G/A	G	A	Schafleitner et al. (2016)
	snpVR00002	5	5,179,402	T/C	T	C	
	snpVR00003	5	5,454,538	T/C	T	C	
	snpVR00004	5	5,622,070	G/A	G	A	
	snpVR00005	5	5,662,479	G/A	G	A	
	snpVR00006	5	5,730,691	G/A	G	A	
	snpVR00007	5	5,953,917	A/T	A	T	
	snpVR00008	5	5,974,663	C/T	C	T	
	snpVR00009	3	10,431,528	T/A	T	A	
	snpVR00010	4	15,255,162	T/G	T	G	

[a]SNPs for bruchid resistance in mung bean are being validated for their selection efficiency in the KASP platform

8.5.2 Mung Bean

Mung bean (*Vigna radiata* (L.) R. Wilczek *var. radiata*), an Asiatic *Vigna* species also known as green gram or moong, is grown in around 7.3 m ha worldwide with an average yield of 721 kg/ha (Nair and Schreinemachers 2020). It is one of the important food and cash crops in the rice-based farming systems of South and Southeast Asia with India and Myanmar together accounting for 60% of global produce of 5.3 m t. Other large producers are China, Indonesia, Thailand, Kenya, and Tanzania. The mung bean yellow mosaic disease (MYMD) and bruchid infestation are major biotic stresses initially focused to develop genomic resources for forward breeding approaches. The RAPD marker OPP 07_{895} was identified to be linked with MYMD resistance using bulk segregant analysis (Dharajiya and Ravindrababu 2019). Two QTLs, i.e., qMYMIV2 and qMYMIV7, with 31.42–37.60% and 29.07–47.36% PV, respectively, were reportedly linked to MYMD resistance (Alam et al. 2014). Four SSR markers, viz., CEDG275, CEDG006, CEDG041, and VES0503, linked to these QTLs could be useful for MAS. Other markers VrD1, CEDG228, CEDG044, and STSbr1 (Singh et al. 2017a, b) and CEDG293, DMB-SSR008, and DMB-SSR059 (Singh et al. 2020) were also reportedly linked with MYMD. Five QTLs with PV that ranged from 10.11 to 20.04% for MYMD resistance were detected on an interspecific recombinant inbred line (RIL) population of mung bean and rice bean. Of these, QTL qMYMV4–1 on LG4 was identified in the same marker interval across years (Mathivathana et al. 2019). The inter-simple sequence repeat, I85420, and ISSR-anchored resistance gene analog markers I42PL-229 and I42PL-222 were successfully used for MAS of powdery mildew (PM) resistance in mung bean. Of these,

I42PL229 was used for negative selection, where I85420 and I42PL-222 were used for positive selection with around 94% selection accuracies when confirmed resistance using detached leaf assay (Chathiranrat et al. 2018). Diagnostic derived cleaved amplified polymorphic sequences (dCAPS1, 2, and 3) and cleaved amplified polymorphic sequences (CAPS) markers (CAPS1, 2, 3, 4, 6, 8, 9, 11, 12, 13, and 14) were reported for resistance to bruchid infestation on LG3, LG4, and LG5 with selection efficiency of over 93% (Schafleitner et al. 2016). Among these, 10 promising markers information used to design SNP markers and their KASP assay to deploy in the breeding program (Table 8.2). These markers are being validated for their selection efficiency using a diverse set of genotypes. The genomic regions qZn-4-3 and qFe-4-1 on LG4 between flanking markers PVBR82-BM210 and qZn-11-2 and qFe-11-1 on LG11 between flanking markers BM141-BM184 were reported for Zn or Fe concentration (Singh et al. 2017a, b). Around 43 SNPs were found to be highly associated with seven seed mineral concentrations traits, including Fe and Zn through genome-wide association study. A total of six genomic regions, one with Fe (five associated SNPs) and five with Zn (7 associated SNPs), were found to be associated with PV ranging from 13 to 22% (Wu et al. 2020).

8.5.3 Black Gram

Black gram is a highly nutritious grain legume crop, mainly grown in the South and Southeast Asian countries, including Afghanistan, Bangladesh, India, Myanmar, Pakistan, Sri Lanka, and Thailand, with India contributing over 70% of global black gram production (Kaewwongwal et al. 2015). The efforts of development linkage map in black gram during the initial years used RFLP and AFLP markers (Chaitieng et al. 2006; Gupta et al. 2008). The efforts were also made to identify and deploy SSR markers available in other crops, such as cowpea (Gupta and Gopalakrishna 2010), mung bean, adzuki bean (Gupta and Gopalakrishna 2009), and common bean (Souframanien and Reddy 2015). The discovery of NGS technologies such as Illumina paired-end sequencing resulted in 17.2 million paired-end reads, and 48,291 transcript contigs (TCS) were used for gene discovery and development of 1840 SSRs that could be used for developing linkage maps and linked molecular markers for target traits (Souframanien and Reddy 2015). In black gram, the efforts on identification of QTLs and molecular markers are limited to MYMD (Souframanien and Gopalakrishna 2006; Maiti et al. 2011; Gupta and Gopalakrishna 2013) and bruchid infestation (Souframanien et al. 2010; Somta et al. 2019). An ISSR marker, ISSR8111357, linked to the MYMD resistance gene with a 6.8 cM distance identified, was sequenced to design a sequence characterized amplified region (SCAR) primer to deploy for MAS (Souframanien and Gopalakrishna 2006). The SSR marker CEDG180 linked to MYMD resistance was also reported (Gupta and Gopalakrishna 2013). Two major QTLs governing resistance to MYMD disease in black gram reported on LG2 and LG10 with 20.90 and 24.90% PVE, respectively (Vadivel et al. 2021). The validation of these QTLs in two other mapping populations identified as qmymv10_60 of LG10 with better selection efficiency could be useful for the MAS/MABC in black gram. Two loci, YR4 and CYR1, were identified associated with resistance to *Mung bean Yellow Mosaic India Virus* (MYMIV) in mung bean, of these CYR1 also co-segregated with MYMIV-resistant F_2, F_3 progenies of black gram (Maiti et al. 2011).

Two QTLs, viz., Cmrae1.1 and Cmrae1.2, were reported for bruchid adult emergence on LG3 and LG4, respectively (Souframanien et al. 2010). However, six QTLs were identified, with two QTLs (Cmrdp1.1 and Cmrdp1.2) on LG 1, three QTLs (Cmrdp1.3, Cmrdp1.4, and Cmrdp1.5) on LG 2, and one QTL (Cmrdp1.6) on LG 10, capturing 8.4 to 16.4% phenotypic variation for developmental period (Souframanien et al. 2010). The draft genome of black gram was sequenced using hybrid genome assembly with Illumina reads and third-generation Oxford Nanopore sequencing technology (Souframanien et al. 2021). It opens tremendous opportunities for the development of marker resources, along with the discovery of QTLs/genes and molecular markers for desirable traits. The genome analysis identified 42,115 genes with a mean coding sequence length of 1131 bp, of which around 80.6% are annotated. Besides, a total of 166,014 SSRs, including 65,180 compound SSRs, were also identified (Jegadeesan et al. 2021). The genome sequence of black gram is expected to provide greater insights and facilitate the identification of genes and QTLs linked to economically important traits for accelerating the genetic gain in black gram. The QTL qCm_PDS2.1 for percent damaged seeds and qVmunBr6.1 (24.32–28.76% PV) and qVmunBr6.2 (15.26–17.37%) for bruchid infestation severity progress mapped on LG 6 in mung bean. Two QTLs, i.e., qVmunBr6.1 and *qVmunBr6.2*, that are new loci for *C. maculatus* resistance in *Vigna* species will be useful for widening the genetic base of bruchid resistance in black gram (Somta et al. 2019). The SSR markers CEDG030 and CEDG248 were successfully used for hybridity test and ingression of MYMD resistance from rice bean to black gram (Sehrawat et al. 2016). Another successful example of the introgression of QTLs for MYMD resistance on LG2 and LG10 from resistant donor Mash 1008 into the popular black gram variety MDU 1 uses the MABC approach. Nine advanced backcross lines were identified with significant superior performances over recurrent parent MDU1 for yield and MYMD resistance (Subramaniyan et al. 2021).

8.6 Future Prospects

Despite the availability of several SNP markers, the public sectors face several challenges compared to private sectors when it comes to accessibility of these platform for implementation. Shared genotyping platform are expected to address several constraints faced by the public breeding program and will enable the implementation of genotyping tools into routine breeding operation. Availability of several LDSG-based trait and QC markers for forward breeding especially for QC and MAS and the MDSG-based medium density SNP panel will be useful for diversity studies, DNA fingerprinting, and MABC for background recovery analysis and GS applications in wheat, potato, and groundnut crops. The *Vigna* species crops were previously considered to be an orphan crop due to the limited availability of genomic resources compared to other legume and cereal crops. However, the recent progress on draft genome sequencing of mung bean (Kang et al. 2014), cowpea (Lonardi et al. 2019), black gram (Jegadeesan et al. 2021), and azuki bean (Kang et al. 2015) would help in accelerating the development of genomic resources and varietal improvement through forward breeding in *Vigna* crops. Collecting and resequencing *Vigna* species from different geographical areas would help researchers investigate allelic variation in beneficial traits that can be mined from wild relatives. These new resources would also open the door to genomic research in other *Vigna* species. More focus should be given to bring the identified markers on a cost-effective genotyping platform, i.e., KASP for their deployment in the breeding program. Utilization of 100-150 SNPs using DArTag panels for QC in potato and sweet potato (polyploids in general) would help streamline the QC implementation in a cost-effective manner. The enhanced precision and selection intensities for different complex target traits using diagnostic cost-effective molecular markers would accelerate the rate of genetic gains in crops and help breeders in developing the market preferred varieties.

References

Agarwal G, Clevenger J, Kale SM, Wang H, Pandey MK, Choudhary D, Yuan M, Wang X, Culbreath AK, Holbrook CC, Liu X, Varshney RK, Guo B (2019) A recombination bin-map identified a major QTL for resistance to tomato spotted wilt virus in peanut (*Arachis hypogaea*). Sci Rep 9(1):1–13. https://doi.org/10.1038/s41598-019-54747-1

Agarwal G, Clevenger J, Pandey MK, Wang H, Shasidhar Y, Chu Y, Fountain JC, Choudhary D, Culbreath AK, Liu X, Huang G, Wang X, Deshmukh R, Holbrook CC, Bertioli DJ, Ozias-Akins P, Jackson SA, Varshney RK, Guo B (2018) High-density genetic map using whole-genome resequencing for fine mapping and candidate gene discovery for disease resistance in peanut. Plant Biotechnol J 16(11):1954–1967. https://doi.org/10.1111/pbi.12930

Agbicodo EM, Fatokun CA, Bandyopadhyay R, Wydra K, Diop NN, Muchero W, Ehlers JD, Roberts PA, Close TJ, Visser RGF, van der Linden CG (2010) Identification of markers associated with bacterial blight resistance loci in cowpea [*Vigna unguiculata* (L.) Walp.]. Euphytica 175(2):215–226. https://doi.org/10.1007/s10681-010-0164-5

Ajibade SR, Weeden NF, Chite SM (2000) Inter simple sequence repeat analysis of genetic relationships in the genus Vigna. Euphytica 111(1):47–55. https://doi.org/10.1023/A:1003763328768

Alam AM, Somta P, Srinives P (2014) Identification and confirmation of quantitative trait loci controlling resistance to mungbean yellow mosaic disease in mungbean [Vigna radiata (L.) Wilczek]. Mol Breed 34:1497–1506

Andargie M, Pasquet RS, Gowda BS, Muluvi GM, Timko MP (2011) Construction of a SSR-based genetic map and identification of QTL for domestication traits using recombinant inbred lines from a cross between wild and cultivated cowpea (*V. unguiculata* (L.) Walp.). Mol Breed 28(3): 413–420. https://doi.org/10.1007/s11032-011-9598-2

Andargie M, Pasquet RS, Muluvi GM, Timko MP (2013) Quantitative trait loci analysis of flowering time related traits identified in recombinant inbred lines of cowpea (*Vigna unguiculata*). Genome 56(5):289–294. https://doi.org/10.1139/gen-2013-0028

Batieno BJ, Danquah E, Tignegre J-B, Huynh B-L, Drabo I, Close TJ, Ofori K, Roberts P, Ouedraogo TJ (2016) Application of marker-assisted backcrossing to improve cowpea (Vignaunguiculata L. Walp) for drought tolerance. Journal of Plant Breeding and Crop Science 8(12):273–286. https://doi.org/10.5897/JPBCS2016.0607

Bera SK, Kamdar JH, Kasundra SV, Dash P, Maurya AK, Jasani MD, Chandrashekar AB, Manivannan N, Vasanthi RP, Dobariya KL, Pandey MK, Janila P, Radhakrishnan T, Varshney RK (2018) Improving oil quality by altering levels of fatty acids through marker-assisted selection of ahfad2 alleles in peanut (Arachis hypogaea L.). Euphytica 214(9):1–15. https://doi.org/10.1007/S10681-018-2241-0/FIGURES/8

Bertioli DJ, Cannon SB, Froenicke L, Huang G, Farmer AD, Cannon EKS, Liu X, Gao D, Clevenger J, Dash S, Ren L, Moretzsohn MC, Shirasawa K, Huang W, Vidigal B, Abernathy B, Chu Y, Niederhuth CE, Umale P et al (2016) The genome sequences of *arachis duranensis* and *arachis ipaensis*, the diploid ancestors of cultivated peanut. Nat Genet 48(4): 438–446. https://doi.org/10.1038/ng.3517

Bertioli DJ, Jenkins J, Clevenger J, Dudchenko O, Gao D, Seijo G, Leal-Bertioli SCM, Ren L, Farmer AD, Pandey MK, Samoluk SS, Abernathy B, Agarwal G, Ballén-Taborda C, Cameron C, Campbell J, Chavarro C, Chitikineni A, Chu Y et al (2019) The genome sequence of segmental allotetraploid peanut *Arachis hypogaea*. Nat Genet 51(5):877–884. https://doi.org/10.1038/s41588-019-0405-z

Bhat RS, Venkatesh Jadhav MP, Patil PV, Shirasawa K (2022) Genomics-assisted breeding for resistance to leaf spots and rust diseases in Peanut. Accelerated Plant Breeding 4:239–278. https://doi.org/10.1007/978-3-030-81107-5_8

Bisht IS, Singh M (2013) Genetic and genomic resources of grain legume improvement: 10. Asian Vigna

Bohar R, Chitkineni A, Varshney RK (2020) Genetic molecular markers to accelerate genetic gains in crops. BioTechniques 69(3). https://doi.org/10.2144/btn-2020-0066

Bomireddy D, Gangurde SS, Variath MT, Janila P, Manohar SS, Sharma V, Parmar S, Deshmukh D, Reddisekhar M, Reddy DM, Sudhakar P, Reddy BVB, Varshney RK, Guo B, Pandey MK (2022) Discovery of major quantitative trait loci and candidate genes for fresh seed dormancy in groundnut. Agronomy 12(2). https://doi.org/10.3390/AGRONOMY12020404

Bonierbale MW, Amoros WR, Salas E, De Jong W (2019) Potato breeding. In: The potato crop: its agricultural, nutritional and social contribution to humankind. Springer International Publishing, pp 163–217. https://doi.org/10.1007/978-3-030-28683-5_6

Boukar O, Belko N, Chamarthi S, Togola A, Batieno J, Owusu E, Haruna M, Diallo S, Umar ML, Olufajo O, Fatokun C (2019) Cowpea (*Vigna unguiculata*): genetics, genomics and breeding. Plant Breed 138(4):415–424. https://doi.org/10.1111/PBR.12589

Bradshaw JE (2017) Review and analysis of limitations in ways to improve conventional potato breeding. Potato Res 60(2):171–193. https://doi.org/10.1007/s11540-017-9346-z

Byrne S, Meade F, Mesiti F, Griffin D, Kennedy C, Milbourne D (2020) Genome-wide association and genomic prediction for fry color in potato. Agronomy 10:90. https://doi.org/10.3390/AGRONOMY10010090

Caruana BM, Pembleton LW, Constable F, Rodoni B, Slater AT, Cogan NOI (2019) Validation of genotyping by sequencing using transcriptomics for diversity and application of genomic selection in tetraploid potato. Front Plant Sci 10. https://doi.org/10.3389/fpls.2019.00670

Caruana BM, Rodoni BC, Constable F, Slater AT, Cogan NOI (2021) Genome enhanced marker improvement for potato virus Y disease resistance in potato. Agronomy 11(5):832. https://doi.org/10.3390/agronomy11050832

CGIAR System Organization (2021) CGIAR 2030 research and innovation strategy: transforming food, land, and water systems in a climate crisis. Montpellier, France: CGIAR System Organization. Available at https://hdl.handle.net/10568/110918. Accessed 20 April 2022

Chaitieng B, Kaga A, Tomooka N, Isemura T, Kuroda Y, Vaughan DA (2006) Development of a black gram [Vigna mungo (L.) hepper] linkage map and its comparison with an azuki bean

[*Vigna angularis* (Willd.) Ohwi and Ohashi] linkage map. Theor Appl Genet 113(7): 1261–1269. https://doi.org/10.1007/S00122-006-0380-5/TABLES/3

Chathiranrat N, Nitisit S, Chaiyapan C, Wansuriwong N, Papan P, Tantasawat PA (2018) Selection of mungbean resistant to powdery mildew in BC_1F_1 progenies based on ISSR and ISSR-RGA markers. Int J Adv Sci Engg Technol 6:2321–9009. http://iraj.in

Chen X, Li H, Pandey MK, Yang Q, Wang X, Garg V, Li H, Chi X, Doddamani D, Hong Y, Upadhyaya H, Guo H, Khan AW, Zhu F, Zhang X, Pan L, Pierce GJ, Zhou G, Krishnamohan KAVS et al (2016) Draft genome of the peanut A-genome progenitor (*arachis duranensis*) provides insights into geocarpy, oil biosynthesis, and allergens. Proc Natl Acad Sci U S A 113(24):6785–6790. https://doi.org/10.1073/PNAS.1600899113

Chen X, Lu Q, Liu H, Zhang J, Hong Y, Lan H, Li H, Wang J, Liu H, Li S, Pandey MK, Zhang Z, Zhou G, Yu J, Zhang G, Yuan J, Li X, Wen S, Meng F et al (2019) Sequencing of cultivated Peanut, *Arachis hypogaea*, yields insights into genome evolution and oil improvement. Mol Plant 12(7):920–934. https://doi.org/10.1016/J.MOLP.2019.03.005

Chu Y, Wu CL, Holbrook CC, Tillman BL, Person G, Ozias-Akins P (2011) Marker-assisted selection to pyramid nematode resistance and the high oleic trait in Peanut. The Plant Genome 4(2):110–117. https://doi.org/10.3835/PLANTGENOME2011.01.0001

Clevenger J, Chu Y, Chavarro C, Agarwal G, Bertioli DJ, Leal-Bertioli SCM, Pandey MK, Vaughn J, Abernathy B, Barkley NA, Hovav R, Burow M, Nayak SN, Chitikineni A, Isleib TG, Holbrook CC, Jackson SA, Varshney RK, Ozias-Akins P (2017) Genome-wide SNP genotyping resolves signatures of selection and Tetrasomic recombination in Peanut. Mol Plant 10(2):309–322. https://doi.org/10.1016/J.MOLP.2016.11.015

Clevenger J, Chu Y, Chavarro C, Botton S, Culbreath A, Isleib TG, Holbrook CC, Ozias-Akins P (2018) Mapping late leaf spot resistance in Peanut (*Arachis hypogaea*) using QTL-seq reveals markers for marker-assisted selection. Front Plant Sci 9. https://doi.org/10.3389/fpls.2018.00083

Cobb JN, Juma RU, Biswas PS, Arbelaez JD, Rutkoski J, Atlin G, Hagen T, Quinn M, Ng EH (2019) Enhancing the rate of genetic gain in public-sector plant breeding programs: lessons from the breeder's equation. Theor Appl Genet 132(3):627–645. https://doi.org/10.1007/S00122-019-03317-0

Crespo-Herrera LA, Crossa J, Huerta-Espino J, Autrique E, Mondal S, Velu G, Vargas M, Braun HJ, Singh RP (2017) Genetic yield gains in CIMMYT's international elite spring wheat yield trials by Modeling the genotype × environment interaction. Crop Sci 57(2):789–801. https://doi.org/10.2135/cropsci2016.06.0553

Crespo-Herrera L, Crossa J, Sci JH-E-C (2018) Genetic gains for grain yield in CIMMYT's semi-arid wheat yield trials grown in suboptimal environments. ScienceopenCom 58(5):1890–1898. https://doi.org/10.2135/cropsci2018.01.0017

Dharajiya DT, Ravindrababu Y (2019) Identification of molecular marker associated with mungbean yellow mosaic virus resistance in mungbean [*Vigna radiata* (L.) Wilczek]. Vegetos 32(4):532–539. https://doi.org/10.1007/s42535-019-00063-y

Dinesh HB, Lohithaswa HC, Viswanatha KP, Singh P, Rao AM (2016) Identification and marker-assisted introgression of QTL conferring resistance to bacterial leaf blight in cowpea (Vigna unguiculata (L.) Walp.). Plant Breed 135(4):506–512

Dodia SM, Joshi B, Gangurde SS, Thirumalaisamy PP, Mishra GP, Narandrakumar D, Soni P, Rathnakumar AL, Dobaria JR, Sangh C, Chitikineni A, Chanda SV, Pandey MK, Varshney RK, Thankappan R (2019) Genotyping-by-sequencing based genetic mapping reveals large number of epistatic interactions for stem rot resistance in groundnut. Theor Appl Genet 132(4): 1001–1016. https://doi.org/10.1007/S00122-018-3255-7

Dreisigacker S, Crossa J, Pérez-Rodríguez P, Montesinos-López OA, Rosyara U, Juliana P, Mondal S, Crespo-Herrera L, Govindan V, Singh RP, Braun H-J (2021) Implementation of genomic selection in the CIMMYT global wheat program, findings from the past 10 years. Crop Breeding, Genetics and Genomics 3(2):e210004. https://doi.org/10.20900/cbgg20210005

Dreisigacker S, Sukumaran S, Guzmán C, He X, Lan C, Bonnett D, Crossa J (2016) Molecular marker-based selection tools in spring bread wheat improvement: CIMMYT experience and prospects, pp 421–474. https://doi.org/10.1007/978-3-319-27090-6_16

EiB-LDSG (2022). Available: https://excellenceinbreeding.org/toolbox/services/low-density-genotyping-services-introduction. Accessed April 20, 2022

EiB-MDSG (2022). Available: https://excellenceinbreeding.org/toolbox/services/mid-density-genotyping-service. Accessed April 20, 2022

Enciso-Rodriguez F, Douches D, Lopez-Cruz M, Coombs J, de los Campos, G. (2018) Genomic selection for late blight and common scab resistance in tetraploid potato (*Solanum tuberosum*). G3 Genes|Genomes|Genetics 8(7):2471–2481. https://doi.org/10.1534/g3.118.200273

Endelman JB, Carley CAS, Bethke PC, Coombs JJ, Clough ME, da Silva WL, De Jong WS, Douches DS, Frederick CM, Haynes KG, Holm DG, Miller JC, Muñoz PR, Navarro FM, Novy RG, Palta JP, Porter GA, Rak KT, Sathuvalli VR et al (2018) Genetic variance partitioning and genome-wide prediction with allele dosage information in autotetraploid potato. Genetics 209(1):77–87. https://doi.org/10.1534/genetics.118.300685

FAOSTAT (2022). Available: http://www.fao.org/faostat/en/#data. Accessed April 20, 2022

Fang J, Chao CCT, Roberts PA, Ehlers JD (2007) Genetic diversity of cowpea [Vigna unguiculata (L.) Walp.] in four West African and USA breeding programs as determined by AFLP analysis. Genet Resour Crop Evol 54:1197–1209

Fatokun CA, Danesh D, Young ND, Stewart EL (1993) Molecular taxonomic relationships in the genus Vigna based on RFLP analysis. Theor Appl Genet 86(1):97–104. https://doi.org/10.1007/BF00223813

Felcher KJ, Coombs JJ, Massa AN, Hansey CN, Hamilton JP, Veilleux RE, Buell CR, Douches DS (2012) Integration of two diploid potato linkage maps with the potato genome sequence. PLoS One 7(4):e36347. https://doi.org/10.1371/journal.pone.0036347

Fulladolsa AC, Navarro FM, Kota R, Severson K, Palta JP, Charkowski AO (2015) Application of marker assisted selection for potato virus Y resistance in the University of Wisconsin Potato Breeding Program. Am J Potato Res 92(3):444–450. https://doi.org/10.1007/s12230-015-9431-2

Garcia Oliveira AL, Ayele L, Uwimana B, Storr S, Tadesse T, Ng EH, Swennen R, Bohar R, Debaene J, Quinn M (2021) Access to genetic analysis accelerates banana breeding in East Africa. Available:https://excellenceinbreeding.org/news/access-genetic-analysis-accelerates-banana-breeding-east-africa. Accessed April 20, 2022

Gemenet DC, Kitavi MN, David M, Ndege D, Ssali RT, Swanckaert J, Makunde G, Craig Yencho G, Gruneberg W, Carey E, Mwanga RO, Andrade MI, Heck S, Campos H (2020a) Development of diagnostic SNP markers for quality assurance and control in sweetpotato [*Ipomoea batatas* (L.) lam.] breeding programs. PLoS One 15(4):e0232173. https://doi.org/10.1371/JOURNAL.PONE.0232173

Gemenet DC, Lindqvist-Kreuze H, De Boeck B, da Silva Pereira G, Mollinari M, Zeng Z-B, Craig Yencho G, Campos H (2020b) Sequencing depth and genotype quality: accuracy and breeding operation considerations for genomic selection applications in autopolyploid crops. Theor Appl Genet 133:3345–3363. https://doi.org/10.1007/s00122-020-03673-2

Gerard GS, Crespo-Herrera LA, Crossa J, Mondal S, Velu G, Juliana P, Huerta-Espino J, Vargas M, Rhandawa MS, Bhavani S, Braun H, Singh RP (2020) Grain yield genetic gains and changes in physiological related traits for CIMMYT's high rainfall wheat screening nursery tested across international environments. Field Crop Res 249:107742. https://doi.org/10.1016/j.fcr.2020.107742

Gupta SK, Gopalakrishna T (2009) Genetic diversity analysis in blackgram (*Vigna mungo* (L.) Hepper) using AFLP and transferable microsatellite markers from azuki bean (*Vigna angularis* (Willd.) Ohwi & Ohashi). Genome 52(2):120–129. https://doi.org/10.1139/G08-107

Gupta SK, Gopalakrishna T (2010) Development of unigene-derived SSR markers in cowpea (Vigna unguiculata) and their transferability to other Vigna species. Genome 53(7):508–523. https://doi.org/10.1139/G10-028/SUPPL_FILE/G10-028SUPPL.PDF

Gupta SK, Gopalakrishna T (2013) Advances in genome mapping in orphan grain legumes of genus *Vigna*. Ind J Genetics Plant Breed 73(1):1. https://doi.org/10.5958/j.0019-5200.73.1.001

Gupta SK, Souframanien J, Gopalakrishna T (2008) Construction of a genetic linkage map of black gram, Vigna mungo (L.) hepper, based on molecular markers and comparative studies. Genome 51(8):628–637. https://doi.org/10.1139/G08-050

Hamilton JP, Hansey CN, Whitty BR, Stoffel K, Massa AN, Van Deynze A, De Jong WS, Douches DS, Buell CR (2011) Single nucleotide polymorphism discovery in elite north american potato germplasm. BMC Genomics 12(1):302. https://doi.org/10.1186/1471-2164-12-302

Han S, Yuan M, Clevenger JP, Li C, Hagan A, Zhang X, Chen C, He G (2018) A snp-based linkage map revealed QTLs for resistance to early and late leaf spot diseases in peanut (*Arachis hypogaea* L.). Front Plant Sci 9:1012. https://doi.org/10.3389/FPLS.2018.01012/BIBTEX

He X, Brar GS, Bonnett D, Dreisigacker S, Hyles J, Spielmeyer W, Bhavani S, Singh RP, Singh PK (2020) Disease resistance evaluation of elite CIMMYT wheat lines containing the coupled *Fhb1* and *Sr2* genes. Plant Dis 104(9):2369–2376. https://doi.org/10.1094/PDIS-02-20-0369-RE

de Herrera MR, Vidalon LJ, Montenegro JD, Riccio C, Guzman F, Bartolini I, Ghislain M (2018) Molecular and genetic characterization of the Ryadg locus on chromosome XI from Andigena potatoes conferring extreme resistance to potato virus Y. Theor Appl Genet 131(9):1925–1938. https://doi.org/10.1007/s00122-018-3123-5

Honsdorf N, Mulvaney MJ, Singh RP, Ammar K, Burgueño J, Govaerts B, Verhulst N (2018) Genotype by tillage interaction and performance progress for bread and durum wheat genotypes on irrigated raised beds. Field Crop Res 216:42–52. https://doi.org/10.1016/j.fcr.2017.11.011

Hu XH, Zhang SZ, Miao HR, Cui FG, Shen Y, Yang WQ, Xu TT, Chen N, Chi XY, Zhang ZM, Chen J (2018) High-density genetic map construction and identification of QTLs controlling oleic and linoleic acid in peanut using SLAFseq and SSRs. Sci Rep 8(1):5479. https://doi.org/10.1038/s41598-018-23873-7

Huang B, Qi F, Sun Z, Miao L, Zhang Z, Liu H, Fang Y, Dong W, Tang F, Zheng Z, Zhang X (2019) Marker-assisted backcrossing to improve seed oleic acid content in four elite and popular peanut (Arachis hypogaea L.) cultivars with high oil content. Breed Sci 69(2):234–243. https://doi.org/10.1270/jsbbs.18107

Hunt LA, Garcia Oliveira AL, William HM, Ng EH, Das B, Kwemoi DB, Asante MD (2021) African breeding programs leap forward by accessing new genotyping data. Available: https://excellenceinbreeding.org/news/african-breeding-programs-leap-forward-accessing-new-genotyping-data. Accessed April 20, 2022

Huynh B-L, Ehlers JD, Ndeve A, Wanamaker S, Lucas MR, Close TJ, Roberts PA (2015) Genetic mapping and legume synteny of aphid resistance in African cowpea (*Vigna unguiculata* L. Walp.) grown in California. Mol Breed 35(1):36. https://doi.org/10.1007/s11032-015-0254-0

Huynh B, Close TJ, Roberts PA, Hu Z, Wanamaker S, Lucas MR, Chiulele R, Cissé N, David A, Hearne S, Fatokun C, Diop NN, Ehlers JD (2013) Gene pools and the genetic architecture of domesticated cowpea. Plant Genome 6(3). https://doi.org/10.3835/plantgenome2013.03.0005

Jadhav MP, Patil MD, Hampannavar M, Venkatesh DP, Shirasawa K, Pasupuleti J, Pandey MK, Varshney RK, Bhat RS (2021) Enhancing oleic acid content in two commercially released peanut varieties through marker-assisted backcross breeding. Crop Sci 61(4):2435–2443. https://doi.org/10.1002/CSC2.20512

Janila P, Pandey MK, Shasidhar Y, Variath MT, Sriswathi M, Khera P, Manohar SS, Nagesh P, Vishwakarma MK, Mishra GP, Radhakrishnan T, Manivannan N, Dobariya KL, Vasanthi RP, Varshney RK (2016) Molecular breeding for introgression of fatty acid desaturase mutant alleles (*ahFAD2A* and *ahFAD2B*) enhances oil quality in high and low oil containing peanut genotypes. Plant Sci 242:203–213. https://doi.org/10.1016/J.PLANTSCI.2015.08.013

Jegadeesan S, Raizada A, Dhanasekar P, Suprasanna P (2021) Draft genome sequence of the pulse crop blackgram [*Vigna mungo* (L.) hepper] reveals potential R-genes. Sci Rep 11(1):1–10. https://doi.org/10.1038/s41598-021-90683-9

Jiang R, Li J, Tian Z, Du J, Armstrong M, Baker K, Tze-Yin Lim J, Vossen JH, He H, Portal L, Zhou J, Bonierbale M, Hein I, Lindqvist-Kreuze H, Xie C (2018) Potato late blight field resistance from QTL dPI09c is conferred by the NB-LRR gene R8. J Exp Bot 69(7):1545–1555. https://doi.org/10.1093/jxb/ery021

Kaewwongwal A, Kongjaimun A, Somta P, Chankaew S, Yimram T, Srinives P (2015) Genetic diversity of the black gram [Vigna mungo (L.) Hepper] gene pool as revealed by SSR markers. Breed Sci 65(2):127–137

Kaga A, Tomooka N, Egawa Y, Hosaka K, Kamijima O (1996) Species relationships in the subgenus Ceratotropis (genus Vigna) as revealed by RAPD analysis. Euphytica 88(1):17–24. https://doi.org/10.1007/BF00029261

Kang YJ, Kim SK, Kim MY, Lestari P, Kim KH, Ha BK, Jun TH, Hwang WJ, Lee T, Lee J, Shim S, Yoon MY, Jang YE, Han KS, Taeprayoon P, Yoon N, Somta P, Tanya P, Kim KS et al (2014) Genome sequence of mungbean and insights into evolution within Vigna species. Nat Commun 5(1):1–9. https://doi.org/10.1038/ncomms6443

Kang YJ, Satyawan D, Shim S, Lee T, Lee J, Hwang WJ, Kim SK, Lestari P, Laosatit K, Kim KH, Ha TJ, Chitikineni A, Kim MY, Ko J-M, Gwag J-G, Moon J-K, Lee Y-H, Park B-S, Varshney RK, Lee S-H (2015) Draft genome sequence of adzuki bean, Vigna angularis. Sci Rep 5(1): 8069. https://doi.org/10.1038/srep08069

Kante M, Lindqvist-Kreuze H, Portal L, David M, Gastelo M (2021) Kompetitive allele specific PCR (KASP) markers for potato: an effective tool for increased genetic gains. Agronomy 11(11):2315. https://doi.org/10.3390/agronomy11112315

Kim SK, Nair RM, Lee J, Lee S-H (2015) Genomic resources in mungbean for future breeding programs. Front Plant Sci 6. https://doi.org/10.3389/fpls.2015.00626

Kolekar RM, Sukruth M, Shirasawa K, Nadaf HL, Motagi BN, Lingaraju S, Patil PV, Bhat RS (2017) Marker-assisted backcrossing to develop foliar disease-resistant genotypes in TMV 2 variety of peanut (*Arachis hypogaea* L.). Plant Breed 136(6):948–953. https://doi.org/10.1111/pbr.12549

Kongjaimun A, Kaga A, Tomooka N, Somta P, Shimizu T, Shu Y, Isemura T, Vaughan DA, Srinives P (2012) An SSR-based linkage map of yardlong bean [*Vigna unguiculata* (L.) Walp. Subsp. unguiculata sesquipedalis group] and QTL analysis of pod length. Genome 55(2):81–92. https://doi.org/10.1139/g11-078

Kumar R, Janila P, Vishwakarma MK, Khan AW, Manohar SS, Gangurde SS, Variath MT, Shasidhar Y, Pandey MK, Varshney RK (2020) Whole-genome resequencing-based QTL-seq identified candidate genes and molecular markers for fresh seed dormancy in groundnut. Plant Biotechnol J 18(4):992–1003. https://doi.org/10.1111/pbi.13266

Lan C, Basnet BR. (2016) Overview of bi-parental QTL mapping and cloning genes in the context of wheat rust. In S. Dreisigacker, Sehgal D, Reyes Jaimez AE, Luna Garrido B, Muñoz Zavala S, Núñez Ríos C, & Mollins JMall S (Eds.), CIMMYT wheat molecular genetics: laboratory protocols and applications to wheat breeding (pp. 39–46). CIMMYT

Lantican MA, Braun HJ, Payne TS, Singh RP, Sonder K, Baum M, van Ginkel M, Erenstein O (2016) Impacts of international wheat improvement research, 1994–2014. International Maize and Wheat Improvement Center (CIMMYT), Mexico, DF, Mexico

Le Thuy CT, Becerra Lopez-Lavalle LA, Vu NA, Hy NH, Nhan PT, Ceballos H, Newby J, Tung NB, Hien NT, Tuan LN, Hung N, Hanh NT, Trang DT, Ha TT, Ham LH, Pham XH, Quynh TN, Rabbi IY, Kulakow PA, Zhang X (2021) Identifying new resistance to cassava mosaic disease and validating markers for the CMD2 locus. Agriculture 11(9):829. https://doi.org/10.3390/agriculture11090829

Li L, Yang X, Cui S, Meng X, Mu G, Hou M, He M, Zhang H, Liu L, Chen CY (2019) Construction of high-density genetic map and mapping quantitative trait loci for growth habit-related traits of peanut (Arachis hypogaea L.). Front Plant Sci 10:745. https://doi.org/10.3389/fpls.2019.00745

Lindqvist-Kreuze H, De Boeck B, Unger P, Gemenet D, Li X, Pan Z, Sui Q, Qin J, Woldegjorgis G, Negash K, Seid I, Hirut B, Gastelo M, De Vega J, Bonierbale M (2021) Global multi-environment resistance QTL for foliar late blight resistance in tetraploid potato with tropical adaptation. G3 Genes|Genomes|Genetics 11(11). https://doi.org/10.1093/g3journal/jkab251

Lindqvist-Kreuze H, Gastelo M, Perez W, Forbes GA, de Koeyer D, Bonierbale M (2014) Phenotypic stability and genome-wide association study of late blight resistance in potato genotypes adapted to the tropical highlands. Phytopathology 104(6):624–633. https://doi.org/10.1094/PHYTO-10-13-0270-R

Lucas MR, Ehlers JD, Roberts PA, Close TJ (2012) Markers for quantitative inheritance of resistance to foliar thrips in cowpea. Crop Sci 52(5):2075–2081

Lonardi S, Muñoz-Amatriaín M, Liang Q, Shu S, Wanamaker SI, Lo S, Tanskanen J, Schulman AH, Zhu T, Luo M, Alhakami H, Ounit R, Hasan AM, Verdier J, Roberts PA, Santos JRP, Ndeve A, Doležel J, Vrána J et al (2019) The genome of cowpea (*Vigna unguiculata* [L.] Walp.). Plant J 98(5):767–782. https://doi.org/10.1111/tpj.14349

Lu Q, Li H, Hong Y, Zhang G, Wen S, Li X, Zhou G, Li S, Liu H, Liu H, Liu Z, Varshney RK, Chen X, Liang X (2018) Genome sequencing and analysis of the peanut B-genome progenitor (*arachis ipaensis*). Front Plant Sci 9. https://doi.org/10.3389/FPLS.2018.00604/FULL

Lucas MR, Ehlers JD, Huynh B-L, Diop N-N, Roberts PA, Close TJ (2013a) Markers for breeding heat-tolerant cowpea. Mol Breed 31(3):529–536. https://doi.org/10.1007/s11032-012-9810-z

Lucas MR, Huynh B-L, da Silva Vinholes P, Cisse N, Drabo I, Ehlers JD, Roberts PA, Close TJ (2013b) Association studies and legume synteny reveal haplotypes determining seed size in *Vigna unguiculata*. Front Plant Sci 4. https://doi.org/10.3389/fpls.2013.00095

Lucas MR, Huynh B-L, Roberts PA, Close TJ (2015) Introgression of a rare haplotype from southeastern Africa to breed California blackeyes with larger seeds. Front Plant Sci 6. https://doi.org/10.3389/fpls.2015.00126

Luo H, Pandey MK, Khan AW, Guo J, Wu B, Cai Y, Huang L, Zhou X, Chen Y, Chen W, Liu N, Lei Y, Liao B, Varshney RK, Jiang H (2019) Discovery of genomic regions and candidate genes controlling shelling percentage using <scp>QTL</scp> −seq approach in cultivated peanut (*Arachis hypogaea* L.). Plant Biotechnol J 17(7):1248–1260. https://doi.org/10.1111/pbi.13050

Maiti S, Basak J, Kundagrami S, Kundu A, Pal A (2011) Molecular marker-assisted genotyping of Mungbean yellow mosaic India virus resistant germplasms of Mungbean and Urdbean. Mol Biotechnol 47(2):95–104. https://doi.org/10.1007/s12033-010-9314-1

Mathivathana MK, Murukarthick J, Karthikeyan A, Jang W, Dhasarathan M, Jagadeeshselvam N, Sudha M, Vanniarajan C, Karthikeyan G, Yang T-J, Raveendran M, Pandiyan M, Senthil N (2019) Detection of QTLs associated with mungbean yellow mosaic virus (MYMV) resistance using the interspecific cross of *Vigna radiata* × *Vigna umbellata*. J Appl Genet 60(3–4):255–268. https://doi.org/10.1007/s13353-019-00506-x

Miedaner T, Korzun V (2012) Marker-assisted selection for disease resistance in wheat and barley breeding. Phytopathology 102(6):560–566. https://doi.org/10.1094/PHYTO-05-11-0157

Mihovilovich E, Aponte M, Lindqvist-Kreuze H, Bonierbale M (2014) An RGA-derived SCAR marker linked to PLRV resistance from *Solanum tuberosum* ssp. andigena. Plant Mol Biol Report 32(1):117–128. https://doi.org/10.1007/s11105-013-0629-5

Morris CF, Kiszonas AM, Murray J, Boehm J, Ibba MI, Zhang M, Cai X (2019) Re-evolution of durum wheat by introducing the hardness and *Glu-D1* loci. Frontiers in Sustainable Food Systems 3. https://doi.org/10.3389/fsufs.2019.00103

Muñoz-Amatriaín M, Mirebrahim H, Xu P, Wanamaker SI, Luo M, Alhakami H, Alpert M, Atokple I, Batieno BJ, Boukar O, Bozdag S, Cisse N, Drabo I, Ehlers JD, Farmer A, Fatokun C, Gu YQ, Guo Y, Huynh B et al (2017) Genome resources for climate-resilient cowpea, an essential crop for food security. Plant J 89(5):1042–1054. https://doi.org/10.1111/tpj.13404

Mwamahonje A, Eleblu JSY, Ofori K, Feyissa T, Deshpande S, Garcia-Oliveira AL, Bohar R, Kigoni M, Tongoona P (2021) Introgression of QTLs for drought tolerance into farmers' preferred sorghum varieties. Agriculture (Switzerland) 11(9). https://doi.org/10.3390/agriculture11090883

Nair R, Schreinemachers P (2020) *Global Status and Economic Importance of Mungbean* (pp. 1–8). https://doi.org/10.1007/978-3-030-20008-4_1

Ndeve AD, Santos JRP, Matthews WC, Huynh BL, Guo Y-N, Lo S, Muñoz-Amatriaín M, Roberts PA (2019) A novel root-knot nematode resistance QTL on chromosome *Vu01* in cowpea. G3 Genes|Genomes|Genetics 9(4):1199–1209. https://doi.org/10.1534/g3.118.200881

Ndjiondjop MN, Semagn K, Zhang J, Gouda AC, Kpeki SB, Goungoulou A, Wambugu P, Dramé KN, Bimpong IK, Zhao D (2018) Development of species diagnostic SNP markers for quality control genotyping in four rice (*Oryza* L.) species. Molecular Breeding 38(11):131. https://doi.org/10.1007/s11032-018-0885-z

Nie X, Sutherland D, Dickison V, Singh M, Murphy AM, De Koeyer D (2016) Development and validation of high-resolution melting markers derived from Ry_{sto} STS markers for high-throughput marker-assisted selection of potato carrying Ry_{sto}. Phytopathology 106(11): 1366–1375. https://doi.org/10.1094/PHYTO-05-16-0204-R

Ogunkanmi L, Ogundipe O, Ng N, Fatokun C (2008) Genetic diversity in wild relatives of cowpea (*Vigna unguiculata*) as revealed by simple sequence repeats (SSR) markers. J Food Agric Environ 6(4):263–268

Omoigui LO, Kamara AY, Moukoumbi YD, Ogunkanmi LA, Timko MP (2017) Breeding cowpea for resistance to *Striga gesnerioides* in the Nigerian dry savannas using marker-assisted selection. Plant Breed 136(3):393–399. https://doi.org/10.1111/pbr.12475

Ongom PO, Fatokun C, Togola A, Salvo S, Oyebode OG, Ahmad MS, Jockson ID, Bala G, Boukar O (2021) Molecular fingerprinting and hybridity authentication in cowpea using single nucleotide polymorphism based Kompetitive allele-specific PCR assay. Front Plant Sci 12. https://doi.org/10.3389/fpls.2021.734117

Ortiz R (2020) Genomic-led potato breeding for increasing genetic gains: achievements and outlook. Crop Breeding, Genetics and Genomics 2:e200010. https://doi.org/10.20900/cbgg20200010

Ottoman RJ, Hane DC, Brown CR, Yilma S, James SR, Mosley AR, Crosslin JM, Vales MI (2009) Validation and implementation of marker-assisted selection (MAS) for PVY resistance (*Ry adg* gene) in a tetraploid potato breeding program. Am J Potato Res 86(4):304–314. https://doi.org/10.1007/s12230-009-9084-0

Padulosi S, Ng NQ (1997) Origin, taxonomy, and morphology of Vigna unguiculata (L.) Walp. In: Singh BB, Mohan Raj DR, Dashiell KE, Jackai LEN (eds) Advances in cowpea research. Japan International Research Center for Agricultural Sciences and International Institute of Tropical Agriculture, Ibadan, Nigeria, pp 1–12

Pandey MK, Agarwal G, Kale SM, Clevenger J, Nayak SN, Sriswathi M, Chitikineni A, Chavarro C, Chen X, Upadhyaya HD, Vishwakarma MK, Leal-Bertioli S, Liang X, Bertioli DJ, Guo B, Jackson SA, Ozias-Akins P, Varshney RK (2017a) Development and evaluation of a high density genotyping 'Axiom_*Arachis*' Array with 58 K SNPs for accelerating genetics and breeding in groundnut. Sci Rep 7(1):1–10. https://doi.org/10.1038/srep40577

Pandey MK, Khan AW, Singh VK, Vishwakarma MK, Shasidhar Y, Kumar V, Garg V, Bhat RS, Chitikineni A, Janila P, Guo B, Varshney RK (2017b) QTL-seq approach identified genomic regions and diagnostic markers for rust and late leaf spot resistance in groundnut (*Arachis hypogaea* L.). Plant Biotechnol J 15(8):927–941. https://doi.org/10.1111/pbi.12686

Pandey MK, Pandey AK, Kumar R, Nwosu CV, Guo B, Wright GC, Bhat RS, Chen X, Bera SK, Yuan M, Jiang H, Faye I, Radhakrishnan T, Wang X, Liang X, Liao B, Zhang X, Varshney RK, Zhuang W (2020) Translational genomics for achieving higher genetic gains in groundnut. Theor Appl Genet 133(5):1679–1702. https://doi.org/10.1007/S00122-020-03592-2

Pandey MK, Bhat RS, Janila P, Guo B, Varshney RK (2017c) Genetic dissection of foliar disease resistance using next-generation sequencing approaches in groundnut. In: 9th international conference advances in arachis through Genomics & Biotechnology (AAGB), pp 14–17

Parmar S, Deshmukh DB, Kumar R, Manohar SS, Joshi P, Sharma V, Chaudhari S, Variath MT, Gangurde SS, Bohar R, Singam P, Varshney RK, Janila P, Pandey MK (2021) Single seed-based high-throughput genotyping and rapid generation advancement for accelerated groundnut genetics and breeding research. Agronomy 11(6):1226. https://doi.org/10.3390/agronomy

Pasupuleti J, Pandey MK, Manohar SS, Variath MT, Nallathambi P, Nadaf HL, Sudini H, Varshney RK (2016) Foliar fungal disease-resistant introgression lines of groundnut (*Arachis hypogaea* L.) record higher pod and haulm yield in multilocation testing. Plant Breed 135(3):355–366. https://doi.org/10.1111/PBR.12358

Pottorff M, Wanamaker S, Ma YQ, Ehlers JD, Roberts PA, Close TJ (2012) Genetic and physical mapping of candidate genes for resistance to fusarium oxysporum f.sp. tracheiphilum race 3 in cowpea [*Vigna unguiculata* (L.) Walp]. PLoS One 7(7):e41600. https://doi.org/10.1371/journal.pone.0041600

Rakosy-Tican E, Thieme R, König J, Nachtigall M, Hammann T, Denes T-E, Kruppa K, Molnár-Láng M (2020) Introgression of two broad-Spectrum late blight resistance genes, Rpi-Blb1 and Rpi-Blb3, from Solanum bulbocastanum dun plus race-specific R genes into potato pre-breeding lines. Front Plant Sci 11. https://doi.org/10.3389/fpls.2020.00699

Ramakrishnan P, Manivannan N, Mothilal A, Mahaingam L, Prabhu R, Gopikrishnan P (2020) Marker assisted introgression of QTL region to improve late leaf spot and rust resistance in elite and popular variety of groundnut (*Arachis hypogaea* L.) cv TMV 2. Australas Plant Pathol 49(5):505–513. https://doi.org/10.1007/s13313-020-00721-9

Razzaq A, Kaur P, Akhter N, Wani SH, Saleem F (2021) Next-generation breeding strategies for climate-ready crops. Front Plant Sci 12:620420. https://doi.org/10.3389/fpls.2021.620420

Roorkiwal M, Bharadwaj C, Barmukh R, Dixit GP, Thudi M, Gaur PM, Chaturvedi SK, Fikre A, Hamwieh A, Kumar S, Sachdeva S, Ojiewo CO, Taran B, Wordofa NG, Singh NP, Siddique KHM, Varshney RK (2020) Integrating genomics for chickpea improvement: achievements and opportunities. Theor Appl Genet 133(5):1703–1720. https://doi.org/10.1007/S00122-020-03584-2

RTB (2019) Making an impact. Annual report 2018. www.rtb.cgiar.org/2018-annual-report

Salifou M, Tignegre JBLS, Tongoona P, Offei S, Ofori K, Danquah E (2016) Introgression of striga resistance gene into farmers' preferred cowpea varieties in Niger. Int J Plant Breed Genet 3(6): 233–240

Schafleitner R, Huang S, Chu S, Yen J, Lin C, Yan M, Krishnan B, Liu M, Lo H, Chen C, Chen LO, Wu D, Bui T-GT, Ramasamy S, Tung C, Nair R (2016) Identification of single nucleotide polymorphism markers associated with resistance to bruchids (*Callosobruchus* spp.) in wild mungbean (*Vigna radiata* var. sublobata) and cultivated V. radiata through genotyping by sequencing and quantitative trait locus analysis. BMC Plant Biol 16(1):159. https://doi.org/10.1186/s12870-016-0847-8

Sehrawat N, Yadav M, Bhat KV, Sairam RK, Jaiwal PK (2016) Introgression of mungbean yellow mosaic virus resistance in *Vigna mungo* (L.)hepper and purity testing of F1 hybrids using SSRs. Turk J Agric For 40(1):95–100

Semagn K, Babu R, Hearne S, Olsen M (2014) Single nucleotide polymorphism genotyping using Kompetitive allele specific PCR (KASP): overview of the technology and its application in crop improvement. Mol Breed 33(1):1–14. https://doi.org/10.1007/s11032-013-9917-x

Semagn K, Beyene Y, Makumbi D, Mugo S, Prasanna BM, Magorokosho C, Atlin G (2012) Quality control genotyping for assessment of genetic identity and purity in diverse tropical maize inbred lines. Theor Appl Genet 125(7):1487–1501. https://doi.org/10.1007/s00122-012-1928-1

Shasidhar Y, Variath MT, Vishwakarma MK, Manohar SS, Gangurde SS, Sriswathi M, Sudini HK, Dobariya KL, Bera SK, Radhakrishnan T, Pandey MK, Janila P, Varshney RK (2020) Improvement of three popular Indian groundnut varieties for foliar disease resistance and high oleic acid using SSR markers and SNP array in marker-assisted backcrossing. Crop J 8(1):1–15. https://doi.org/10.1016/J.CJ.2019.07.001

Simon MV, Benko-Iseppon A-M, Resende LV, Winter P, Kahl G (2007) Genetic diversity and phylogenetic relationships in *Vigna* Savi germplasm revealed by DNA amplification fingerprinting. Genome 50(6):538–547. https://doi.org/10.1139/G07-029

Simpson CE, Starr JL, Church GT, Burow MD, Paterson AH (2003) Registration of 'NemaTAM' Peanut. Crop Sci 43(4):1561–1561. https://doi.org/10.2135/cropsci2003.1561

Singh CM, Pratap A, Gupta S, Biradar RS, Singh NP (2020) Association mapping for mungbean yellow mosaic India virus resistance in mungbean (*Vigna radiata* L. Wilczek). 3Biotech 10(2): 33. https://doi.org/10.1007/s13205-019-2035-7

Singh N, Mallick J, Sagolsem D, Mandal N, Bhattacharyya S (2017a) Mapping of molecular markers linked with MYMIV and yield attributing traits in mungbean. Indian J Genet Plant Breed 78(1):118. https://doi.org/10.5958/0975-6906.2018.00014.7

Singh RP, Herrera-Foessel S, Huerta-Espino J, Singh S, Bhavani S, Lan C, Basnet BR (2014) Progress towards genetics and breeding for minor genes based resistance to Ug99 and other rusts in CIMMYT high-yielding spring wheat. J Integr Agric 13(2):255–261. https://doi.org/10.1016/S2095-3119(13)60649-8

Singh V, Yadav RK, Yadav NR, Yadav R, Malik RS, Singh J (2017b) Identification of genomic regions/genes for high iron and zinc content and cross transferability of SSR markers in mungbean (*Vigna radiata* L.). Legume Res Intern J 40(6). https://doi.org/10.18805/lr.v40i04.9006

Sinha P, Bajaj P, Pazhamala LT, Nayak SN, Pandey MK, Chitikineni A, Huai D, Khan AW, Desai A, Jiang H, Zhuang W, Guo B, Liao B, Varshney RK (2020) Arachis hypogaea gene expression atlas for fastigiata subspecies of cultivated groundnut to accelerate functional and translational genomics applications. Plant Biotechnol J 18(11):2187–2200. https://doi.org/10.1111/PBI.13374

Slater AT, Cogan NOI, Hayes BJ, Schultz L, Dale MFB, Bryan GJ, Forster JW (2014) Improving breeding efficiency in potato using molecular and quantitative genetics. Theor Appl Genet 127(11):2279–2292. https://doi.org/10.1007/s00122-014-2386-8

Sneller C, Ignacio C, Ward B, Rutkoski J, Mohammadi M (2021) Using genomic selection to leverage resources among breeding programs: consortium-based breeding. Agronomy 11(8): 1555. https://doi.org/10.3390/AGRONOMY11081555

Somta P, Chen J, Yundaeng C, Yuan X, Yimram T, Tomooka N, Chen X (2019) Development of an SNP-based high-density linkage map and QTL analysis for bruchid (Callosobruchus maculatus F.) resistance in black gram (*Vigna mungo* (L.) hepper). Sci Rep 9(1):1–9. https://doi.org/10.1038/s41598-019-40669-5

Sood S, Bhardwaj V, Chourasia KN, Kaur RP, Kumar V, Kumar R, Sundaresha S, Bohar R, Garcia-Oliveira AL, Singh RK, Kumar M (2022) KASP markers validation for late blight, PCN and PVY resistance in a large germplasm collection of tetraploid potato (*Solanum tuberosum* L.). Sci Hortic 295. https://doi.org/10.1016/j.scienta.2021.110859

Souframanien J, Gopalakrishna T (2006) ISSR and SCAR markers linked to the mungbean yellow mosaic virus (MYMV) resistance gene in blackgram [*Vigna mungo* (L.) hepper]. Plant Breed 125(6):619–622. https://doi.org/10.1111/j.1439-0523.2006.01260.x

Souframanien J, Gupta SK, Gopalakrishna T (2010) Identification of quantitative trait loci for bruchid (Callosobruchus maculatus) resistance in black gram [*Vigna mungo* (L.) hepper]. Euphytica 176(3):349–356. https://doi.org/10.1007/S10681-010-0210-3/FIGURES/2

Souframanien J, Reddy KS (2015) De novo assembly, characterization of immature seed transcriptome and development of genic-SSR markers in black gram [*Vigna mungo* (L.) hepper]. PLoS One 10(6):e0128748. https://doi.org/10.1371/journal.pone.0128748

Stefańczyk E, Plich J, Janiszewska M, Smyda-Dajmund P, Sobkowiak S, Śliwka J (2020) Marker-assisted pyramiding of potato late blight resistance genes *Rpi-rzc1* and *Rpi-phu1* on di- and tetraploid levels. Mol Breed 40(9):89. https://doi.org/10.1007/s11032-020-01169-x

Stich B, Van Inghelandt D (2018) Prospects and potential uses of genomic prediction of key performance traits in tetraploid potato. Front Plant Sci 9. https://doi.org/10.3389/FPLS.2018.00159/FULL

Subramaniyan R, Kulanthaivel V, Narayana M, Angamuthu M, Kothandaraman SV (2021) Marker-assisted backcross breeding for enhancing Mungbean yellow mosaic virus (MYMV) disease resistance in blackgram [*Vigna mungo* (L.) hepper] cv MDU 1. Physiol Mol Plant Pathol 116:101732. https://doi.org/10.1016/j.pmpp.2021.101732

Sverrisdóttir E, Byrne S, Sundmark EHR, Johnsen HØ, Kirk HG, Asp T, Janss L, Nielsen KL (2017) Genomic prediction of starch content and chipping quality in tetraploid potato using genotyping-by-sequencing. Theor Appl Genet 130(10):2091–2108. https://doi.org/10.1007/S00122-017-2944-Y

Takahashi Y, Tomooka N (2020) Taxonomy of mungbean and its relatives (pp. 27–41). https://doi.org/10.1007/978-3-030-20008-4_3

Tiwari JK, Siddappa S, Singh BP, Kaushik SK, Chakrabarti SK, Bhardwaj V, Chandel P (2013) Molecular markers for late blight resistance breeding of potato: an update. Plant Breed 132(3): 237–245. https://doi.org/10.1111/pbr.12053

Uitdewilligen JGAML, Wolters AMA, D'hoop BB, Borm TJA, Visser RGF, van Eck HJ (2013) A next-generation sequencing method for genotyping-by-sequencing of highly heterozygous autotetraploid potato. PLoS One 8(5). https://doi.org/10.1371/JOURNAL.PONE.0062355

Vadivel K, Manivannan N, Mahalingam A, Satya VK, Vanniarajan C, Ragul S (2021) Identification and validation of quantitative trait loci of Mungbean yellow mosaic virus disease resistance in Blackgram [*Vigna mungo* (L). Hepper]. Legume Res Int J 1:1–7. https://doi.org/10.18805/LR-4459

Varshney RK, Pandey MK, Bohra A, Singh VK, Thudi M, Saxena RK (2019) Toward the sequence-based breeding in legumes in the post-genome sequencing era. Theor Appl Genet 132(3):797–816. https://doi.org/10.1007/S00122-018-3252-X/FIGURES/1

Varshney RK, Pandey MK, Janila P, Nigam SN, Sudini H, Gowda MVC, Sriswathi M, Radhakrishnan T, Manohar SS, Nagesh P (2014) Marker-assisted introgression of a QTL region to improve rust resistance in three elite and popular varieties of peanut (*Arachis hypogaea* L.). Theor Appl Genet 127(8):1771–1781. https://doi.org/10.1007/S00122-014-2338-3/FIGURES/3

Velásquez AC, Mihovilovich E, Bonierbale M (2007) Genetic characterization and mapping of major gene resistance to potato leafroll virus in *Solanum tuberosum* ssp. andigena. Theor Appl Genet 114(6):1051–1058. https://doi.org/10.1007/S00122-006-0498-5

Vishwakarma MK, Kale SM, Sriswathi M, Naresh T, Shasidhar Y, Garg V, Pandey MK, Varshney RK (2017) Genome-wide discovery and deployment of insertions and deletions markers provided greater insights on species, genomes, and sections relationships in the genus arachis. Front Plant Sci 8:2064. https://doi.org/10.3389/FPLS.2017.02064/BIBTEX

Vos PG, Uitdewilligen JGAML, Voorrips RE, Visser RGF, van Eck HJ (2015) Development and analysis of a 20K SNP array for potato (*Solanum tuberosum*): an insight into the breeding history. Theor Appl Genet 128(12):2387–2401. https://doi.org/10.1007/s00122-015-2593-y

Wageningen FSC (2016) Multi-level mapping and exploration of wheat production and consumption and their potential contribution to alleviation of poverty, malnutrition and gender inequality. Final Report WHEAT Competitive Grant. Wageningen University Food Security Center Study, The Netherlands

Whitworth JL, Novy RG, Hall DG, Crosslin JM, Brown CR (2009) Characterization of broad Spectrum potato virus Y resistance in a *Solanum tuberosum* ssp. andigena-derived population and select breeding clones using molecular markers, grafting, and field inoculations. Am J Potato Res 86(4):286–296. https://doi.org/10.1007/s12230-009-9082-2

Wu X, Islam ASMF, Limpot N, Mackasmiel L, Mierzwa J, Cortés AJ, Blair MW (2020) Genome-wide SNP identification and association mapping for seed mineral concentration in mung bean (Vigna radiata L.). Front Genet 11. https://doi.org/10.3389/fgene.2020.00656

Xiong H, Shi A, Mou B, Qin J, Motes D, Lu W, Ma J, Weng Y, Yang W, Wu D (2016) Genetic diversity and population structure of cowpea (*Vigna unguiculata* L. Walp). PLoS One 11(8): e0160941. https://doi.org/10.1371/journal.pone.0160941

Xu P, Wu X, Wang B, Hu T, Lu Z, Liu Y, Qin D, Wang S, Li G (2013) QTL mapping and epistatic interaction analysis in asparagus bean for several characterized and novel horticulturally important traits. BMC Genet 14(1):4. https://doi.org/10.1186/1471-2156-14-4

Xu Y, Li P, Zou C, Lu Y, Xie C et al (2017) Enhancing genetic gain in the era of molecular breeding. J Exp Bot 68(11):2641–2666. https://doi.org/10.1093/jxb/erx135

Yeri SB, Bhat RS (2016) Development of late leaf spot and rust resistant backcross lines in Jl 24 variety of groundnut (*Arachis hypogaea* L.). Electron J Plant Breed 7(1):37. https://doi.org/10.5958/0975-928X.2016.00005.3

Yin D, Ji C, Ma X, Li H, Zhang W, Li S, Liu F, Zhao K, Li F, Li K, Ning L, He J, Wang Y, Zhao F, Xie Y, Zheng H, Zhang X, Zhang Y, Zhang J (2018) Genome of an allotetraploid wild peanut *Arachis monticola*: a de novo assembly. GigaScience 7(6):1–9. https://doi.org/10.1093/GIGASCIENCE/GIY066

Zhao C, Qiu J, Agarwal G, Wang J, Ren X, Xia H, Guo B, Ma C, Wan S, Bertioli DJ, Varshney RK, Pandey MK, Wang X (2017) Genome-wide discovery of microsatellite markers from diploid progenitor species, *A. duranensis* and *A. Ipaensis*, and their application in cultivated peanut (*a. hypogaea*). Front Plant Sci 8:1209. https://doi.org/10.3389/FPLS.2017.01209/BIBTEX

Zhao Y, Ma J, Li M, Deng L, Li G, Xia H, Zhao S, Hou L, Li P, Ma C, Yuan M, Ren L, Gu J, Guo B, Zhao C, Wang X (2020) Whole-genome resequencing-based QTL-seq identified *AhTc1* gene encoding a R2R3- MYB transcription factor controlling peanut purple testa colour. Plant Biotechnol J 18(1):96–105. https://doi.org/10.1111/pbi.13175

Zhang X, Zhang J, He X, Wang Y, Ma X, Yin D (2017) Genome-wide association study of major agronomic traits related to domestication in peanut. Front Plant Sci 8:1611. https://doi.org/10.3389/fpls.2017.01611

Zhuang W, Chen H, Yang M, Wang J, Pandey MK, Zhang C, Chang W-C, Zhang L, Zhang X, Tang R, Garg V, Wang X, Tang H, Chow C-N, Wang J, Deng Y, Wang D, Khan AW, Yang Q et al (2019) The genome of cultivated peanut provides insight into legume karyotypes, polyploid evolution and crop domestication. Nat Genet 51(5):865–876. https://doi.org/10.1038/s41588-019-0402-2

Chapter 9
Genomic Selection in Crop Improvement

H. V. Veerendrakumar, Rutwik Barmukh, Priya Shah, Deekshitha Bomireddy, Harsha Vardhan Rayudu Jamedar, Manish Roorkiwal, Raguru Pandu Vasanthi, Rajeev K. Varshney, and Manish K. Pandey

Abstract A boost in the crop improvement rate is essential for accomplishing a sustainable food supply and other demands of rapid population growth. Genomic selection (GS), a very promising breeding strategy used effectively in animal breeding, is now used in crop improvement. GS offers a reduced duration of breeding cycles by rapidly selecting better genotypes. Several empirical and simulated research on GS and their implications on agricultural production enhancement

H. V. Veerendrakumar · D. Bomireddy
International Crops Research Institute for the Semi-Arid Tropics (ICRISAT), Hyderabad, India

S. V. Agricultural College, Acharya N.G Ranga Agricultural University (ANGRAU), Tirupati, India

R. Barmukh · P. Shah
International Crops Research Institute for the Semi-Arid Tropics (ICRISAT), Hyderabad, India

Department of Genetics, Osmania University, Hyderabad, India

H. V. R. Jamedar
International Crops Research Institute for the Semi-Arid Tropics (ICRISAT), Hyderabad, India

Agricultural College, Acharya N.G Ranga Agricultural University (ANGRAU), Bapatla, India

M. Roorkiwal
Khalifa Center for Genetic Engineering and Biotechnology, United Arab Emirates University, Al-Ain, United Arab Emirates
e-mail: mroorkiwal@uaeu.ac.ae

R. P. Vasanthi
S. V. Agricultural College, Acharya N.G Ranga Agricultural University (ANGRAU), Tirupati, India

R. K. Varshney
Murdoch's Centre for Crop and Food Innovation, State Agricultural Biotechnology Centre, Food Futures Institute, Murdoch University, Murdoch, WA, Australia
e-mail: rajeev.varshney@murdoch.edu.au

M. K. Pandey (✉)
Center of Excellence in Genomics and Systems Biology (CEGSB), International Crops Research Institute for the Semi-Arid Tropics (ICRISAT), Hyderabad, India
e-mail: Manish.Pandey@icrisat.org

M. K. Pandey et al. (eds.), *Frontier Technologies for Crop Improvement*, Sustainable Agriculture and Food Security, https://doi.org/10.1007/978-981-99-4673-0_9

have lately been published. We briefly discuss the GS methodology, its present position, the GS advantages over alternative methods of breeding, commonly used prediction models of GS, and factors interfering with the prediction accuracy of GS to provide a comprehensive grasp of the technology. In addition, the integration of speed breeding and other modern techniques for increasing the effectiveness and speed of GS are discussed.

9.1 Introduction

Plant breeding programme aims to develop genotypes improved for desirable traits to fulfil the requirements of key stakeholders. The breeder needs to explore a large genetic landscape to identify the superior genotypes and the material required to surpass the capacity of breeding programs (Chenu 2015). Plant breeding may be viewed as a number game in which breeding plans are meant to enhance the likelihood of identifying genotypes with acceptable combinations of traits with minimal resources (Brown et al. 2008). The assessment, which includes multiple phenotyping processes aimed to quantify the heritable genetic value of selection genotypes, is a critical component of the breeding scheme (Lynch and Walsh 1998).

For a characteristic such as yield, a set of individuals identified for high heritable traits is often assessed in multi-environmental trials meant to resemble the target population of environments (TPE), where the product is predicted to perform (Chenu 2015). Throughout most breeding programmes, such assessment processes need large amount of resources and span several years (Brown et al. 2008). Several approaches and technologies have arisen over the last three decades to overcome these constraints and boost the efficiency of breeding programmes, due largely to developments in the characterisation of DNA polymorphisms and computational capacity (Xu et al. 2020). Among them, methods that can identify the phenotypic performance by using molecular information (marker-assisted selection) (Cobb et al. 2019) and GS (Meuwissen et al. 2001) are suitable tools that help modern breeding programmes to get maximum outcome from limited resources. In contrast to traditional marker-assisted selection (MAS), current genomic prediction algorithms account for both minor and major QTLs, capturing the large amount of genetic diversity in a trait. The GS original idea was initially proposed in the field of animal breeding by estimating the marker effects for two generations (Meuwissen et al. 2001). Genotyping based on NGS has increased the genomic estimated breeding values (GEBVs) accuracy of prediction over other platforms developed in cereals and also in other crops. In this way, the dream of GS in crop plants has come true. To get the most benefits out of GS, these marker techniques should be used along with high-throughput phenotyping to get higher genetic gain from complex traits.

GS is rapidly becoming the favourable strategy for accelerating breeding by utilising genetic markers. Prediction models for GS are built by regressing observed phenotypes in a training population (TP) on markers that are genotyped on same population (Meuwissen et al. 2001). Best individuals are found in future generations by utilising these models, which are based purely on the genetic profile. For GS to be successful, genetic markers should be densely and widely dispersed across the genome, with a high probability that every quantitative trait locus (QTL) is really in linkage disequilibrium (LD) with at least one marker (de Resende and de Assis 2010). As the marker number increases, so does the prediction ability of model (Meuwissen and Goddard 2010). As a result, a low-cost, high-density, adaptable, and precise genotyping platform is required for the implementation of GS effectively. Many genome sequencing approaches in the earlier days have recorded a vast number of useful SNPs (single-nucleotide polymorphisms), and to reduce the subsequent challenging problem of high cost incurred characterising those, numerous DNA chip-based SNP genotyping systems are advanced and have become an extremely renowned platform of genetic variation analysis across the genome (Maresso and Broeckel 2008). All of these systems rely largely upon annealing single or several oligonucleotides near a point close to the variable site known before, followed by detection of extension reaction of the attached nucleotide on a chip. DNA chip techniques, including Axiom Affymetrix and Illumina Infinium, provide large-scale, simpler SNP analysis for a large number of loci (Crossa et al. 2013). The emergence of such platforms has laid the foundation for the development of GS as a suitable molecular breeding strategy in species having genotyping tool accessibility.

9.2 Basics of GS

GS is applicable to minor gene governed traits, and its accuracy of prediction is more time efficient than the usual phenotypic selection (Spindel and Iwata 2018). The main advantage of genomic prediction is that it may be used in making decisions for the effective selection of breeding material at different levels (generations) of the breeding program. As a result, efficient genomic prediction integration necessitates more grip over the breeding strategy and its various factors. The breeding strategy is often in the breeder's head, and converting this information into a systematic structure is an important aspect in thoroughly designing various methods (Cobb et al. 2019). Genomic prediction is indeed a continuous long-term commitment requiring breeding programme, and switching to an optimal GS breeding strategy is not always feasible. As a result, the breeding team and specialists must design a transition strategy that outlines precise ways to achieve the objectives (Bartholomé et al. 2021).

Optimal GS strategies are seldom simply evolutions of existing breeding methods. In general, pedigree breeding is used in the majority of conventional breeding systems in self-pollinated plants (Guimarães 2009), but GS is very much suitable to recurrent selection methods relying on superior by superior crosses to improve complex traits. Due to the greater LD among QTL and markers, undetectable or lower population structure, and greater similarity among genotypes, a well-structured breeding programme with clearly defined elite germplasm and a low population size (Ne 40) seems to be more likely to benefit from prediction using GS (Bartholomé et al. 2021).

Just like with traditional marker-assisted selection (MAS), the GS will be using genome-wide markers for estimating the breeding values of individual genotypes rather than individual markers. The genomically determined breeding values could then be used for effective trait selection. A plethora of algorithms have been developed to model GEBV. In consideration of both dimensions (columns) and instances, modelling GEBVs generates a massive genotype (rows). The right feature combinations can let you figure out which phenotype is being represented. Preparing a solid TP sample is seen to be a practical strategy for dealing with such complicated genetic data (Purbarani et al. 2017). GEBV is measured on a set of beneficial loci present in each BP's genome, and it gives a direct assessment of each individual's likelihood of having a better phenotype (i.e. higher breeding value). Newer parent choices are made using the GEBVs. This shortens the breeding cycle since there is no need to phenotype quantitative parameters like yield and its components in successive generations. The validation population (VP) is a third set of individuals that undergoes genotyping and phenotyping. The GEBVs are calculated for VP, and the correlation between it and the phenotype is used to assess the precision of GS model (Bassi et al. 2015).

The genetic gain expected per unit time from GS is

$$\Delta G = \mathrm{ir\sigma A/T}$$

i–intensity of selection,
r–accuracy of selection,
σ_A–the square root value of the additive variance,
T–time required to complete one cycle of breeding (Falconer and Mackay 1996).

Following the first study of Meuwissen et al., the researchers Bernardo and Yu first established the effect of GS on crop breeding (Bernardo and Yu 2007). The researchers utilised a computer-based simulation to show that the use of an entire set of genotyping markers produced better breeding value prediction accuracy than using only a couple of markers that were significantly associated with QTL. Years later, the first genomic-enabled predictions in the actual crop breeding conditions were demonstrated, indicating that strong genomic predictions may be obtained in a variety of corn and wheat data sets (De Los Campos et al. 2009). This study was the first to use pedigree and genetic relationship information to make wheat predictions, which researchers applied in different non-parametric and parametric statistical models. Following these initial discoveries, a significant number of scientific research into forecast accuracy in a different crop species have been conducted and published; some of them are mentioned in Table 9.1.

Higher genetic advance per generation can be attained if the reduction in breeding cycle duration by GS compensates for the drop in selection accuracy, assuming equal selection intensities and genetic variation for both GS and phenotypic selection (PS). Considering assumptions about breeding cycle lengths, selection accuracies,

Table 9.1 GS in different crops

Sl. No.	Crop	Traits	Model	Population	Markers	Reference
1.	Wheat	Plant height (PH), grain yield (GY), heading date (HD), test weight (TW)	RR-BLUP	816 breeding lines	21,643 SNPs	(Xu 2013)
2.	Wheat	Quality traits and agro-morphological traits	G-BLUP	329 genotypes	7748 SNPs	(Ward et al. 2019)
3.	Wheat	Fusarium head blight, Septoria tritici blotch, plant height, heading date	RR-BLUP	1120 winter	5628 SNPs	(Herter et al. 2019)
4.	Wheat	Grain yield, kernel number per spike, kernel weight per spike, thousand kernel weight	RR-BLUP	156 RILs, 239 lines,100 DHs	1188, 5661, and 2780 SNPs	(Lozada et al. 2019)
5.	Wheat	Grain yield	G-BLUP	1100 lines	27,000 SNPs	(Belamkar et al. 2018)
6.	Wheat	Gluten index, protein content, and alveograph	G-BLUP, RR-BLUP, RKHS, BayesA, B, BL	324 lines	9752 SNPs	(Haile et al. 2018)
7.	Buck wheat	Nodes, stem length, flower clusters, primary branches, selection index	G-BLUP	92 lines	14,598–50,000 markers	(Yabe et al. 2018)
8.	Wheat	Fusarium head blight resistance	RR-BLUP	273	19,992 SNPs	(Arruda et al. 2016)
9.	Wheat	Grain yield, flour yield, softness equivalence, fusarium head blight index	RR-BLUP, BL, RF	470 lines	4858 SNPs	(Hoffstetter et al. 2016)
10.	Wheat	Drought and heat stress	G-BLUP	8416 and 2403	40,000 DArTs	(Crossa et al. 2016)
11.	Wheat	Stem rust resistance	BRR and G-BLUP	374 lines	18,653 GBS	(Rutkoski et al. 2013)
12.	Wheat	Quality of grain	BayesCπ, RR-BLUP	174, 209	574 DArTs and 399 various markers	(Heffner et al. 2011)
13.	Maize	Starch, oil content, and protein	G-BLUP	257 inbreds	48,814 SNPs	(Guo et al. 2014)

(continued)

Table 9.1 (continued)

Sl. No.	Crop	Traits	Model	Population	Markers	Reference
14.	Maize	Resistance ear rot	RR-BLUP	238 lines	23,154 DArTs	(Dos Santos et al. 2016)
15.	Maize	Drought stress	G-BLUP	3273 lines	58,731 SNPs	(Zhang et al. 2015)
16.	Maize	Grain yield, silking-anthesis interval, and anthesis date	RKHS and GBLUP	504 DH and 296 inbreds	158,281 and 235,265 SNPs	(Crossa et al. 2013)
17.	Maize	Kernel rows, kernels per row, and ear length	G-BLUP	635, 5 DH populations	16,741 SNPs	(Riedelsheimer et al. 2013)
18.	Maize	10 agro-morphological traits	RR-BLUP, BLUP	441 hybrids and 294 RILs	261 SSRs	(Guo et al. 2013)
19.	Maize	Lignin content, biomass, female flowering, plant height, starch, and sugar	RR-BLUP	285 diverse inbred lines	56,110 SNPs &130 metabolites	(Riedelsheimer et al. 2013)
20.	Rice	Flowering, florets per panicle, protein, height	G-BLUP	405 lines	36,901 SNPs	(Isidro et al. 2015)
21.	Rice	Panicle weight, days to flowering; grain yield, height	BL, G-BLUP, LASSO, RR-BLUP, BRR	343 lines	8336 SNPs	(Grenier et al. 2015)
22.	Rice	Plant height, grain yield, head rice percentage, milling yield, and grain chalkiness percentage	GK and G-BLUP	327 indica and 309 japonica	Japonica: 44598 and indica: 92430 SNP	(Monteverde et al. 2018)
23.	Barley	12 malting traits	RR-BLUP	Winter: 148 Spring: 116–294	Winter:4359 SNPs Spring:4095	(Schmidt et al. 2016)
24.	Barley	Grain protein content, plant height, amylase activity, malt extract, grain yield	BayesA and B, G-BLUP	2 datasets, 140 and150 DH lines	223 RFLP/107 RFLP/AFLP	(Lorenzana and Bernardo 2009)
25.	Barley	Resistance to fusarium head blight	BayesCπ, BL, RR-BLUP	691 lines	3072 SNPs	(Lorenz et al. 2012)
26.	Rye	Protein content and grain yield		219 and 201 lines		

			Multi-trait RR-BLUP		584 and 394 DArTs	(Schulthess et al. 2016)
27.	Rye	Total pentosan content, grain yield, starch content, and plant height	RR-BLUP	220 (2 sets)	1048 DArTs	(Wang et al. 2014)
28.	Sorghum	Condensed tannins, polyphenols, total antioxidant capacity, flavonoids	BayesB, GBLUP, BL, Bayesian RR	114 genotypes	61,976 SNPs	(Habyarimana and Lopez-Cruz 2019)
29.	Pearl millet	Plant height, grain weight, grain yield, days to flowering	RR-BLUP	37 inbreds, 320 hybrids	306, 15, 463, 32 SNPs	(Liang et al. 2018)
30.	Soybean	Protein, yield, oil	RR-BLUP	483 elite lines	5403 SNPs	(Stewart-Brown et al. 2019)
31.	Soybean	15 amino acids content	RR-BLUP	249 accessions	23,279 SNPs	(Qin et al. 2019)
32.	Groundnut	Total yield/plant, days to 50% flowering, hundred seed weight, days to maturity, rust, late leaf spot, oleic acid, pods/plant, and shelling %	BGLR	340 lines	13,355 SNPs	(Pandey et al. 2020)
33.	Chickpea	Early vigour score, grain yield, seed number, and hundred seed weight	RR-BLUP, BRR, and BL	119 and 13, respectively, Indian and Australian	144,777 SNPs	(Li et al. 2018)
34.	Cowpea	Seed size, flowering time (FT), maturity	SVR, RRBLUP, RKHS	305 RILs	51,128 SNPs	(Olatoye et al. 2019)
35.	n	Seed weight, grain yield, lodging susceptibility, and onset of flowering	BL, G-BLUP, RR-BLUP	306 genotypes	6058 SNPs	(Annicchiarico et al. 2019)

and selection intensities, GS can outperform PS in terms of genetic advance per year (Schaeffer 2006; Shengqiang et al. 2009; Wong and Bernardo 2008). Furthermore, when evaluating traits with a long generation time, GS becomes easier or cheaper than PS, enabling more number of candidates to be characterised for a unit cost, boosting higher selection intensity. As a result, GS is currently utilised in all crops; some of the crops in which GS is used is mentioned in Table 9.1.

9.3 Methodology of GS

The main principle of GS is to estimate breeding values of genotypes under the test purely depending on genotypic data with the usage of statistical models designed on TP (Meuwissen et al. 2001). TP is a group of individuals having a lot of genotypic and the phenotypic data that can be used to get GS model parameters. Following that, using genotypic data, these GS methods are used to calculate breeding values called as GEBV of breeding population (BP) genotypes. A TP is a group of individuals who are linked and have known ancestors, such as half-sibs and closely related groups. The BP is composed of TP descendants or aristocratic lineages with strong ties to the TP. To forecast genetic values of a BP for various traits, allelic similarities with loci connected to the phenotype in the TP is employed. As a result, the amount of genetic resemblance between BP and TP in LD of markers with trait loci determines GS (Edwards et al. 2019).

9.3.1 Designing Training Population (TP)

The overall design of TP is crucial for the success of GS because it adds to higher accuracy of prediction in BP, allowing actual individuals to be selected in present breeding activities (Zhang et al. 2017). The TP should be made up of individuals from single biparental family or germplasm collection accessions. The BP makeup is the most important issue in TP creation; therefore, the BP must always be described first; then, the TP design should be done, which is focused on lowering phenotyping/genotyping costs and improving the accuracy of candidate prediction. First of all, the size of TP, its composition, and its relation with BP will act as key elements to estimate the GS's accuracy of prediction. One of the most challenging aspects to optimise is the selection of individuals to be included in the TP, yet it is critical to achieve excellent accuracy of prediction. Generally, the most accurate GS methods with a TP of high population size are all closely related to BP, and no population structure (Isidro et al. 2015).

Crop breeders must make it a priority to design ad hoc TP for every BP, and the ideal TP will be made up of half-sibs or full-sibs of BP. Maintaining the similarity is crucial around the GS cycles. BP individuals may accrue some genetic variation, and gene frequencies may alter to the point that the TP deviates from BP with each round of recombination and selection. As a result, the plant breeder must be well prepared to continuously update its TP every cycle (Heffner et al. 2010) or use closed recurrent selection strategies, such as crossing only half-sib or full-sibs. The first option has been extensively researched and is now in use due to the relative ease of execution; nevertheless, as stated previously, it has substantial downsides in selection accuracy. A hybrid strategy integrating these two alternatives was utilised to create the breeding schemes in the second and third rounds of recurrent GS to give better accuracies. Inter-mating full-sibs with half-sibs further ensures a quick progression towards inbreeding as well as the fixing of advantageous genes. Moreover, because the TP stays linked to the BP during closed recurrent selection cycles, multiple morphological scorings for such BP may be accumulated through time and space, boosting the accuracy of prediction (Heffner et al. 2010). The level of LD in TP and BP must be always similar, and it has been shown that more LD results in better predictions (Bassi et al. 2016). Insufficient marker density may lead to an unnatural exaggeration of LD, which, when paired with the near homozygosity of later filial generations (i.e. F6 and later), leads to a considerable reduction in the accuracy of prediction (Shengqiang et al. 2009). As a consequence, if BP and TP are not at all from same parents or have different levels of inbreeding, the density of marker should be raised to account for the larger effective size of the population and rate of recombination (Heffner et al. 2010). However, several TP designs can be feasible for particular breeding conditions, like the following:

1. TP and BP are segregants developed from same cross: genotypes from a biparental cross are chosen for both BP and TP. All genotypes (progenies) in a cross are genotyped, but only a portion of population is phenotyped to act as TP required for deciding a model for genomic prediction; this model is then used to forecast the genetic value of genotypes (BP), which are not phenotyped. Furthermore, the trained model may be utilised to forecast future rounds of selection in populations formed by the intercrossing the family's chosen genotypes (Combs and Bernardo 2013). This type of TP design was widely investigated in several breeding programmes, including huge biparental families and doubled haploids. Numerous investigations on rice (Cui et al. 2020), wheat (Miedaner et al. 2013; Thavamanikumar et al. 2015), rye (Wang et al. 2014), and maize (Zhang et al. 2015) have been conducted. The benefit of within-family genomic predictions would be that a high accuracy of prediction may be obtained with a limited marker number as well as a small population. Because of the large LD found in segregating lines of early hybridisation cycle, improved precision is feasible here (Zhao et al. 2012). This is comparable to the utility of biparental populations QTL mapping. The demerits of this type of TP design include the high cost associated with genotyping a larger number of genotypes in highly segregating generations as well as phenotyping data from replicates and trials at multi-locations, as well as the non-fixation of trait-related alleles in populations, which may impact the identification of an effective GS model.
2. Population lines for TP and BP that include both related as well as unrelated individuals: Prediction models derived from mono biparental populations offer

minimal practical use, except in a few breeding systems. Plant breeders might benefit more from TP schemes that integrate data both from related and unrelated groups. TP must be generated by combining progenies from multiple pedigrees and genetic lineages with varying degrees of relatedness, such as half-sibs, full-sibs, and other people with related heritage. From a large number of studies, we came to know that, when TP is not related to BP, the precision of GS decreases significantly. In hybrid wheat, genomic accuracy of prediction for resistance to disease was substantially higher for sets that are related (0.65–0.92, compared to unrelated sets (0.06–0.43)) (Gowda et al. 2014). The findings also reveal that aggregating a large number of population to forecast a specific target family yields improved prediction accuracies if the TP families have one parent in common with the target population (BP). Families with one parent included, in addition to the family-specific TP, may aid to improve prediction accuracy when solely compared to the family-specific TP, especially for small-sized target families (Schulz-Streeck et al. 2012). Furthermore, it has been proposed that using high-density markers may increase the predictive performance in unrelated families via exchanging marker data among families (Hickey et al. 2014). The essential benefit of TP design is that it would be ideally suited to implementing GS in current breeding strategies, as it comprises both closely related and distantly related genotypes. In general, TPs made up of exclusively unrelated individuals to the BP have very poor to nil accuracy of prediction.

3. Breeding and training genotypes are drawn out from a wide germplasm's: GS predicts the variability of germplasm in addition to measuring the breeding value of individuals from the successful breeding approach. There are large collections of numerous accessions, making it difficult to choose few best genotypes through phenotyping all collections of gene bank. High-throughput genotyping methods have allowed the genotyping of large number of germplasm, enabling the breeding value of germplasm accessions to be predicted (Jarquín et al. 2014). GS has been shown to be effective in harnessing germplasm potential in a variety of agricultural species, including sorghum (Fernandes et al. 2018), wheat (Daetwyler et al. 2014), soybean (Qin et al. 2019), sugar beet (Würschum et al. 2013), and lentil (Haile et al. 2018). Even with low-density genotyping techniques and a well-represented selection population in TP, it was revealed that GS could be used to efficiently unleash the potential of larger germplasms. In spring wheat, 1163 genotypes were phenotyped for resistance against *Puccinia striiformis* that causes stripe rust and then genotyped with a 9 K SNP array, and multiple genomic prediction techniques were tested. The results showed that increasing the marker density and TP size improved the accuracy of prediction. Whenever the size of TP was increased from 210 to 959, the accuracy of prediction increased from 0.50 to 0.63, by an average of 1% improvement for each 50 individuals added towards the size of TP. Beyond an SNP marker density of 1 per 3.2 cM, there was no further improvement in prediction accuracy observed. However, when subpopulations were created based on kinship and structure analysis, prediction accuracy increased. In one subpopulation, it ranged from 0.75 to 0.79, while in another, it was between 0.51 and 0.58. These

variations in prediction accuracy were linked to the genetic relatedness among accessions within each subpopulation, highlighting the importance of genetic relationships between the TP and selection population for germplasm collection decisions (Muleta et al. 2017). Utilising GS, that TP structure may be used to discover promising accessions with higher GEBVs out of huge number of accessions. It is one of several alternative methods for utilising essential gene bank data, as phenotyping the entire collection is a time-consuming process due to a number of practical challenges.

9.4 Statistical Tools and Models Adopted in GS

Many approaches in GS have progressed in recent decades, such as general approaches and their extensions. General GS approaches rely on additive models, but their accuracies might differ due to differences in assumptions and algorithms in regard to complicated trait variations. These general approaches can be enhanced by integrating multiple variates or non-additive effects.

9.4.1 Prediction Methods for Additive Genetic Effects

To generate reliable GEBVs of individuals for selection, GS utilises correlations of a huge number of markers across the entire genome with phenotype. For whole genome regression, though, the marker number (k) is typically much greater than the number of observations (n). Degrees of freedom are inadequate to simultaneously assess the effects of all markers, which is enhanced by multicollinearity (Neves et al. 2012). This ordinary least squares approach will be invalid if k marker effects are evaluated simultaneously. For GS, several approaches, such as GBLUP, machine learning, and Bayes, are being used to overcome these difficulties. The GBLUP and Bayesian techniques consider marker effects to be random effects, and the fundamental model is (Meuwissen et al. 2001)

$$y = X\beta + Z\alpha + \varepsilon$$

where

y–a vector of phenotypes,
β–a vector of non-genetic fixed effects, for which a flat prior is often used,
X–an incidence matrix for the fixed effects β.
α–a vector of random regression coefficients of all the marker effects,
Z–an $n \times k$ genotypic matrix for markers.
ε–a vector of residuals.

9.4.1.1 GS Based on a Single Trait

In the literature, many single-trait GS (STGS) approaches are being suggested. We'll go through some of the most common methods in this section.

Linear Regression Model

Our major goal in GS is designing a model to connect between genotypic and phenotypic variables in order to predict GEBVs and choose desirable individuals. Simple linear regression is the simplest model for analysing this relationship

$$Y_i = \mu + \sum_{j=1}^{P} X_{ij}\beta_j + e_i$$

where μ is an intercept,

X_{ij} is the genotype of the ith individual i = 1to n of the jth marker j = 1 to p,
β_j is effect of marker,
e_i is i^{th} individual's associated random residuals.
This may be expressed as follows:

$$Y = X\,\beta + e'$$

where

Y is phenotypic character observation vector,
X is a design matrix,
β is a vector of marker effects that are unknown, and,
e is a random residual.

The advantage of employing a linear model is that it is a relatively basic model with straightforward inference because it is statistically sound. However, when the fundamental criteria of normality, linearity, explanatory variable independence, and $p < n$ are met, this model performs effectively.

Ridge Regression

Multi-collinearity has been discovered among markers. As a result, the individual's GEBV that is estimated becomes inaccurate (Hoerl and Kennard 1970). RR was determined to be a preferable option in this case. RR obtains a smaller variance estimation of b, but as a price to pay estimator, it becomes biased. One more benefit of RR is that it could be utilised in cases where $p >$ n problem. Instead of reducing the sum of the squared residuals, just like in linear regression models, RR minimises (μ, β) = (Y-Xβ)' (Y-Xβ) + λβ' β, where k = 0 is a regularisation parameter that controls the intensity of the penalty. The higher the value of k, the more shrinkage variables there are. In this scenario, the estimate of regression coefficients, or marker effects, could be provided by

$$\beta = (X' X + \lambda I)\ X'\ Y$$

Best Linear Unbiased Prediction

Henderson for the first time presented the theory of mixed random effect model, and that has been widely utilised in conventional animal breeding programmes earlier (Henderson et al. 1959). In the situation of imbalanced data, BLUP is commonly utilised. Because of its versatility and adaptability, BLUP is used not just in animal breeding programs but also in crop improvement (Bernardo 1994). BLUP has the added benefit of being able to accommodate any family information. Hayes and Goddard were the first to use BLUP in GS (Meuwissen et al. 2001). The model is as follows:

$$Y = X\beta + Zm + e'$$

where β is a $p \times 1$ fixed effect vector that needs to be calculated and m is a random effect that is also a parameter of interest, that is, $m \sim N(0, G)$ and $e \sim N(0, R)$. BLUE is the fixed effect β estimator, whereas BLUP is the random effect m estimator. The main downside of BLUP is that it necessitates complicated statistical calculations, which demands enough computational resources. However, as computer power improves, this problem becomes limited.

Least Absolute Shrinkage and Selection Operator (LASSO)

RR may utilise LASSO for GS to overcome the restrictions of linear regression. This is an RR variation that is generated by changing the penalty function in RR, i.e., assigning a linear penalty rather than quadratic penalty. As a result of this, the influence of certain least important markers is reduced, and the effects of the less important markers are set to zero, thereby solving the $p > n$ problem. Tibshirani was the first to develop LASSO. It is written as

$$\beta\ (\text{lasso}) = \text{argmin}\ (Y - X\ \beta)'\ (Y - X\ \beta) + \lambda\beta$$

Bayesian Methods

In a Bayesian method, we must assume a prior distribution of such model's parameters based on previous knowledge and experience. The previous distributions of such model variables are combined well with likelihood function to get the posterior distribution. The posterior distribution is used to get model parameter inferences. Hayes and Goddard (2001) applied the Bayesian technique for the first time in GS, utilising the previously mentioned linear regression model. The parameters of the model are computed using the prior and posterior distributions of the variables. Bayesian approaches include BayesA, BayesB, BayesCp, and BayesDp. The assumptions of such prior distribution of model parameters, model variance, and other factors differ across different approaches.

BayesA implies that the variance of every marker site has the same prior, whereas BayesB considers that the location of all markers doesn't contribute to overall genetic variation. In comparison to BayesA, the BayesB technique is more realistic for GS. Other Bayes variants have been created to address the shortcomings of BayesA and BayesB. BayesC assumes and utilises the same variance for all markers, whereas BayesD, a scaled parameter, is computed rather than given by the user. BayesCp and BayesDp are derived from BayesC and BayesD, respectively, with the probability p determined for low-effect SNPs.

Support Vector Machine (SVM)

The methods given in this chapter are parametric in nature. These approaches always need the data to be subjected to a number of assumptions. However, parametric model assumptions do not always hold. Parametric approaches perform poorly in such situations. In this scenario, nonparametric approaches may perform better. They believe that the response and predictors have an unknown connection. In GS, nonparametric approaches, such as neural network (NN), SVM, and RKHS (reproducing Kernel Hilbert space), have been employed (Budhlakoti et al. 2019). SVM is a machine learning approach. Supervised learning was the principle on which it was based. It creates a separating hyperplane with the goal of classifying data into distinct categories. It is based on the maximum separation hyperplane idea. The SVM approach is used extensively in support vector regression.

9.4.1.2 Multi-Trait-Based GS

The models outlined before are based on single attribute's genetic information. However, we now have access to data on various traits. We lose lot of details related to association among many traits if we employ approaches based on single factors, since they fit the method by evaluating each attribute separately. Multivariate-based approaches have been developed to use this kind of info in the model. A number of multivariate regression-based methods have been developed. Multivariate methods are indeed extensions of basic regression models, in which users regress two or more responses ($q > 1$) upon p predictors instead of one response onto p predictors. Consider a straightforward multivariate regression model.

$$Y = X\beta + e$$

Here,

X is the marker's $n \times p$ matrix,
Y is the $n \times q$ matrix,
β is the dimensional vector p of coefficients of regression, and,
e is the random residual.

Multivariate Regression with Covariance Estimation

By applying a LASSO-like penalty to β and Ω while accounting for correlated errors, this method minimises the parameter number that are to be estimated (Rothman et al. 2010). Based on $\Omega = [\omega_{j'j}]$ and $\sum_{jk} |b_{jk}|$, the LASSO constraint (Tibshirani 1996) on the entries of b, two penalties are employed to construct a sparse estimator of b. The following is the form in which the function can be written

$$(\beta^\wedge, \Omega^\wedge) = \min_{\beta,\Omega} \left\{ f(\beta, \Omega) + \lambda_1 \sum_{i \neq j} |\omega_{j'j}| \, \lambda_2 \sum_{j=1}^{p} \sum_{k=1}^{q} |b_{jk}| \right\}$$

where $\lambda 1 \geq 0$ and $\lambda 2 \geq 0$ are tuning parameters.

We impose a related penalty for off-diagonal entries of the inverse error covariance matrix O like in the case of LASSO because (1) it makes sure a solution to the issue of ($q > n$), i.e. more response than that of the total count of sample population, and (2) it reduces the amount of parameters in O and has been found to be effective whenever the number of response parameters is greater.

Multivariate Mixed-Model-Based Approach

This is simply a multivariate version of the univariate mixed model technique. Multivariate BLUP is another name for it. Because a mixed model incorporates random along with fixed variables, it employs the covariance structure for random effects in the multivariate situation. It employs the very same model as that of the mixed-model-based method, namely, BLUP, with the exception that Y is a response matrix rather than a vector.

Conditional Gaussian Graphical Models

cGGMs are multivariate linear regression models that have been reparametrised (Chiquet et al. 2016). It makes use of the predictor-response variable covariance structure. Using multivariate regression parameterisation, many statistical formulations may be generated. Because partial covariance reflects crucial relationships among variables, these models are valuable. Multivariate penalised techniques, like univariate penalised methods, resemble regularisation. By applying penalty-based approaches, the multivariate frame takes use of the sparsity within the predictors. Partial Gaussian graphical model is another name for it.

9.5 Factors Influencing GS Predictions

The use of GS in ordinary crop breeding programmes is dependent on the precision of prediction; hence, it is crucial to use cross-validation to check a trained model for high accuracy of prediction. A second genotyped and also phenotyped population, termed as the validation population (VP) or BP, is employed for this. Based on TP phenotypic data, marker effects throughout the particular genome are computed. To measure the model's accuracy of prediction, comparison of GEBVs is calculated from VP with its true breeding values (TBV). A trained model of TP and VP genotypic data is used to construct the GEBV for VP. Correlation between TBV and calculated GEVBs based on VP's phenotypic data is being used to assess the validity of the GS model. Cross-validation is commonly used in TP to train and design the optimal model of prediction that can be used to calculate BP's GEBV (Pérez-Cabal et al. 2012). The Pearson product moment correlation among calculated TBVs and GEBVs is being used to determine predictive performance. GS accuracy is influenced by below mentioned variables:

- Genetic relatedness (Duangjit et al. 2016; Endelman et al. 2014).
- Efficient population size (Poland and Rife 2012; Zhao et al. 2012).
- Structure of population (Isidro et al. 2015).
- Genetic inheritance of the trait (Heffner et al. 2009; Ornella et al. 2012).
- DNA marker characteristic and density (Zhao et al. 2012).
- The distribution and level of LD between markers and genomic sequences correlated with the desired trait (Rajsic et al. 2016).
- Statistical models that are frequently used to calibrate any best-fitted model (Heslot et al. 2012).
- Gene effects (Akdemir 2013).
- Genotype environment interactions (Rajsic et al. 2016).

An investigation of corn doubled haploid lines found that when full-sibs were replaced with half-sibs, accuracy of prediction dropped by 42% (Riedelsheimer et al. 2013). To increase accuracy of genomic prediction, BP and TP must be close enough to exchange long-range haplotypes (Lorenz and Smith 2015). The size of TP is also a significant component in achieving improved prediction accuracy. More genotypic and phenotypic data might be accessible for precise assessment of the genetic components impacting the characteristics' expression, which would enhance the accuracy of genetic effect estimation (T. Guo et al. 2013), especially for low inherited characteristics (Lian et al. 2014). Since GS intends to eliminate the necessity of phenotyping as well as the costs associated with it, determining the ideal TP size to produce relevant accuracy of prediction is critical when using it in a breeding programme. Raising TP size, boosted accuracy of prediction for many characters, these prediction accuracies plateaued around 700 lines (Cericola et al. 2017). Likewise, many studies with respect to the effect of size of TP on genomic predictions have been published (Liu et al. 2018).

Based on one consensus finding, the optimal TP size required for improved prediction accuracy ranged from less than one hundred full-sibs or thousands of half-sibs of a BP to hundreds of unrelated individuals. As a consequence, enhancing relatedness of TP by incorporating populations that are strongly related could be preferred to increasing size of TP through including distantly associated groups (Brandariz and Bernardo 2019). However, in long-term GS, the usage of strongly related individuals may cause the response of selection to be delayed (Moeinizade et al. 2019). As a result, the TP-BP interaction must be carefully considered in order to achieve better genetic gains and effective GS implementation in breeding operations. Another major element controlling prediction accuracy in GS investigations is trait heritability. The typical GS model's prediction accuracy has a positive relationship with heritability (Hayes et al. 2009). Previous research has shown that as trait heritability rises, accuracy of prediction improves. For example, under wet and water-stressed situations in maize, heritability of grain yield, a complex trait, was significantly less than less complex characteristics like days to anthesis and height of the plant.

Furthermore, for each of the three attributes, heritability mean values during watered situations were greater than those under water-constrained situations in the investigated populations. In watered situations, heritability mean values for plant height, anthesis date, and grain yield were 0.59, 0.55, and 0.38, respectively, in comparison with 0.37, 0.47, and 0.27 in stressed conditions (Zhang et al. 2017). In peanut, though, it has recently been established that there's no correlation between heredity and trait accuracy of prediction in genomic investigations (Pandey et al. 2020). Among eleven agronomic traits included for the GS research, plant height had the maximum heritability (92.3%), while the main branches/plant had the lowest (78.7%). However, employing four GS models (Pandey et al. 2020), the accuracies of prediction of these two characteristics were 0.56 and 0.64, respectively, indicating that there is no link between the heritability of a characteristic and its accuracy of prediction. GS for sucrose solvent retention (heritability: 0.45) and flour amount of protein (heritability: 0.56) in wheat found that the accuracy of prediction for the two traits was 0.74 and 0.64, respectively. As a result, GS might be a useful technique for accelerating genetic gains, especially for traits with low heritability.

Another essential factor that influences the accuracy of prediction is density of markers that varies with the type of population, trait of interest, and plant species. A few markers are needed for biparental populations and self-pollinated crops, and more markers are needed for crops that undergo cross-pollinated and also for natural populations (Juliana et al. 2019; Liu et al. 2018). In theory, genome-wide high-density markers assure nearly perfect LD among at least one marker with each QTL, leading to greater accuracy of prediction, but in fact, once the optimum density (markers) is reached, there is no further genetic gain in accuracy of prediction significantly (Wang et al. 2018). To obtain same levels of accuracy of prediction, GS in such a bi-parental population uses few markers (hundreds) in comparison with vast number of markers in a multiple family population (Crossa et al. 2014; Technow et al. 2012). In a research on quality of grain in bi-parental populations of wheat, GS accuracy of prediction plateaued at a lower density of markers (Hoffstetter et al. 2016), and a similar trend was reported in predicting the performance of rice hybrids (Wang et al. 2017). According to study, validated functional markers might potentially be utilised as fixed effects in the model to increase accuracy of prediction (Xu et al. 2020).

Furthermore, factual investigations have demonstrated that the statistical models used have an impact on the accuracy of GS prediction (Daetwyler et al. 2010; Resende et al. 2012). In comparison to the GS models, which examine the genetic relationship matrices among individuals of BP and TP, simpler models with the hypothesis that no link exists across genotypes result in lower estimates of genetic variation (Cericola et al. 2017). RR-BLUP, BayesA, GBLUP, RKHS, Bayesian LASSO, BayesC, and BayesB were used to examine the genomic prediction of six maize variables (yield per plant, 100-kernel weight, plant height, ear height, ear diameter, and ear length). For complex characteristics with lower heritability, additive-dominance and RKHS models demonstrated improved accuracy of prediction (Liu et al. 2018). Accurate phenotyping of TP for model construction could also improve the accuracy of prediction (Voss-Fels et al. 2019). As a result, precise phenotyping is critical for training the prediction model while implementing GS, since the risk of GEI is higher in plants than in cattle (Jonas and De Koning 2013).

The cost-benefit ratio of generating a new variety in crop improvement programmes is determined by factors like phenotyping and genotyping expenses, which are determined by the nature and heritability of traits, as well as the size of the TP and BP. The GS method was found to be advantageous for variables with less than 0.1% heritability when the size of TP was >400 and the effective chromosomal segments (Me) was >100 (Rajsic et al. 2016). GS also offers economic benefits for characteristics with heritability less than 0.25 and effective chromosomal segments less than 100, as far as phenotypic expenses per genotype remain less than the expenditures for genotyping. For example, the break-even cost ratio for resistance to common beans bacterial blight (heritability = 0.24) was significantly lower than that of maize tryptophan and lysine content (heritability = 0.96) that showed the heritability consistency. As a result, if phenotyping is less expensive or not complex, traditional selection may be highly cost-efficient for breeding tryptophan and lysine concentration in maize. Furthermore, when assessing total performance, GS was determined to be financially efficient (Rajsic et al. 2016). Same patterns in cost efficiency of GS were seen in other investigations (Heffner et al. 2010; Wong and Bernardo 2008).

When utilising GS in breeding, the heritability of characteristics and the size of TP must be considered. One of the current obstacles to the effectiveness of GS is the massive price of genotyping. Breeders' interest is in using GS on fixed materials of later generations, such as preliminary yield trials, although the genetic gain is not as significant as with PS. As a result, breeders have been slow to include GS into ordinary varietal development efforts, particularly in research that is government-funded. When we include unrelated families in the TPs, the expense rises even more. A large-sized population of ~20,000 requires nearly 10,000 markers to achieve a higher prediction accuracy of 0.7 (Hickey et al. 2014). Yet, GS's financial efficiency can be improved if genotyping expenses continue to decline and predictions made in one generation are utilised to influence decisions in the following generations (Rajsic et al. 2016).

Moreover, having an open-source GS breeding network, wherein high-throughput systems and infrastructure—including high-throughput phenotyping (HTP), high-throughput genotyping (HTG), and effective models for prediction—are built and allowed to share among scientific organisations and private industries, will lead to significant savings and facilitate the most GS applications in crop improvement programmes.

9.6 Part Strategy of GS

9.6.1 Two-Part Strategy

The aims of developing inbred lines are (i) new inbred's identification and (ii) parent's identification for subsequent breeding cycles. Traditional breeding programmes would be reorganised into two different components under this two-part strategy: a product development (PD) component to produce as well as evaluate inbred lines and a population improvement (PI) component to boost the number of favourable alleles via rapid recurrent genomic selection. Most breeding strategies that produce inbred lines involve crossing to generate new germplasm and thereafter selfing to derive new inbred lines (Bernardo 2014). Instead, doubled haploid technique can be utilised to generate inbred lines quickly (Forster et al. 2007). These newly obtained inbred lines are then phenotyped for one or even more cycles prior to the final selection in order to achieve one or both of the previously mentioned goals of product development and germplasm enhancement. Within this context, genomic selection may be used to find promising lines earlier, lowering cycle time and boosting genetic gain per generation (Heffner et al. 2009). The practice of using inbred lines as parents might be avoided totally with the introduction of GS (Heffner et al. 2009). Strategies based on this concept have been detailed for crops that are simple to cross (Bernardo 2009; Bernardo and Yu 2007) as well as for those which are difficult to cross due to the self-pollination behaviour (Bernardo

2010). The two-part technique is an extension of previous methods and attempts to optimise the capacity of genomic selection over an entire breeding programme (Gaynor et al. 2017). The population improvement component of the two-part strategy employs fast recurrent selection via GS. The idea is to shorten the breeding cycle time in order to enhance genetic gains per year. Each phase of population improvement starts with a large number of genetically different plants. These plants are genotyped, and the finest ones are involved in intercrossing to generate a new generation. The procedure is repeated. Thus, in two-part strategy, population improvement is just a recurrent genomic selection scheme. To assure a consistent supply of enhanced germplasm, a part of the seed generated in few or all cycles is passed towards the product development component. The product development aspect of the two-part approach is primarily dedicated on producing inbreds to release as inbred varieties or hybrid parents. This component's structure is similar to existing breeding programmes and may thus be designed flexibly to fit current or newer breeding program. This design flexibility in the product development aspect also enables different ways to implement at GS. The fundamental distinction between the two-part strategy's product development aspect and conventional breeding programmes is that lines are not selected for subsequent breeding cycles, as this is managed by population improvement component. Furthermore, certain phenotyped plants must be genotyped as part of the product development component. This is required for revising the GS training population utilised in population improvement component aspect as well as for using GS in product development component. The product development component aspect of the two-part approach leads population improvement component by facilitating the development and revision of the training set. The two-part breeding method produced the largest genetic gain, and all genomic selection techniques generated genetic gain greater than conventional breeding.

9.6.2 Multi-Part Strategy

The multi-part approach is an expansion of the two-part strategy, in which exotic germplasm is brought into the PI component via pre-breeding bridges. The pre-breeding bridges take up the role of breaking the linkage between favourable and unfavourable alleles to lower the performance gap between elite germplasm and exPVP. Introgression into the PI component began a year after the last bridge was established. Individuals from the previous year's final breeding cycle were randomly separated into males and females. Individuals from the exPVP breeding programmes or a prior bridge were introduced into the female population when germplasm exchange occurred. Germplasm was pushed back from the PI component to improve pre-breeding performance, with the goal of decreasing the performance gap between exPVP and elite germplasm. Individuals that had been excluded from the PI component were returned into the male population. The number of crossings made between males and females for every bridging cycle was maximised. GS was used to pick the best individuals for the following breeding cycle. When germplasm exchange occurred, outstanding individuals were advanced to the next stage of the breeding programme. After all bridges were constructed, the PI component was introduced. Each cycle of the PI component began with a large number of lines under the genotyping constraints. These individuals were evenly and randomly divided into males and females. When no germplasm exchange occurred, 50 males and 50 females were chosen. When germplasm exchange occurred, 50 females and 50 males were chosen, minus the number of males to be introgressed. The results of the multi-part strategy imply that it has the ability to improve quantitative (polygenic) traits while also giving a tool for avoiding suboptimal convergence of long-term genetic gain (Breider et al. 2022).

9.7 Advantage of GS over Other Breeding Methods Using MAS

In the last century, classical plant breeding together with agronomic practices has developed several high-yielding nutrient responsive varieties and achieved notable gains in terms of production and productivity. However, with diversified food consumption patterns and to address demands of the ever-increasing population, there is a need for increase in crop improvement (Krishnappa et al. 2021). Food and nutritional security in verge of the rapid population growth can be attained by increased yield potential and reduced crop yield gaps. For major food crops, 0.8–1.2% of the current crop improvement rate is short of 2.4% required for meeting food demands of the 10 billion population predicted over the next 30 years (Hickey et al. 2019; Ray et al. 2013). Globally, plant scientists are striving hard to achieve this required genetic improvement rate by developing high yielding, more nutritious, biotic, and abiotic resistant crops, peculiarly during the degrading land and water resources due to the scenario climate change. This indeed is a challenging task to the plant scientists to develop climate smart crops.

Most of the agriculturally important grain yield, stress adaptive and plant growth traits are controlled by quantitative and genes with minor effects leading to higher epistatic interactions (Mackay 2003). Conventional breeding methods are less precise and reliable for multi-genic, low heritable traits as they are highly influenced by environmental cues and G × E interactions, which make their improvement difficult. Besides, these methods require large land and maintenance of large breeding populations, which are often laborious, time-consuming, and cost-prohibitive. It also requires hybridisation of genetically distinct parents and continuous PS over successive generations, which indeed require 5–12 years for a variety development. This necessitated the development of efficient and rapid selection methods to address the restrictions of traditional breeding methods, as shown in Fig. 9.1.

Extensive use of DNA markers over the last two decades has enabled the usage of MAS in crop improvement programmes as it requires minimum phenotypic information for the indirect selection of traits of interest (Collard and Mackill 2008). In MAS, plants with desirable alleles are selected by using the markers related to the

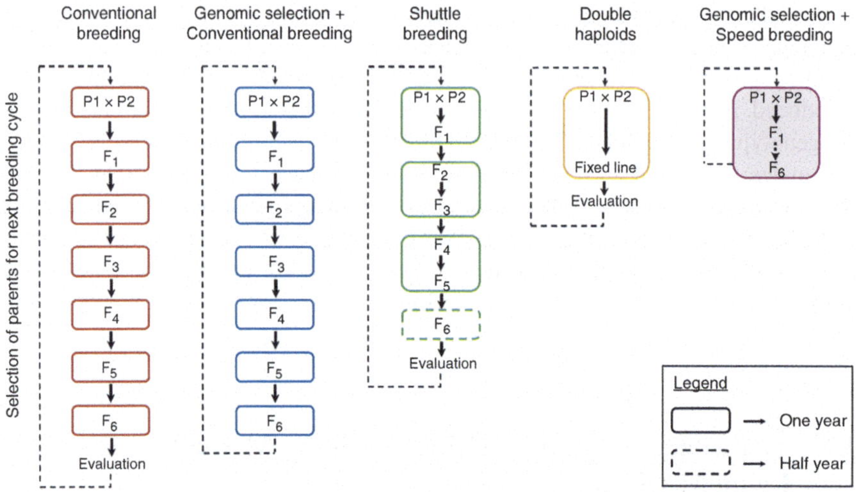

Fig. 9.1 Depicts comparison of breeding cycle length for conventional breeding versus modern strategies that exploit GS, shuttle breeding, double haploids, and speed breeding

desired trait, which makes it efficient only for the traits governed by few major effect QTLs. For the complex traits regulated by several minor QTLs, this approach is inferior to the traditional phenotypic selection as the effect of QTL estimation is biased through linkage or association mapping (Zhao et al. 2014). Initially, marker-assisted backcross breeding (MABB) was used for introgressing one or few genes (from donor) with large genetic effect into the background of adapted cultivars (recipient) with the main aim of recovering the genome content of the recipient. Later on, as emphasis was made only on foreground selection, marker-assisted recurrent selection (MARS) turned out as an alternative strategy for integrating multiple favourable alleles within the same population from different sources (Rai et al. 2018). MAS and MARS are being used for stacking multiple genes (gene pyramiding) into widely adapted cultivars to correct the drawbacks associated with it or for introgressing novel genes (Rana et al. 2019). As of today, the success of gene pyramiding has been shown mainly for major effect genes and is found inefficient for complex agronomic traits governed by multiple genes with minor (very few) effects. These strategies are also constrained mainly because of the usage of low-marker density systems. But, advancement in genome sequencing technologies has made GS a powerful selection tool to tackle these complex traits. In GS, an individual's genetic worth is estimated based on the information from a vast set of markers distributed across the genome instead of a few markers as in MAS. By integrating phenotypic and genotypic data of TP, GS develops a prediction model to derive GEBVs of all the base population individuals that substantially increases the selection efficiency (Poland and Rife 2012). Thus, GEBVs help in selecting the better performing individuals that can be used either as parents in hybridisation programmes or for generation advancement. Based on realistic assumptions of selection intensities, selection accuracies, and generation time, GS has a comparative advantage over conventional PS and MAS in enhancing the genetic gain per year

(Crossa et al. 2017; Heffner et al. 2010). Furthermore, GS is cheaper and easier to evaluate difficult traits like insect resistance, where more individuals can be assessed in a given time and cost (Bhat et al. 2016). Hence, GS enables faster development of crop varieties than PS with increased genetic gains, selection intensity, efficiency, and reduced duration, hence saving resources and time (Desta and Ortiz 2014).

Major consideration for applying GS in crop plants is its cost-benefit ratio. Next-generation sequencing (NGS) technologies have provided accessibility to cost-effective, high-throughput genome-wide markers for model and non-model crops, which made implementation of GS successful mainly in large breeding populations (Poland and Rife 2012). In many crop species, GEBV estimates through NGS-based genotyping are more accurate than other established genotyping platforms. GS is efficient in capturing many additive and small effect loci, which might not be covered by MAS (Heffner et al. 2009). Deploying genome-wide dense markers, GS increases the precision of detecting the markers in LD with desired QTLs of interest (Bekele et al. 2014). Upgradation in high-throughput phenotyping techniques and continuous downswing in sequencing cost will enable GS in accelerating the genetic gains from complex traits required for crop improvement in the near future.

9.8 Limitations of GS

Despite all the positive sides of GS, there are some constraints in situations where the GS is not so good at performance. Let us know the things where this GS performs poorly.

1. To assess the efficiency of genetic prediction, accuracy alone will not be sufficient. Almost the majority of the studies assessed genetic selection using prediction accuracy. Even though accuracy is a crucial factor in establishing prediction model efficacy, it never shows which are all the individuals selected in good faith by using various methodologies. The realised selection differential is arguably a superior criterion for comparing different genomic prediction systems, since breeders examine numerous traits together to advance material, making trait evaluation individually less relevant. Finally, it was correctly pointed out that the phenotype is the final predictor of the genuine breeding value and, like a GEBV, does have an error variance (Bassi et al. 2016).
2. The accuracy of within-family prediction is not fully considered. There has been no comprehensive investigation of prediction accuracy of within-family utilising numerous biparental families or parental information as the training set. Indeed, with the exception of research involving a single biparental household, findings on within-family accuracy are sparse. This is also evident in the hybrid literature, where most studies concentrate on predicting particular hybrid combinations instead of assessing general combining ability in a group of new males or females. This is an essential consideration when adopting genomic prediction

because improved within-family accuracy may help to accelerate genetic advancement while optimising the fraction of inbreeding within the population. Differences among crosses are well predicted since the model accounts for both within and between family variables (Edwards et al. 2019).

9.9 Speed GS High-Throughput Genotyping

In order to meet the growing demand for food, it is required to double the crop production by 2050. Continuous advancement in genotyping and phenotyping technology offers great potential for enhancing genetic gain in crop plants (Phillips 2010) and improves the efficiency of breeders. In GS, to train a prediction model, the phenotypes are utilised. Through high-throughput phenotyping, a large number of lines can be phenotypes more rapidly and accurately, and the precise selection of the best progeny can be made. Similarly, advancement in the high-throughput genotyping have led to the generation of valuable genomic information in a fasten and cost-effective way that has eased the development and study of the large mapping population (McMullen et al. 2009).

Speed breeding in completely contained, controlled environment like growth chambers can speed up the development of superior varieties, such as phenotyping mature plant features, mutant research, and transformation. Additional illumination in a greenhouse leads to faster generation advancement through single seed descent (SSD) and adaptability to plant growth activities at a larger scale. Speed breeding significantly reduces generation time and speeds up the crop breeding process. Durum wheat, Spring wheat, barley, pea, and chickpea could now produce nearly up to six generations per year, especially in comparison to 2–3 under regular glasshouse situations, and rapeseed (*Brassica napus*) could now produce nearly four generations per year (Watson et al. 2018) (Fig. 9.1).

Selection by breeding (SB) helps to resolve difficulties related to the double haploid (DH) technique, like poor germination percentage, reduced vigour, and deformed development (Ferrie 2006). Recombinant inbred lines (RILs) formed during numerous generations of autogamy may be preferred to DH for genetic mapping purposes because of the many meiotic events that take place all through repetitive fertilisation as well as the associated increased recombination frequency. Likewise, SSD can create and assess segregating generations quickly under SB circumstances (Sinha et al. 2021), saving a lot of time over the conventional pedigree breeding approach (Jähne et al. 2020). SB methodologies may be used in many crops and integrated with other current techniques of crop breeding like high-throughput genotyping, genome editing, and GS to accelerate crop development. In orphan crops, SB approaches might be utilised to cut down the breeding cycles and to hasten research. High-throughput phenotyping is one of the major advances in agricultural research in the twenty-first century that has the ability to overcome long-standing barriers to crop improvement advancement. Carrying out high-throughput phenotyping under SB circumstances opens new possibilities for discovering and incorporating favourable features while conserving resources (Al-Tamimi et al. 2016). Because of their precision and convenience of use, high-throughput phenotyping platforms (HTPPs) have sparked a lot of interest (Furbank and Tester 2011). The HTPPs feature completely automated facilities in greenhouses with regulated environmental conditions, as well as remote sensing technology that allows for exact assessment of crop growth and performance. Recently, efforts have been made to develop low-cost HTPP approaches to widen its adoption in breeding programs. Under SB circumstances that include increased planting density, temperature control, and prolonged photoperiod, targeting proxy variables like seminal root number and angle of seedlings permitted quick selection for superior root architecture of mature plants (Richard et al. 2015). SB has been used to assess different stages of plant breeding operations. In spring wheat, GS was paired with SB to maximise genetic gain for complex traits (Voss-Fels et al. 2019). SB was employed for certain trait phenotyping of carried in TP of wheat, as well as the development and phenotyping of selection candidates. Indirect selection in SB settings with targeted populations was found to predict plant height and blooming time with accuracy comparable to direct field selection. In comparison to field phenotyping directly, SB allows for a genetic gain at a higher rate (Watson et al. 2018).

Similarly, imaging technology allowed for the acquisition of field plot photos at a rate of 7400 plots per hour based on wheat colour features (Walter et al. 2019). When compared to terrestrial sensing, the technology using automated aerial vehicles demonstrated a substantial correlation with enhanced grain output. For phenotyping plant characteristics, several sensing methods have been identified, including proximal (remote) sensing and imaging, laboratory studies of samples, and near-infrared reflectance spectroscopy (NIRS) analysis with the harvestable portion of the crop. The approach chosen is determined by the kind of trait as well as the period of evaluation. Remote sensing techniques could be used to conduct in situ screening for a broad range of breeding objectives, such as yield potential, tolerance to abiotic and biotic limiting circumstances, and even quality-related traits. Several different sorts of properties, ranging from green biomass to photosynthetic transpirative gas exchange, quality features, and even grain production prediction, may be measured using remote sensing techniques in diverse environmental conditions (Weber et al. 2012).

NIRS is generally utilised in breeding for a wide range of feed and food quality parameters. In fact, NIRS could be used to evaluate for drought tolerance, nutrient efficiency, and other breeding/gene discovery objectives. Because of the employment of proximal sensing with VIS-NIR and far-infrared light in imaging formats, the measurement process has been upscaled: for example, from analysing a single plot towards dissecting an entire trial made up of several plots, providing the picture has an adequate resolution (pixels). Additionally, the aerial HTPPs have enabled to take measurements of all plots simultaneously in the trial, and this has made the phenotyping process to overcome the largest limitation, time, and allows rapid characterisation of several plots within a short duration. The use of these approaches in the GS would provide an overall understanding of the role of genetic factors in

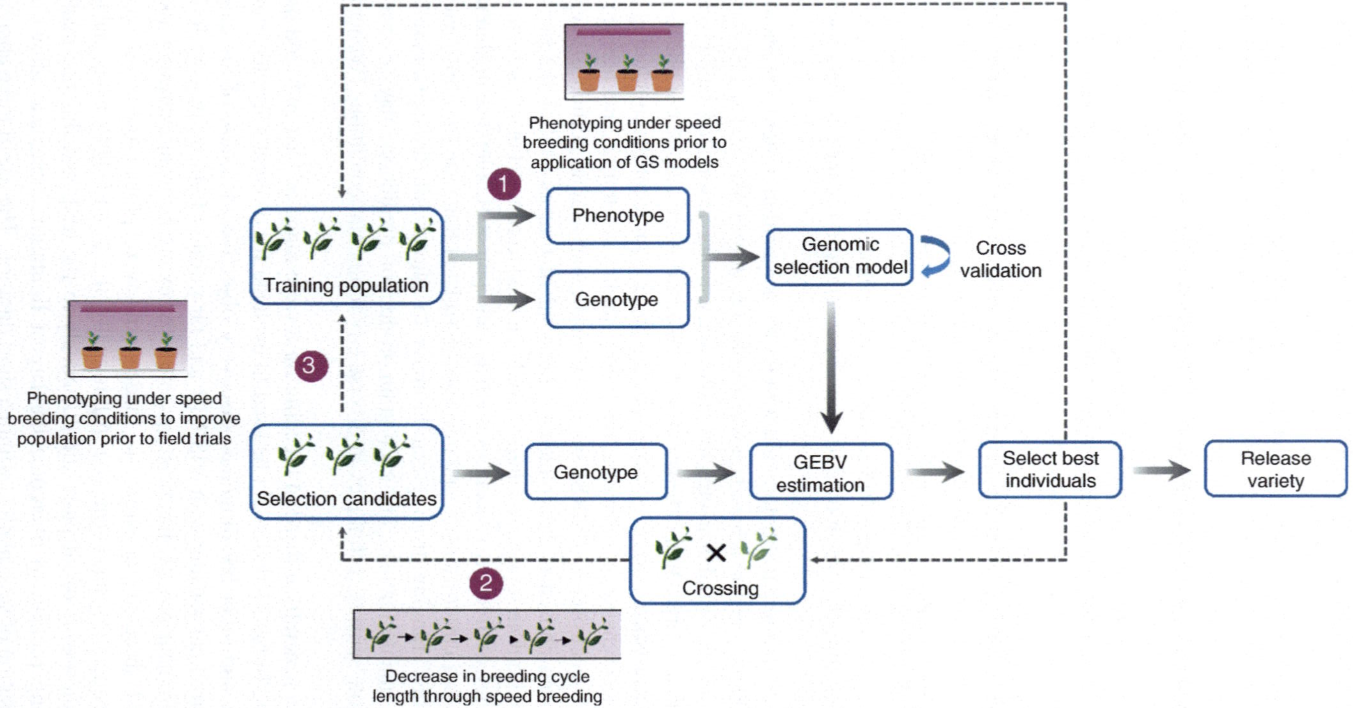

Fig. 9.2 Represents crop breeding cycle employing GS and points within the cycle wherein speed breeding can be applied to increase the rate of genetic gain

determining crop performance, the stages where SB can be efficiently applied in GS are mentioned in Fig. 9.2.

The high-throughput genotyping platforms (HTGP) and high-throughput phenotyping platforms (HTPP) allow effective GS and enhance the success rate of breeding programs (Cobb et al. 2019). The ease in availability of high-throughput genotyping platforms, such as high-density, mid-density, and low-density platforms, has increased the precision and accelerated the genetic gain in crops with large and complex genetic makeup. To estimate GEBVs of (BP), phenotyping and genotyping are very critical for the identification of the appropriate gene and GS model to be utilised. Hence, the combination of these high-throughput technologies with appropriate genetic diversity and analytical tools, along with databases, would lead to the new variety development having better yield, quality, and resistance to stresses.

Over the last decade, the dependence on phenotypic selection has gradually shifted to a greater reliance upon genotypic-based approaches for plant selection, enabled feasible in part by NGS-based sequencing platforms (Bhat et al. 2020; Pandey et al. 2016). The advancements in the HTPP and HTGP have been successful in enhancing the accuracy of genomic prediction and mapping of genes. Platforms such as AgriSeq, DartTag, and RiCa are widely used in the GS. The HTGP technology has improved genome-wide genotyping throughput, cost-effectiveness, and speed (Getachew et al. 2020). Before the advent of NGS-based marker genotyping, the generation of markers was costly and time-consuming, especially for the GS, which was limited by the number of markers that could be tested economically (Bhat et al. 2016). As a result, to forecast the presence or absence of agriculturally beneficial characteristics, only markers in crucial genomic regions were used (Varshney et al. 2014). The HTGP has enabled cost and time effective genotyping with precise identification of the desirable genotypes. Hence, it enables the development of markers, which further caters the GS.

9.10 Conclusion

Plant breeders can use GS to forecast genomic-estimated breeding values of individuals by employing markers that span the whole genome. However, the best way to implement GS is still up for dispute. Predictions within the breeding cycle in the breeding programme can provide high selection accuracies, but selections across the breeding cycle might suffer from a poor association between the training and test populations, making predictions less accurate. More research on predicting distantly related individuals is required. Due to the lack of precision in prediction ahead of several breeding cycles, lower accuracies can be predicted when GS is paired with the usage of untested parents. The best answer for using GS in plant breeding initiatives can be a mix of diverse approaches. Pedigree information might be used in GS to improve forecast accuracy and provide breeding values for non-genotyped lines. The characteristics and relationships of individuals influence the size of the TP and marker set, which should be examined separately before implementing GS in a breeding programme. GS is commonly used to estimate individual additive genetic value while ignoring non-additive genetic variation, which reflects the performing power of a line as a parent. Future research will look at the assessment of total genetic value, which would be ideal for variety marketing. We can conclude that within generation, GS is now an attractive and realistic alternative, with expenditures in genotyping being recovered through improved selection decisions, reduced phenotyping, and a reduction in the number of candidates retained in the breeding programme. Because forecast accuracies in such systems may be poor, cross-breeding cycle GS, and in particular the use of untested parents, has to be examined further. We also come to the conclusion that plant breeders may benefit more from employing pedigree data, as well as combined pedigree-genomic data than they really do now.

References

Akdemir D (2013) Locally epistatic genomic relationship matrices for genomic association, prediction and selection

Al-Tamimi N, Brien C, Oakey H, Berger B, Saade S, Ho YS, Schmöckel SM, Tester M, Negraõ S (2016) Salinity tolerance loci revealed in rice using high-throughput non-invasive phenotyping. Nat Commun 7(1):1–11. https://doi.org/10.1038/ncomms13342

Annicchiarico P, Nazzicari N, Pecetti L, Romani M, Russi L (2019) Pea genomic selection for Italian environments. BMC Genomics 20(1). https://doi.org/10.1186/S12864-019-5920-X

Arruda MP, Lipka AE, Brown PJ, Krill AM, Thurber C, Brown-Guedira G, Dong Y, Foresman BJ, Kolb FL (2016) Comparing genomic selection and marker-assisted selection for fusarium head blight resistance in wheat (Triticum aestivum L.). Mol Breed 36(7). https://doi.org/10.1007/S11032-016-0508-5

Bartholomé J, Parthiban TP, Cobb JN (2021) Genomic prediction : progress and perspectives for rice improvement. Methods Mol Biol 2467:569–617

Bassi FM, Bentley AR, Charmet G, Ortiz R, Crossa J (2015) Breeding schemes for the implementation of genomic selection in wheat (Triticum spp.). Plant Sci 242:23–36. https://doi.org/10.1016/J.PLANTSCI.2015.08.021

Bassi FM, Bentley AR, Charmet G, Ortiz R, Crossa J (2016) Breeding schemes for the implementation of genomic selection in wheat (Triticum spp.). Plant Sci 242:23–36. https://doi.org/10.1016/J.PLANTSCI.2015.08.021

Bekele WA, Fiedler K, Shiringani A, Schnaubelt D, Windpassinger S, Uptmoor R, Friedt W, Snowdon RJ (2014) Unravelling the genetic complexity of sorghum seedling development under low-temperature conditions. Plant, Cell and Environment 37(3):707–723. https://doi.org/10.1111/PCE.12189/SUPINFO

Belamkar V, Guttieri MJ, Hussain W, Jarquín D, El-basyoni I, Poland J, Lorenz AJ, Baenziger PS (2018) Genomic selection in preliminary yield trials in a winter wheat breeding program. G3: Genes, Genomes, Genetics 8(8):2735–2747. https://doi.org/10.1534/G3.118.200415

Bernardo R (1994) Prediction of maize single-cross performance using RFLPs and information from related hybrids. Crop Sci 34(1):20–25. https://doi.org/10.2135/CROPSCI1994.0011183X003400010003X

Bernardo R (2009) Genomewide selection for rapid introgression of exotic germplasm in maize. Crop Sci 49(2):419–425. https://doi.org/10.2135/CROPSCI2008.08.0452

Bernardo R (2010) Genomewide selection with minimal crossing in self-pollinated crops. Crop Sci 50(2):624–627. https://doi.org/10.2135/CROPSCI2009.05.0250

Bernardo R (2014) Process of plant breeding. Essentials Plant Breed 57(2):9–13. https://www.nhbs.com/essentials-of-plant-breeding-book

Bernardo R, Yu J (2007) Prospects for genomewide selection for quantitative traits in maize. Crop Sci 47(3):1082–1090. https://doi.org/10.2135/CROPSCI2006.11.0690

Bhat JA, Ali S, Salgotra RK, Mir ZA, Dutta S, Jadon V, Tyagi A, Mushtaq M, Jain N, Singh PK, Singh GP, Prabhu KV (2016) Genomic selection in the era of next generation sequencing for complex traits in plant breeding. Front Genet 7(DEC):1–11. https://doi.org/10.3389/fgene.2016.00221

Bhat JA, Deshmukh R, Zhao T, Patil G, Deokar A, Shinde S, Chaudhary J (2020) Harnessing high-throughput phenotyping and genotyping for enhanced drought tolerance in crop plants. J Biotechnol 324:248–260. https://doi.org/10.1016/J.JBIOTEC.2020.11.010

Brandariz SP, Bernardo R (2019) Small ad hoc versus large general training populations for genomewide selection in maize biparental crosses. Theor Appl Genet 132(2):347–353. https://doi.org/10.1007/S00122-018-3222-3

Breider I, Gaynor R, Gorjanc G, Thorn S, Pandey MK, Varshney RK, Hickey J (2022) A multi-part strategy for introgression of exotic germplasm into elite plant breeding programs using genomic selection. Research Square. https://doi.org/10.21203/RS.3.RS-1246254/V1

Brown J, Caligari P, Peter DS (2008) An introduction to plant breeding. John Wiley & Sons, p 209

Budhlakoti N, Mishra DC, Rai A, Lal SB, Chaturvedi KK, Kumar RR (2019) A comparative study of single-trait and multi-trait genomic selection. J Comput Biol 26(10):1100–1112. https://doi.org/10.1089/CMB.2019.0032

Cericola F, Jahoor A, Orabi J, Andersen JR, Janss LL, Jensen J (2017) Optimizing training population size and genotyping strategy for genomic prediction using association study results and pedigree information. A case of study in advanced wheat breeding lines. PLoS One 12(1): e0169606. https://doi.org/10.1371/JOURNAL.PONE.0169606

Chenu K (2015) Characterizing the crop environment–nature, significance and applications. In: Crop physiology: applications for genetic improvement and agronomy: second edition, pp 321–348. https://doi.org/10.1016/B978-0-12-417104-6.00013-3

Chiquet J, Mary-Huard T, Robin S (2016) Structured regularization for conditional gaussian graphical models. Statistics and Computing 27(3):789–804. https://doi.org/10.1007/S11222-016-9654-1

Cobb JN, Biswas PS, Platten JD (2019) Back to the future: revisiting MAS as a tool for modern plant breeding. Theor Appl Genet 132(3):647–667. https://doi.org/10.1007/S00122-018-3266-4/FIGURES/12

Collard BCY, Mackill DJ (2008) Marker-assisted selection: an approach for precision plant breeding in the twenty-first century. Philos Trans R Soc B Biol Sci 363(1491):557–572. https://doi.org/10.1098/RSTB.2007.2170

Combs E, Bernardo R (2013) Accuracy of Genomewide selection for different traits with constant population size, heritability, and number of markers. The Plant Genome 6(1). https://doi.org/10.3835/PLANTGENOME2012.11.0030

Crossa J, Beyene Y, Semagn K, Pérez P, Hickey JM, Chen C, de Campos GL, Burgueño J, Windhausen VS, Buckler E, Jannink JL, Cruz MAL, Babu R (2013) Genomic prediction in maize breeding populations with genotyping-by-sequencing. G3: Genes, Genomes, Genetics 3(11):1903–1926. https://doi.org/10.1534/G3.113.008227

Crossa J, Jarquín D, Franco J, Pérez-Rodríguez P, Burgueño J, Saint-Pierre C, Vikram P, Sansaloni C, Petroli C, Akdemir D, Sneller C, Reynolds M, Tattaris M, Payne T, Guzman C, Peña RJ, Wenzl P, Singh S (2016) Genomic prediction of gene bank wheat landraces. G3: Genes, Genomes, Genetics 6(7):1819–1834. https://doi.org/10.1534/G3.116.029637

Crossa J, Pérez-Rodríguez P, Cuevas J, Montesinos-López O, Jarquín D, de los Campos G, Burgueño J, González-Camacho JM, Pérez-Elizalde S, Beyene Y, Dreisigacker S, Singh R, Zhang X, Gowda M, Roorkiwal M, Rutkoski J, Varshney RK (2017) Genomic selection in plant breeding: methods, models, and perspectives. Trends Plant Sci 22(11):961–975. https://doi.org/10.1016/j.tplants.2017.08.011

Crossa J, Pérez P, Hickey J, Burgueño J, Ornella L, Cerón-Rojas J, Zhang X, Dreisigacker S, Babu R, Li Y, Bonnett D, Mathews K (2014) Genomic prediction in CIMMYT maize and wheat breeding programs. Heredity 112(1):48–60. https://doi.org/10.1038/HDY.2013.16

Cui Y, Li R, Li G, Zhang F, Zhu T, Zhang Q, Ali J, Li Z, Xu S (2020) Hybrid breeding of rice via genomic selection. Plant Biotechnol J 18(1):57–67. https://doi.org/10.1111/PBI.13170

Daetwyler HD, Bansal UK, Bariana HS, Hayden MJ, Hayes BJ (2014) Genomic prediction for rust resistance in diverse wheat landraces. Theor Appl Genet 127(8):1795–1803. https://doi.org/10.1007/S00122-014-2341-8

Daetwyler HD, Pong-Wong R, Villanueva B, Woolliams JA (2010) The impact of genetic architecture on genome-wide evaluation methods. Genetics 185(3):1021–1031. https://doi.org/10.1534/GENETICS.110.116855

De Los Campos G, Naya H, Gianola D, Crossa J, Legarra A, Manfredi E, Weigel K, Cotes JM (2009) Predicting quantitative traits with regression models for dense molecular markers and pedigree. Genetics 182(1):375–385. https://doi.org/10.1534/GENETICS.109.101501

Desta ZA, Ortiz R (2014) Genomic selection: genome-wide prediction in plant improvement. Trends Plant Sci 19(9):592–601. https://doi.org/10.1016/j.tplants.2014.05.006

Dos Santos JPR, Pires LPM, de Castro Vasconcellos RC, Pereira GS, Von Pinho RG, Balestre M (2016) Genomic selection to resistance to Stenocarpella maydis in maize lines using DArTseq markers. BMC Genet 17(1). https://doi.org/10.1186/S12863-016-0392-3

Duangjit J, Causse M, Sauvage C (2016) Efficiency of genomic selection for tomato fruit quality. Mol Breed 36(3). https://doi.org/10.1007/S11032-016-0453-3

Edwards SMK, Buntjer JB, Jackson R, Bentley AR, Lage J, Byrne E, Burt C, Jack P, Berry S, Flatman E, Poupard B, Smith S, Hayes C, Gaynor RC, Gorjanc G, Howell P, Ober E, Mackay IJ, Hickey JM (2019) The effects of training population design on genomic prediction accuracy in wheat. Theor Appl Genet. https://doi.org/10.1007/S00122-019-03327-Y

Endelman JB, Atlin GN, Beyene Y, Semagn K, Zhang X, Sorrells ME, Jannink JL (2014) Optimal design of preliminary yield trials with genome-wide markers. Crop Sci 54(1):48–59. https://doi.org/10.2135/CROPSCI2013.03.0154

Falconer DS, Mackay TFC (1996) Introduction to quantitative genetics. In: Trends in genetics, vol 12, 4th edn. Prentice Hall

Fernandes SB, Dias KOG, Ferreira DF, Brown PJ (2018) Efficiency of multi-trait, indirect, and trait-assisted genomic selection for improvement of biomass sorghum. Theor Appl Genet 131(3):747–755. https://doi.org/10.1007/S00122-017-3033-Y

Ferrie AMR (2006) Doubled haploid production in nutraceutical species: a review. Euphytica 158(3):347–357. https://doi.org/10.1007/S10681-006-9242-0

Forster BP, Heberle-Bors E, Kasha KJ, Touraev A (2007) The resurgence of haploids in higher plants. Trends Plant Sci 12(8):368–375. https://doi.org/10.1016/J.TPLANTS.2007.06.007

Furbank RT, Tester M (2011) Phenomics–technologies to relieve the phenotyping bottleneck. Trends Plant Sci 16(12):635–644. https://doi.org/10.1016/J.TPLANTS.2011.09.005

Gaynor RC, Gorjanc G, Bentley AR, Ober ES, Howell P, Jackson R, Mackay IJ, Hickey JM, Gaynor RC, Gorjanc G, Hickey JM, Institute R, Dick R, Bent-Ley AR, Ober ES, Howell P, Jackson R, Mackay IJ, Bingham J (2017) A two-part strategy for using genomic selection to develop inbred lines. Crop Sci 57(5):2372–2386. https://doi.org/10.2135/CROPSCI2016.09.0742

Getachew, T., Haile, A., Rekik, M., & Rischkowsky, B. (2020). A genetic database tool for data capture in small ruminant community-based breeding programs

Gowda M, Zhao Y, Würschum T, Longin CF, Miedaner T, Ebmeyer E, Schachschneider R, Kazman E, Schacht J, Martinant JP, Mette MF, Reif JC (2014) Relatedness severely impacts accuracy of marker-assisted selection for disease resistance in hybrid wheat. Heredity 112(5):552–561. https://doi.org/10.1038/HDY.2013.139

Grenier C, Cao TV, Ospina Y, Quintero C, Châtel MH, Tohme J, Courtois B, Ahmadi N (2015) Accuracy of genomic selection in a Rice synthetic population developed for recurrent selection breeding. PLoS One 10(8):e0136594. https://doi.org/10.1371/JOURNAL.PONE.0136594

Guimarães EP (2009) Rice breeding. Cereals 99–126. https://doi.org/10.1007/978-0-387-72297-9_2

Guo T, Li H, Yan J, Tang J, Li J, Zhang Z, Zhang L, Wang J (2013) Performance prediction of F1 hybrids between recombinant inbred lines derived from two elite maize inbred lines. Theor Appl Genet 126(1):189–201. https://doi.org/10.1007/S00122-012-1973-9

Guo Z, Tucker DM, Basten CJ, Gandhi H, Ersoz E, Guo B, Xu Z, Wang D, Gay G (2014) The impact of population structure on genomic prediction in stratified populations. Theor Appl Genet 127(3):749–762. https://doi.org/10.1007/S00122-013-2255-X

Habyarimana E, Lopez-Cruz M (2019) Genomic selection for antioxidant production in a panel of sorghum bicolor and s. bicolor × s. halepense lines. Genes 10(11). https://doi.org/10.3390/GENES10110841

Haile JK, N'Diaye A, Clarke F, Clarke J, Knox R, Rutkoski J, Bassi FM, Pozniak CJ (2018) Genomic selection for grain yield and quality traits in durum wheat. Mol Breed 38(6). https://doi.org/10.1007/S11032-018-0818-X

Hayes BEN, Goddard ME (2001) The distribution of the effects of genes affecting quantitative traits in livestock. Genet Sel Evol 33:1–21

Hayes BJ, Visscher PM, Goddard ME (2009) Increased accuracy of artificial selection by using the realized relationship matrix. Genet Res 91(1):47–60. https://doi.org/10.1017/S0016672308009981

Heffner EL, Jannink J-L, Sorrells ME (2011) Genomic selection accuracy using multifamily prediction models in a wheat breeding program. Plant Genome J 4(1):65. https://doi.org/10.3835/PLANTGENOME.2010.12.0029

Heffner EL, Lorenz AJ, Jannink JL, Sorrells ME (2010) Plant breeding with genomic selection: gain per unit time and cost. Crop Sci 50(5):1681–1690. https://doi.org/10.2135/CROPSCI2009.11.0662

Heffner EL, Sorrells ME, Jannink JL (2009) Genomic selection for crop improvement. Crop Sci 49(1):1–12. https://doi.org/10.2135/CROPSCI2008.08.0512

Henderson CR, Kempthorne O, Searle SR, von Krosigk CM (1959) The estimation of environmental and genetic trends from records subject to culling. Biometrics 15(2):192. https://doi.org/10.2307/2527669

Herter CP, Ebmeyer E, Kollers S, Korzun V, Miedaner T (2019) An experimental approach for estimating the genomic selection advantage for fusarium head blight and Septoria tritici blotch in winter wheat. Theor Appl Genet 132(8):2425–2437. https://doi.org/10.1007/s00122-019-03364-7

Heslot N, Yang HP, Sorrells ME, Jannink JL (2012) Genomic selection in plant breeding: a comparison of models. Crop Sci 52(1):146–160. https://doi.org/10.2135/CROPSCI2011.06.0297

Hickey JM, Dreisigacker S, Crossa J, Hearne S, Babu R, Prasanna BM, Grondona M, Zambelli A, Windhausen VS, Mathews K, Gorjanc G (2014) Evaluation of genomic selection training population designs and genotyping strategies in plant breeding programs using simulation. Crop Sci 54(4):1476–1488. https://doi.org/10.2135/CROPSCI2013.03.0195

Hickey LT, Hafeez AN, Robinson H, Jackson SA, Leal-Bertioli SCM, Tester M, Gao C, Godwin ID, Hayes BJ, Wulff BBH (2019) Breeding crops to feed 10 billion. Nat Biotechnol 37(7): 744–754. https://doi.org/10.1038/S41587-019-0152-9

Hoerl AE, Kennard RW (1970) Ridge regression: applications to nonorthogonal problems. Technometrics 12(1):69. https://doi.org/10.2307/1267352

Hoffstetter A, Cabrera A, Huang M, Sneller C (2016) Optimizing training population data and validation of genomic selection for economic traits in soft winter wheat. G3: Genes, Genomes, Genetics 6(9):2919–2928. https://doi.org/10.1534/G3.116.032532

Isidro J, Jannink JL, Akdemir D, Poland J, Heslot N, Sorrells ME (2015) Training set optimization under population structure in genomic selection. Theor Appl Genet 128(1):145–158. https://doi.org/10.1007/S00122-014-2418-4

Jähne F, Hahn V, Würschum T, Leiser WL (2020) Speed breeding short-day crops by LED-controlled light schemes. Theor Appl Genet 133(8):2335–2342. https://doi.org/10.1007/S00122-020-03601-4/FIGURES/3

Jarquín D, Crossa J, Lacaze X, Du Cheyron P, Daucourt J, Lorgeou J, Piraux F, Guerreiro L, Pérez P, Calus M, Burgueño J, de los Campos, G. (2014) A reaction norm model for genomic selection using high-dimensional genomic and environmental data. Theor Appl Genet 127(3): 595–607. https://doi.org/10.1007/S00122-013-2243-1

Jonas E, De Koning DJ (2013) Does genomic selection have a future in plant breeding? Trends Biotechnol 31(9):497–504. https://doi.org/10.1016/J.TIBTECH.2013.06.003

Juliana P, Poland J, Huerta-Espino J, Shrestha S, Crossa J, Crespo-Herrera L, Toledo FH, Govindan V, Mondal S, Kumar U, Bhavani S, Singh PK, Randhawa MS, He X, Guzman C, Dreisigacker S, Rouse MN, Jin Y, Pérez-Rodríguez P et al (2019) Improving grain yield, stress resilience and quality of bread wheat using large-scale genomics. Nat Genet 51(10):1530–1539. https://doi.org/10.1038/S41588-019-0496-6

Krishnappa G, Savadi S, Tyagi BS, Singh SK, Mamrutha HM, Kumar S, Mishra CN, Khan H, Gangadhara K, Uday G, Singh G, Singh GP (2021) Integrated genomic selection for rapid improvement of crops. Genomics 113(3):1070–1086. https://doi.org/10.1016/J.YGENO.2021.02.007

Li Y, Ruperao P, Batley J, Edwards D, Khan T, Colmer TD, Pang J, Siddique KHM, Sutton T (2018) Investigating drought tolerance in chickpea using genome-wide association mapping and genomic selection based on whole-genome resequencing data. Front Plant Sci 9:190. https://doi.org/10.3389/FPLS.2018.00190/BIBTEX

Lian L, Jacobson A, Zhong S, Bernardo R (2014) Genomewide prediction accuracy within 969 maize biparental populations. Crop Sci 54(4):1514–1522. https://doi.org/10.2135/CROPSCI2013.12.0856

Liang Z, Gupta SK, Yeh CT, Zhang Y, Ngu DW, Kumar R, Patil HT, Mungra KD, Yadav DV, Rathore A, Srivastava RK, Gupta R, Yang J, Varshney RK, Schnable PS, Schnable JC (2018) Phenotypic data from inbred parents can improve genomic prediction in pearl millet hybrids. G3 (Bethesda, Md) 8(7):2513–2522. https://doi.org/10.1534/G3.118.200242

Liu X, Wang H, Wang H, Guo Z, Xu X, Liu J, Wang S, Li WX, Zou C, Prasanna BM, Olsen MS, Huang C, Xu Y (2018) Factors affecting genomic selection revealed by empirical evidence in maize. Crop J 6(4):341–352. https://doi.org/10.1016/J.CJ.2018.03.005

Lorenz AJ, Smith KP (2015) Adding genetically distant individuals to training populations reduces genomic prediction accuracy in barley. Crop Sci 55(6):2657–2667. https://doi.org/10.2135/CROPSCI2014.12.0827

Lorenz AJ, Smith KP, Jannink JL (2012) Potential and optimization of genomic selection for fusarium head blight resistance in six-row barley. Crop Sci 52(4):1609–1621. https://doi.org/10.2135/CROPSCI2011.09.0503

Lorenzana RE, Bernardo R (2009) Accuracy of genotypic value predictions for marker-based selection in biparental plant populations. Theor Appl Genet 120(1):151–161. https://doi.org/10.1007/S00122-009-1166-3

Lozada DN, Mason RE, Sarinelli JM, Brown-Guedira G (2019) Accuracy of genomic selection for grain yield and agronomic traits in soft red winter wheat. BMC Genet 20(1):1–12. https://doi.org/10.1186/S12863-019-0785-1/TABLES/2

Lynch M, Walsh B (1998) Chapter 27 BT–genetics and analysis of quantitative traits. *Genetics and analysis of quantitative traits.* papers2://publication/uuid/6E5E388C-7C1E-49F6-937F-13FB4B8E5A68

Mackay TFC (2003) The genetic architecture of quantitative traits. Annu Rev Genet 35:303–339. https://doi.org/10.1146/ANNUREV.GENET.35.102401.090633

Maresso K, Broeckel U (2008) Genotyping platforms for mass-throughput genotyping with SNPs, including human genome-wide scans. Adv Genet 60:107–139. https://doi.org/10.1016/S0065-2660(07)00405-1

McMullen MD, Kresovich S, Villeda HS, Bradbury P, Li H, Sun Q, Flint-Garcia S, Thornsberry J, Acharya C, Bottoms C, Brown P, Browne C, Eller M, Guill K, Harjes C, Kroon D, Lepak N, Mitchell SE, Peterson B et al (2009) Genetic properties of the maize nested association mapping population. Science 325(5941):737–740. https://doi.org/10.1126/SCIENCE.1174320/SUPPL_FILE/MCMULLEN.SOM.PDF

Meuwissen T, Goddard M (2010) Accurate prediction of genetic values for complex traits by whole-genome resequencing. Genetics 185(2):623–631. https://doi.org/10.1534/GENETICS.110.116590

Meuwissen THE, Hayes BJ, Goddard ME (2001) Prediction of total genetic value using genome-wide dense marker maps. Genetics 157(4):1819–1829. https://doi.org/10.1093/GENETICS/157.4.1819

Miedaner T, Zhao Y, Gowda M, Longin CFH, Korzun V, Ebmeyer E, Kazman E, Reif JC (2013) Genetic architecture of resistance to Septoria tritici blotch in European wheat. BMC Genomics 14(1):1–8. https://doi.org/10.1186/1471-2164-14-858

Moeinizade S, Hu G, Wang L, Schnable PS (2019) Optimizing selection and mating in genomic selection with a look-ahead approach: an operations research framework. G3 Genes|Genomes|Genetics 9(7):2123–2133. https://doi.org/10.1534/G3.118.200842

Monteverde E, Rosas JE, Blanco P, Pérez de Vida F, Bonnecarrère V, Quero G, Gutierrez L, McCouch S (2018) Multienvironment models increase prediction accuracy of complex traits in advanced breeding lines of Rice. Crop Sci 58(4):1519–1530. https://doi.org/10.2135/CROPSCI2017.09.0564

Muleta KT, Bulli P, Rynearson S, Chen X, Pumphrey M (2017) Loci associated with resistance to stripe rust (Puccinia striiformis f. sp. tritici) in a core collection of spring wheat (Triticum aestivum). PLoS One 12(6). https://doi.org/10.1371/JOURNAL.PONE.0179087

Neves HHR, Carvalheiro R, Queiroz SA (2012) A comparison of statistical methods for genomic selection in a mice population. BMC Genet 13. https://doi.org/10.1186/1471-2156-13-100

Olatoye MO, Hu Z, Aikpokpodion PO (2019) Epistasis detection and modeling for genomic selection in cowpea (Vigna unguiculata. L. Walp.). Front Genet 10(JUN):677. https://doi.org/10.3389/FGENE.2019.00677/BIBTEX

Ornella L, Singh S, Perez P, Burgueño J, Singh R, Tapia E, Bhavani S, Dreisigacker S, Braun H-J, Mathews K, Crossa J (2012) Genomic prediction of genetic values for resistance to wheat rusts. The Plant Genome 5(3). https://doi.org/10.3835/PLANTGENOME2012.07.0017

Pandey MK, Chaudhari S, Jarquin D, Janila P, Crossa J, Patil SC, Sundravadana S, Khare D, Bhat RS, Radhakrishnan T, Hickey JM, Varshney RK (2020) Genome-based trait prediction in multi-environment breeding trials in groundnut. Theor Appl Genet 133(11):3101–3117. https://doi.org/10.1007/S00122-020-03658-1/TABLES/5

Pandey MK, Roorkiwal M, Singh VK, Ramalingam A, Kudapa H, Thudi M, Chitikineni A, Rathore A, Varshney RK (2016) Emerging genomic tools for legume breeding: current status and future prospects. Front Plant Sci 7(MAY2016):455. https://doi.org/10.3389/FPLS.2016.00455/BIBTEX

Pérez-Cabal MA, Vazquez AI, Gianola D, Rosa GJM, Weigel KA (2012) Accuracy of genome-enabled prediction in a dairy cattle population using different cross-validation layouts. Front Genet 3(FEB):27. https://doi.org/10.3389/FGENE.2012.00027/BIBTEX

Phillips RL (2010) Mobilizing science to break yield barriers. Crop Sci 50:S-99. https://doi.org/10.2135/CROPSCI2009.09.0525

Poland JA, Rife TW (2012) Genotyping-by-sequencing for plant breeding and genetics. *The plant.* Genome 5(3). https://doi.org/10.3835/PLANTGENOME2012.05.0005

Purbarani SC, Wasito I, Kusuma I (2017) Adaptive genetic algorithm for reliable training population in plant breeding genomic selection. In: 2016 international conference on advanced computer science and information systems, ICACSIS 2016, pp 556–563. https://doi.org/10.1109/ICACSIS.2016.7872803

Qin J, Shi A, Song Q, Li S, Wang F, Cao Y, Ravelombola W, Song Q, Yang C, Zhang M (2019) Genome wide association study and genomic selection of amino acid concentrations in soybean seeds. Front Plant Sci 10:1445. https://doi.org/10.3389/FPLS.2019.01445/BIBTEX

Rai N, Bellundagi A, Kumar PKC, Kalasapura Thimmappa R, Rani S, Sinha N, Krishna H, Jain N, Singh GP, Singh PK, Chand S, Prabhu KV (2018) Marker-assisted backcross breeding for improvement of drought tolerance in bread wheat (Triticum aestivum L. em Thell). Plant Breed 137(4):514–526. https://doi.org/10.1111/PBR.12605

Rajsic P, Weersink A, Navabi A, Peter Pauls K (2016) Economics of genomic selection: the role of prediction accuracy and relative genotyping costs. Euphytica 210(2):259–276. https://doi.org/10.1007/S10681-016-1716-0/TABLES/4

Rana MM, Takamatsu T, Baslam M, Kaneko K, Itoh K, Harada N, Sugiyama T, Ohnishi T, Kinoshita T, Takagi H, Mitsui T (2019) Salt tolerance improvement in Rice through efficient SNP marker-assisted selection coupled with speed-breeding. Int J Mol Sci 20:2585. https://doi.org/10.3390/IJMS20102585

Ray DK, Mueller ND, West PC, Foley JA (2013) Yield trends are insufficient to double global crop production by 2050. PLoS One 8(6):e66428. https://doi.org/10.1371/JOURNAL.PONE.0066428

de Resende MDV, de Assis TF (2010) Seleção Recorrente Recíproca entre Populações Sintéticas Multi- Espécies (SRR-PSME) de Eucalipto. Pesquisa Florestal Brasileira 0(57):57. https://pfb.cnpf.embrapa.br/pfb/index.php/pfb/article/view/57

Resende MDV, Resende MFR, Sansaloni CP, Petroli CD, Missiaggia AA, Aguiar AM, Abad JM, Takahashi EK, Rosado AM, Faria DA, Pappas GJ, Kilian A, Grattapaglia D (2012) Genomic selection for growth and wood quality in eucalyptus: capturing the missing heritability and accelerating breeding for complex traits in forest trees. New Phytol 194(1):116–128. https://doi.org/10.1111/J.1469-8137.2011.04038.X

Richard CAI, Hickey LT, Fletcher S, Jennings R, Chenu K, Christopher JT (2015) High-throughput phenotyping of seminal root traits in wheat. Plant Methods 11(1):1–11. https://doi.org/10.1186/S13007-015-0055-9/TABLES/2

Riedelsheimer C, Endelman JB, Stange M, Sorrells ME, Jannink JL, Melchinger AE (2013) Genomic predictability of interconnected biparental maize populations. Genetics 194(2):493–503. https://doi.org/10.1534/GENETICS.113.150227

Rothman N, Garcia-Closas M, Chatterjee N, Malats N, Wu X, Figueroa JD, Real FX, Van Den Berg D, Matullo G, Baris D, Thun M, Kiemeney LA, Vineis P, De Vivo I, Albanes D, Purdue MP, Rafnar T, Hildebrandt MAT, Kiltie AE et al (2010) A multi-stage genome-wide association study of bladder cancer identifies multiple susceptibility loci. Nature Genetics 42(11):978–984. https://doi.org/10.1038/ng.687

Rutkoski JE, Poland J, Jannink JL, Sorrells ME (2013) Imputation of unordered markers and the impact on genomic selection accuracy. G3: Genes, Genomes, Genetics 3(3):427–439. https://doi.org/10.1534/G3.112.005363

Schaeffer LR (2006) Strategy for applying genome-wide selection in dairy cattle. J Anim Breed Genet 123(4):218–223. https://doi.org/10.1111/J.1439-0388.2006.00595.X

Schmidt M, Kollers S, Maasberg-Prelle A, Großer J, Schinkel B, Tomerius A, Graner A, Korzun V (2016) Prediction of malting quality traits in barley based on genome-wide marker data to assess the potential of genomic selection. Theor Appl Genet 129(2):203–213. https://doi.org/10.1007/S00122-015-2639-1

Schulthess AW, Wang Y, Miedaner T, Wilde P, Reif JC, Zhao Y (2016) Multiple-trait- and selection indices-genomic predictions for grain yield and protein content in rye for feeding purposes. Theor Appl Genet 129(2):273–287. https://doi.org/10.1007/S00122-015-2626-6

Schulz-Streeck T, Ogutu JO, Karaman Z, Knaak C, Piepho HP (2012) Genomic selection using multiple populations. Crop Sci 52(6):2453–2461. https://doi.org/10.2135/CROPSCI2012.03.0160

Shengqiang Z, Dekkers JCM, Fernando RL, Jannink JL (2009) Factors affecting accuracy from genomic selection in populations derived from multiple inbred lines: a barley case study. Genetics 182(1):355–364. https://doi.org/10.1534/GENETICS.108.098277

Sinha R, Fritschi FB, Zandalinas SI, Mittler R (2021) The impact of stress combination on reproductive processes in crops. Plant Sci 311:111007. https://doi.org/10.1016/J.PLANTSCI.2021.111007

Spindel J, Iwata H (2018) Genomic selection in Rice breeding. In: Rice genomics, genetics and breeding, pp 473–496. https://doi.org/10.1007/978-981-10-7461-5_24

Stewart-Brown BB, Song Q, Vaughn JN, Li Z (2019) Genomic selection for yield and seed composition traits within an applied soybean breeding program. G3 (Bethesda, Md.) 9(7):2253–2265. https://doi.org/10.1534/G3.118.200917

Technow F, Riedelsheimer C, Schrag TA, Melchinger AE (2012) Genomic prediction of hybrid performance in maize with models incorporating dominance and population specific marker effects. Theor Appl Genet 125(6):1181–1194. https://doi.org/10.1007/S00122-012-1905-8

Thavamanikumar S, Dolferus R, Thumma BR (2015) Comparison of genomic selection models to predict flowering time and spike grain number in two hexaploid wheat doubled haploid populations. G3: Genes, Genomes, Genetics 5(10):1991–1998. https://doi.org/10.1534/g3.115.019745

Tibshirani R (1996) Regression shrinkage and selection via the lasso. J Royal Statistical Soc: Series B (Methodological) 58(1):267–288. https://doi.org/10.1111/J.2517-6161.1996.TB02080.X

Varshney RK, Terauchi R, McCouch SR (2014) Harvesting the promising fruits of genomics: applying genome sequencing technologies to crop breeding. PLoS Biol 12(6):e1001883. https://doi.org/10.1371/JOURNAL.PBIO.1001883

Voss-Fels KP, Cooper M, Hayes BJ (2019) Accelerating crop genetic gains with genomic selection. Theor Appl Genet 132(3):669–686. https://doi.org/10.1007/S00122-018-3270-8

Walter JDC, Edwards J, McDonald G, Kuchel H (2019) Estimating biomass and canopy height with LiDAR for field crop breeding. Front Plant Sci 10:1145. https://doi.org/10.3389/FPLS.2019.01145/BIBTEX

Wang X, Li L, Yang Z, Zheng X, Yu S, Xu C, Hu Z (2017) Predicting rice hybrid performance using univariate and multivariate GBLUP models based on North Carolina mating design II. Heredity 118(3):302–310. https://doi.org/10.1038/HDY.2016.87

Wang X, Xu Y, Hu Z, Xu C (2018) Genomic selection methods for crop improvement: current status and prospects. Crop J 6(4):330–340. https://doi.org/10.1016/j.cj.2018.03.001

Wang Y, Mette MF, Miedaner T, Gottwald M, Wilde P, Reif JC, Zhao Y (2014) The accuracy of prediction of genomic selection in elite hybrid rye populations surpasses the accuracy of marker-assisted selection and is equally augmented by multiple field evaluation locations and test years. BMC Genomics 15(1):1–12. https://doi.org/10.1186/1471-2164-15-556/FIGURES/4

Ward BP, Brown-Guedira G, Tyagi P, Kolb FL, van Sanford DA, Sneller CH, Griffey CA (2019) Multienvironment and multitrait genomic selection models in unbalanced early-generation wheat yield trials. Crop Sci 59(2):491–507. https://doi.org/10.2135/CROPSCI2018.03.0189

Watson A, Ghosh S, Williams MJ, Cuddy WS, Simmonds J, Rey MD, Asyraf Md Hatta M, Hinchliffe A, Steed A, Reynolds D, Adamski NM, Breakspear A, Korolev A, Rayner T, Dixon LE, Riaz A, Martin W, Ryan M, Edwards D, Hickey LT et al (2018) Speed breeding is a powerful tool to accelerate crop research and breeding. Nature Plants 4(1):23–29. https://doi.org/10.1038/s41477-017-0083-8

Weber N, Halpin C, Curtis Hannah L, Jez JM, Kough J, Parrott W (2012) Editor's choice: crop genome plasticity and its relevance to food and feed safety of genetically engineered breeding stacks. Plant Physiol 160(4):1842–1853. https://doi.org/10.1104/PP.112.204271

Wong CK, Bernardo R (2008) Genomewide selection in oil palm: increasing selection gain per unit time and cost with small populations. Theor Appl Genet 116(6):815–824. https://doi.org/10.1007/S00122-008-0715-5

Würschum T, Reif JC, Kraft T, Janssen G, Zhao Y (2013) Genomic selection in sugar beet breeding populations. BMC Genet 14. https://doi.org/10.1186/1471-2156-14-85

Xu S (2013) Genetic mapping and genomic selection using recombination breakpoint data. Genetics 195(3):1103–1115. https://doi.org/10.1534/GENETICS.113.155309/-/DC1

Xu Y, Liu X, Fu J, Wang H, Wang J, Huang C, Prasanna BM, Olsen MS, Wang G, Zhang A (2020) Enhancing genetic gain through genomic selection: from livestock to plants. Plant Communications 1(1):100005. https://doi.org/10.1016/j.xplc.2019.100005

Yabe S, Hara T, Ueno M, Enoki H, Kimura T, Nishimura S, Yasui Y, Ohsawa R, Iwata H (2018) Potential of genomic selection in mass selection breeding of an allogamous crop: an empirical study to increase yield of common buckwheat. Front Plant Sci 9:276. https://doi.org/10.3389/FPLS.2018.00276/BIBTEX

Zhang X, Pérez-Rodríguez P, Burgueño J, Olsen M, Buckler E, Atlin G, Prasanna BM, Vargas M, Vicente FS, Crossa J (2017) Rapid cycling genomic selection in a multiparental tropical maize population. G3 (Bethesda, Md.) 7(7):2315–2326. https://doi.org/10.1534/G3.117.043141

Zhang X, Pérez-Rodríguez P, Semagn K, Beyene Y, Babu R, López-Cruz MA, San Vicente F, Olsen M, Buckler E, Jannink JL, Prasanna BM, Crossa J (2015) Genomic prediction in biparental tropical maize populations in water-stressed and well-watered environments using low-density and GBS SNPs. Heredity 114(3):291–299. https://doi.org/10.1038/HDY.2014.99

Zhao Y, Gowda M, Liu W, Würschum T, Maurer HP, Longin FH, Ranc N, Reif JC (2012) Accuracy of genomic selection in European maize elite breeding populations. Theor Appl Genet 124(4):769–776. https://doi.org/10.1007/S00122-011-1745-Y/TABLES/3

Zhao Y, Mette MF, Gowda M, Longin CFH, Reif JC (2014) Bridging the gap between marker-assisted and genomic selection of heading time and plant height in hybrid wheat. Heredity 112(6):638–645. https://doi.org/10.1038/HDY.2014.1

Chapter 10
Genetic Engineering: A Powerful Tool for Crop Improvement

Mamta Bhattacharjee, Swapnil Meshram, Jyotsna Dayma, Neha Pandey, Naglaa Abdallah, Aladdin Hamwieh, Nourhan Fouad, and Sumita Acharjee

Abstract Rising population, changing climatic conditions, and various biotic and abiotic stresses are contributors to lowering crop yields. This, in turn, has augmented the number of people suffering from malnutrition. The applications of genetic engineering including genome editing are important as it can complement modern breeding activities to mitigate the effects of changing environment and boost crop production. The genetically modified (GM) crops thus offer one or more advantageous attributes, such as herbicide resistance, tolerance against pests and pathogens, and nutritional enhancement. The discovery of the natural ability of *Agrobacterium tumefaciens* to transfer a segment of its DNA (T-DNA) into the host was one of the breakthroughs of the twentieth century. It marked the beginning of achieving successful genetic transformation in a wide range of plants. Further, with the advent of technologies like zinc-finger nucleases (ZFNs), transcription activator-like effector nucleases (TALENs), and clustered regularly interspaced short palindromic repeats (CRISPR)/Cas, it has been possible to overcome the limitations of conventional breeding techniques. The synergism of scientific skills with sophisticated technologies resulted in many successful GM crops that were resistant to insects, pests, and weeds and enriched in micronutrients like vitamins and various minerals. Although not all GM crops have been commercialized, a few like soybean, papaya, maize, cotton, common bean, sweet potato, cowpea, etc. are practising. Recently, genome-edited crops are also approved for commercialization. The technology holds immense promise to achieve UN's sustainable development goals (SDGs) to fight hunger, attain food security, enhance nutrition, and promote sustainable agriculture.

Keywords Genetic engineering · GM crops · Food security · Biofortification

---

M. Bhattacharjee · S. Meshram · J. Dayma · N. Pandey · S. Acharjee (✉)
Department of Agricultural Biotechnology, Assam Agricultural University, Jorhat, Assam, India

N. Abdallah
Faculty of Agriculture, Cairo University, Giza, Egypt

A. Hamwieh · N. Fouad
International Center for Agricultural Research in the Dry Areas, ICARDA, Cairo, Giza, Egypt

M. K. Pandey et al. (eds.), *Frontier Technologies for Crop Improvement*, Sustainable Agriculture and Food Security, https://doi.org/10.1007/978-981-99-4673-0_10

10.1 Introduction

Genetic engineering has played a critical role in crop improvement either by improving the pre-existing traits or by introducing new desirable traits to improve crop production. GM crops offer one or more beneficial attributes, such as herbicide resistance, tolerance against pest/pathogens, and nutritional improvement (Kumar et al. 2020). Some of the famous instances of GM crops showcasing the potential to avert the challenges in agriculture include Bt cotton for insect tolerance and golden rice for improved vitamin A content (Qaim and Kouser 2013). Adoption of GM crops has revealed that the use of GM crops could boost the crop yield by 22% and lower the use of pesticides by 37% (Taheri et al. 2017). GM crops are the ones whose genomes have been modified such that economically important traits of the plants could be improved along with their yield. In this process, plants have been produced by inserting specific segments of foreign DNA or nucleic acid into their genome using the transformation methods like direct gene transfer or by *Agrobacterium*-mediated transformation (Griffiths et al. 2005). Such crops are referred to as transgenic crops, and the gene that has been inserted is known as a transgene (Kumar et al. 2020). The development of genetic transformation techniques marked the beginning of exponential growth in the field of plant research and offered a major advantage over conventional plant breeding technologies as compatibility was no longer a requirement. This breakthrough can be credited to the natural ability of *Agrobacterium tumefaciens* to insert its T-DNA into the host. The achievements in this field extend from the development of the first transgenic antibiotic-resistant tobacco and petunia (Fraley et al. 1983; Herrera-Estrella et al. 1983) to the commercialization of glyphosate-resistant soybean, bromoxynil herbicide-resistant cotton, and Bt maize (James 1998). The new editing techniques, like zinc-finger nucleases (ZFNs), transcription activator-like effector nucleases (TALENs), and clustered regularly interspaced short palindromic repeats (CRISPR)/Cas, have further utilized site-specific nucleases that have helped in addressing many concerns related to the unpredictability and inefficiency associated with conventional mutagenesis and transgenesis methods (Kumar et al. 2020). Thus, it is evident that traditional transgenic technologies along with genome-editing tools technology can prove useful in not only boosting agricultural productivity but also reducing dependency on agrochemicals, minimizing the environmental footprint of agriculture (Kumar et al. 2020).

Attaining food security is one of the prime concerns for any country. However, the growing population, diminishing arable land, and changing climate have widened the gap between population growth and food security (Islam and Karim 2019). Therefore, technological innovations in agriculture sector are necessary to sustainably increase production and decrease food losses (UNCTAD 2017). GM crops could positively impact food security by increasing the availability as well as improving the quality of food and influencing the farmers' socio-economically (Juma 2011). The recent global pandemic has triggered a serious social and economic crisis, presenting profound threats to nutritional status and food security around the world, especially in countries with low income (FAO 2020). Of particular concern is the rise in malnutrition, occurring due to changes in the availability, accessibility, and affordability of nutritious foods during this unprecedented situation. More women and children are suffering from malnutrition due to the declining quality of their diet, which is one of the worst repercussions of this outbreak (Osendarp et al. 2021; The World Bank Report 2021). In the uncertain circumstances provoked by the COVID-19 pandemic, where it is expected to decrease financial security and exacerbate all forms of malnutrition, wise adoption of GM crops can play a pivotal role in nutritional improvement (Wesseler and Purnhagen 2020; Gbashi et al. 2021). Furthermore, benefits like better agricultural yield, farm profits, and reduction of post-harvest losses could contribute to reducing food insecurity (Gbashi et al. 2021). The unanticipated chain of events during the pandemic has caused a shortage of workforce, restrictions on transportation, limited market operations, and the crumbling of the food supply chain, increasing our dependency on domestic food systems. Thus, the current situation necessitates the enabling of the environment, improving the resilience and nutrition sensitivity of the local food systems. Additionally, the threat to the food security scenario across the world has also encouraged people to accept and appreciate the need for every nation to adopt technologies that can support the farmers to boost yields with scarce resources. A pandemic like the current one may have an extensive and long-term influence on the agriculture and food industry.

This chapter highlights the importance of GM crops in agriculture by emphasizing the traits like herbicide tolerance, insect resistance, virus resistance, and biofortification. Different techniques utilized for the development of GM crops have also been exemplified in the text. Furthermore, the chapter also focused on the commercially available GM crops and the benefits of GM crops for addressing issues like food insecurity.

10.2 Pandemic and GM Crops

Like any other industry, the sudden outbreak of the COVID-19 pandemic had severely affected the agricultural sector. The situation of a pandemic can be regarded as the wake-up call for all the key stakeholders associated with this industry to reflect on the prevailing strategies such that the loopholes could be recognized. Just like human health, plant health must also be protected and secured by adopting preventive measures (Lamichhane and Reay-Jones 2021). During the time of Covid-19, the imposition of quarantine measures along with the lockdown had affected the mobilization of farmers and other people leading to a shortage of labourers as well as farm operators, especially in countries whose agricultural system is labour-intensive. This had not only affected the planting of the seasonal crops but also upsurge the loss of crops due to biotic and abiotic stress as the farmers were unable to carry out interventions like mechanical weeding or spraying pesticides. Moreover, activities like capacity development initiatives and pest management activities were severely affected in the agricultural sector due to the pandemic (FAO 2020).

In addition, equipment shortages during the time of the COVID-19 pandemic have also affected timely crop protection. For instance, a shortage in the fogging equipment was reported due to their utilization in disinfecting the areas for reducing the spread of the corona virus. Additionally, shortages in the respiratory protective equipment used by the pesticides handlers had also been reported in the United States. Under such conditions, not only did the health risk of the applicator increase but suboptimal applications of the pesticides had also upsurged the crop damage (Lamichhane and Reay-Jones 2021).

The outbreak of the pandemic has made it clear that it is crucial to make the agricultural system more resilient. In this context, one of the most important strategies that can be used is to develop GM crops that are self-sustainable and has the potential to overcome biotic and abiotic threats. Therefore, by using herbicide-resistant, virus-resistant, and drought-resistant GM crops, it could be hoped to minimize the loss of the crops and food insecurity despite measures like lockdown and social distancing. Therefore, the development of GM crops must be augmented so that crop yields could be improved irrespective of the pandemic. Furthermore, the outbreak of the Covid-19 pandemic had played a crucial role in increasing the burden of the malnutrition epidemic (Littlejohn and Finlay 2021). Further, Gastélum-Estrada et al. (2021) revealed that nutrients, such as vitamin C, D, and selenium, have the potential to reduce the risk as well as amelioration of COVID-19. These micronutrients are present in small quantities in most foods; however, the natural concentrations are not at par in meeting the needs of the human body. Henceforth, for meeting the immune-modulating needs, biofortification must be done for the foods like chickpea, tomato, wheat, and others such that their consumption could help in enhancing the immune system of humans.

10.3 Abiotic Stress and GM Crops

A deleterious profusion of various environmental factors, which can be summarised as abiotic stress, such as flood, heat, cold, drought, etc., imposes a detrimental impact on the overall growth of the crops and also on the grain yield of the crop plants. The plants cope up themselves with the existing obliterated condition by changing the physiological mechanism and altering the signalling cascades, regulatory proteins, and modification of the antioxidant defence system ultimately to limitise the cellular homeostatic condition. The changes that the plant made to adapt with the changing environmental regarding the abiotic stresses ultimately helps to minimise the loss of the plant along with its yield; also the near-optimal conditions facilitate them for better growth and development. At the molecular level, several arrays of genes get expressed variably due to the abiotic stresses, and also several disruptions of normal functioning can be seen due to this factor.

Many of the reported deficits of the crop plants, like cold, heat, and water-deficit of rice, water-deficit of maize, etc., have been studied through mitigation of the effects of the bacterial cold shock proteins (*csp*), as mentioned in the study of Castiglioni et al. (2008). The whole study summarises that maintaining the RNA stability and protein translation proved to be effective in the maintenance of the cellular functions during dehydration stress conditions, where *cspA* and *cspB* gene were used as a bacterial RNA chaperone extracted, respectively, from *E. coli* and a soil bacterium *B. subtilis*. The phenotypic behaviour of the transgenic maize was normal under adequate-watered condition but was showing better adaptation in water scarcity condition. Here, the RNA chaperone helps in stabilising the mis-folded RNA structures.

Another example of drought stress tolerance can be sugarcane, which was approved by Indonesia as commercial cultivation in 2013 (Waltz 2014). The gene *betA* extracted from *E. coli* and *Rhizobium meliloti* codes for choline dehydrogenase and catalyses formation of the osmoprotectant compound glycinebetaine, which helps in adapting to the water stress (Takabe et al. (1998); Khan et al. (2009); Singh et al. (2015)). It was inferred by another study of Chen and Murata (2002) and Singh et al. (2015) that the accumulation of the compatible solutes designated as the osmoprotectants, which comprise non-reducing sugars like fructan, trehalose, mannitol, and sorbitol, along with the proline and glycinebetaine, aids the survival of the plants under osmotic stress. In 2016, Nahar et al. have stated that increased titre of glycinebetaine helps in stabilising the enzymes and the protein structures, thus maintaining the cellular integrity during the stress condition. According to Waltz (2014), these transgenic sugarcane plants can produce 10–30% higher sugar than that of the non-transgenic plants under drought conditions in field trial.

Heat as an abiotic stress is also vulnerable to the crop plants, and as a by-product of this particular stress, abundant number of reactive oxygen species (ROS), such as hydrogen peroxide and superoxide, are produced inside the plant body, which hampers in the growth and development and declines the yield of the crop. Scavenging the ROS and restricting it to denature the enzymes and damaging the internal cellular components are straightforward approaches to tolerate the detrimental effects of heat stress (Chaitanya et al. 2002). Overexpression of the cytosolic ascorbate peroxidise (*cAPX*) gene to enhance tolerance to heat stress is already reported in transgenic apple and tomato, which seems to withstand a temperature of 40 degree Celsius in field condition as documented by Wisniewski et al. (2002) and Wang et al. (2005), respectively.

SAMDC (S-adenosyl-l-methionine decarboxylase) is one of the regulatory target enzymes in polyamine biosynthesis, and a related study on a tomato plant for enhancing polyamines production was done, where Chen and Xiong (2009) overexpressed *SAMDC* cDNA isolated from *Saccharomyces cerevisiae* and the transgenic lines was found to produce about 1.7–2.4-fold higher levels of spermidine and spermine along with enhanced antioxidant enzyme activity for better protection of membrane lipid peroxidation when compared to wild-type plants, which ultimately can withstand a temperature up to 38 degree Celsius. It is quite obvious that proline works as an important osmoprotectant, which protects cells from damage under heat stress, and according to Boston et al. (1996), HSPs facilitate correct protein folding, assembly, and translocation and provide stability to the integral proteins and cell membranes under heat stress; therefore, in 2014, Song et al., in his experiment, had overexpressed *CgHSP70* gene in chrysanthemum. The result in the transgenic lines showed increased peroxidase activity and higher proline content along with reduced malondialdehyde content.

Salinity or salt stress stands out to be the prevalent and one of the important abiotic stresses in the crop plants, which is engulfing around 20% of the agricultural land according to Rengasamy (2005). Here, *osmotin* comes out to be the important pathogenesis-related protein, which is solely for the plants to manage various abiotic stresses. Chilli plants are vulnerable to the salinity stress, and on the same hand, they are also very inept to genetic transformations and tissue culture, which also prevents the scope of genetic transformation in the chilli plants to withstand the salinity stress, but it has been made possible by Subramanyam et al. (2011), when he had successfully made the chilli pepper (*Capsicum annum L.* cv. Aiswarya 2103) by ectopic expression of tobacco *osmotin* gene via *Agrobacterium tumefaciens*-mediated gene transfer technique. The transgenic pepper plants also showed increased levels of chlorophyll, proline, glycine betaine, ascorbate peroxidase (*APX*), superoxide dismutase (*SOD*), glutathione reductase (*GR*), and relative water content (*RWC*) in biochemical analysis and survived in salinity level up to 300 mM NaCl concentration. Also previously in 2008, Husaini and Abdin have overexpressed the tobacco osmotin gene in strawberry plants (*Fragaria ananassa* Duch.), which also gave a positive result to withstand the salinity stress.

10.4 Biotic Stress and GM Crops

The drastically declining yield and productivity of the crops due to biotic is a major concern faced by the agricultural sector across the globe. Some of the biotic constraints responsible for crop destruction include infestation of crops by pests, diseases, weeds, and herbicides. This is the reason that has led to the emergence of conventional breeding methods, such that better crop varieties could be developed with superior traits. This was the only and most reliable method that could lead to crop protection, improvement, as well as quality management. However, the limitations of the conventional methods led to the development of genetic engineering techniques. This, in turn, allowed the scientists to tailor the plant varieties for expressing economically important traits (Tohidfar and Khosravi 2015; Parmar et al. 2017). Unlike abiotic stress, biotic stresses cause their hosts to directly face deprivation of nutrients thus leading to a reduction in the vigour and often the death of the plant. This is the major reason that causes pre-harvest and post-harvest losses in the field (Singla and Krattinger 2016). Additionally, the limited availability of agricultural resources along with insufficient nutritious food has increased the issue of food insecurity and malnutrition across the globe (Chrispeels and Sadava 2003). Thus, it is imperative to minimise the damage caused by the different stresses and to improve the quality of the food being produced for consumption. This, in turn, will resolve half of the food issues that are being faced by the rising population of the world.

Here, in this chapter, emphasis has been led to illuminate the ways genetic engineering could aid in reducing the biotic stresses (insect, weeds, and pathogens) and improving the nutrition profile of the crops (biofortification).

10.4.1 Herbicide Resistance

Herbicides are widely utilised in the farmlands across the globe for increasing crop yield along with farm labour efficiency (Schütte et al. 2017). However, a serious concern associated with the usage of herbicide is that other than affecting the target crops, it could also drift and run off towards the non-target plants. This, in turn, has deteriorated the ecological adaptability of the plants. Thus, during the development of GM crops, researchers had focused on involving the herbicide-resistant trait (Table 10.1). GM crops having resistance against herbicides like glyphosate and glufosinate-ammonium have been developed by plant scientists. Out of these two herbicides, the underlying mechanism of herbicide glyphosate is to inhibit 5-enolpyruvylshikimate-3-phosphate synthase (EPSPS). This enzyme is involved in carrying out the biosynthesis of aromatic acids and phenolics via the shikimate pathway in plants (OCED 1999). As a result, protein synthesis, production of the phenolics, defence molecules, salicylic acid, and lignin derivatives all get impacted. On the other hand, glufosinate ammonia is the racemic mixture of D- and L-isomers of phosphinothricin (PPT) and is found to inhibit the activity of glutamine synthetase via L-isomer. This, in turn, leads to the accretion of a lethal level of ammonia, ultimately causing death in plants (OCED 1999). Thus, the mode of action for herbicide-tolerant crops to get rid of the herbicides (either at the tissue or cellular level of the target plant) occurs in two ways, i.e. (i) the modification/alteration of an enzyme or other herbicidal target in plants in order to render its insensitive to the action of herbicide (ii) and the use of enzyme/enzyme system to degrade or detoxify the herbicide before they can act on the plant (Botterman and Leemans 1988). Some examples of the GM crops that are resistant to glyphosate herbicide include alfalfa, canola, maize, and soybean (Kumar et al. 2020). The molecular mechanism that has been used in the development of the above-mentioned crops is by creating modification at the target site of the herbicide such that the herbicide is unable to bind to its target. In most of the cases, the resistance mechanism is based on the heterologous expression of glyphosate-insensitive forms of *epsps* (epsps derived from *A. tumefaciens* strain CP4, mutant version of maize epsps, chemically synthesized gene similar to *epsps*grg23 gene of *Arthrobacter globiformis*) (Barry et al. 1997; Padgette et al. 1995).

In addition, herbicide-degrading enzymes were explored to generate herbicide resistance in plants. Such enzymes are extracted from bacteria residing in soil and water. One such gene is *atzA*, which is plasmid-borne. This gene is present in the

Table 10.1 Summary of 32 researches published between 1999 and 2020 using 18 techniques related to GM for enhancing herbicide tolerance in 11 crops

Technique and crop	Year
Base-editing-mediated gene evolution	
Rice	2020
BE	
Rice	2017, 2018
CRISPR-Cas9	
Maize	2020
Rice	2019
Tomato	2018
CRISPR-Cas9-based cytosine base editing	
Rice	2016
CRISPR-Cpf1	
Rice	2015
CRISPR/Cas9 SDN1	
Linum	2016
Tomato	2019
Watermelon	2018
CRISPR/Cas9 SDN2	
Rice	2016, 2017
Soybean	2015
CRISPR/Cas9 SDN3	
Cassava	2018
CRISPR/Cas9, TALENs SDN2	
Potato	2016
CRISPR/Cas-mediated base editing	
Maize	2019
Oilseed rape	2018
CRISPRnCas9-RT	
Maize	2020
Rice	2020
Meganuclease SDN3	
Cotton	2013
ODM	
Canola	2015
Maize	1999, 2000
Rice	2004
SDN1, SDN2, SDN3	
Maize	2015
SDN1, SDN2, SDN4	
Maize	2016
SDN3	
Soybean	2017

(continued)

Table 10.1 (continued)

Technique and crop	Year
TALENs SDN2	
Rice	2015
ZFN SDN3	
Maize	2013

gram-negative bacteria *Pseudomonas* sp. strain ADP. The gene encodes atrazine chlorohydrolase that catalyses the hydrolytic dechlorination of atrazine (Wang et al. 2005). Other genes that have been introduced into the plants for the development of the transgenic lines include soybean cytochrome P450 monooxygenase (Siminszky et al. 1999), glutamylcysteine synthetase (Gullner et al. 2001), *bar*, and *pat* genes (Lutz et al. 2001). Apart from this, GM crops have also been developed that are resistant to two herbicides. One of the GM plants that have been developed by introducing the genes tolerant to bensulfuron-methyl (BM) and glufosinate herbicides is rice AHAS (acetohydroxyacid synthase). The development of dual herbicide resistance confers the advantage that the chance of the evolution of weed resistance to herbicide within a short period decreases (Green and Castle 2010). Additionally, photoperiod-sensitive genic male-sterile transgenic rice has been developed that is tolerant to glyphosate and glufosinate herbicides (Deng et al. 2014). Other examples include the development of herbicide-tolerant tobacco by the introduction of the chimeric psbA gene [resistant to atrazine] (Cheung et al. 1988) and protoporphyrinogen oxidase [resistant to diphenyl ether] (Lermontova and Grimm 2000).

10.4.2 Insect Resistance

Among the various biotic stresses, the effect of insects has always been a major concern for farmers (Manosathiyadevan et al. 2017). This is because the quantity of crops damaged by insects is equal to 42% of the direct calorie consumed by humans worldwide every year. The rate of damage increases with the increase in the temperature of the climate. Studies have revealed that a warmer climate increases the metabolic rate of insects (Petersen et al. 2000). Thus, it leads to an augmentation of the food consumption rate among the insects. Apart from this, the population growth rate of insects also increases with the change in temperature (Deutsch et al. 2008). This had necessitated the development of plant lines that would be resistant to insects. In this regard, *Bt* genes that confer resistance to *Bacillus thuringiensis* have been extensively studied and introduced in economically important plants, like maize, cotton, and others. The GM crops express *Cry* genes that encode *Cry* protein. The *Cry* protein is responsible for pore formation in the cell membranes of the insect midgut, which finally leads to paralysis and death in the insects (Labbé et al. 2017). Some of the examples of the crops that have been made insect resistant with the help of the *Bt* gene, include maize, rice, cotton, and potato. GM maize has been developed with the integration of *Cry1Ab*, *Cry1Ac,* or *Cry9C* for protecting against *Ostrinia nubilalis* (Lepidoptera: European corn borer and Crambidae) and *Sesamia nonagrioides* (Lepidoptera: Noctuidae, Mediterranean corn borer). Further, *Cry1F* gene had been used for protecting maize against *Spodoptera frugiperda* (Lepidoptera: Noctuidae). On the other hand, the development of the GM maize with *Cry3Bb, Cry34Ab*, and *Cry35Ab* has helped the plant to get protected from rootworms belonging to the genus *Diabrotica* (Coleoptera: Chrysomelidae). For protecting cotton plants from getting affected by *Heliothis virescens, Pectinophora gossypiella, Helicoverpa zea, and Helicoverpa armigera,* GM cotton had been developed by inserting either *Cry1Ac* or by the stacking of *Cry1Ac* and *Cry1Ab* (Romeis et al. 2008). Apart from this, other genes for insect resistance that have been introduced into GE crops include Vip proteins (source of isolation: *B. thuringiensis* and *B. cereus*), lectins (source of isolation: plants like *Nicotiana tabacum* and *Oryza sativa)*, and protease inhibitors (source of isolation: plants, bacteria, and fungus). Vip gene encodes vegetative insecticidal proteins and has been successfully inserted for the development of transgenic cotton and maize (Chakroun et al. 2016).

Likewise, lectins interact with various glycoproteins or glycan structures, which can interfere with a variety of physiological functions in insects (Macedo et al. 2015). Orysata (agglutinin, a type of lectin isolated from the seedlings of *Oryza sativa*) was successfully expressed in transgenic tobacco and found to exhibit insecticidal activity against beet armyworm, the green peach aphid, and the pea aphid (Al Atalah et al. 2014). Similarly, protease inhibitors (PIs) work by inhibiting the proteolytic enzymes present in the guts of insects preventing them to procure the amino acids necessary for their growth and development (Broadway and Duffey 1986). For instance, PIs like trypsin inhibitor (Hilder et al. 1987) and potato protease inhibitor II (Duan et al. 1996; Majeed et al. 2011) were successfully explored to impart insect resistance in tobacco, rice, and cotton, respectively (Jagdish et al. 2020; Quilis et al. 2014; Dunse et al. 2010).

10.4.3 Virus Resistance

Plant viruses are another biotic factor that has been causing severe crop losses across the globe. As per the reports from FAO, 20% to 40% of the global crop production is lost to pests (FAO 2019). Every year, plant diseases cost the global economy approximately $220 billion. The expeditious distribution of the global plant virus diseases is the outcome of the rapid expansion of the international trade in the plant as well as plant produce (Jones 2021). This, in turn, has facilitated the introduction of virus diseases in parts of the world where they were not present earlier. All this has occurred for three main reasons. Firstly, trade globalisation that involves international agreements over free trade has led to the transfer of crops from one continent to another. Secondly, lower subsidies for developed countries had let the developing countries expand trade in international crop produce. Lastly, a well-developed and efficient transportation system with loosening plant quarantine facilities has further facilitated crop trading (Jones 2009). Thus, the movement of the plants from their domestication centre to other regions for monoculture had also led to the emergence of new virus diseases. In addition to this, the rapid change in climatic conditions has also been found to be one of the major reasons that have made it difficult to manage virus-caused pandemics (Jones and Naidu 2019). To get rid of the virus-mediated destruction, transgenic plants resistant to the virus are being developed at a swift pace by plant researchers. In this context, the gene silencing mechanism has been used for the development of transgenic plants, where genetic constructs having similar sequence identity as that of the pathogen genes are used for eliciting RNAi (RNA interference) in plants for silencing the gene. This method, where a particular gene in the pathogen is silenced, has been used to transform papaya with the coat protein gene of the papaya ring spot virus (PRSV) (Ferreira et al. 2002; Gonsalves et al. 2004). This method has also been successfully applied for developing GM cassava (Odipio et al. 2014), summer squash (Klas et al. 2011), and soybean (Abbas et al. 2021). The other methods that have been used for introducing virus resistance in plants are by genetically modifying targets of pathogen virulence factors so that host resistance could be increased without the insertion of any transgene or any exogenous biochemical pathway in the plants (Vincelli 2016). Apart from this, attempts have been made to develop virus-resistant plants to transform plants with satellite RNAs (satRNA). These are the small RNAs that depend upon the helper virus for replication and get encapsidated in the particles of the helper virus (Tien and Wu 1991). The transgenic expression of satRNA has helped in decreasing the virus systems by integrating DNA-copy of the responsible satRNA. The attempt had been successfully made in tobacco plants where a DNA copy of satRNA of cucumber mosaic virus (CMV) was introduced which helped in restricting the replication of CMV (Kundu and Mandal 2001). Similarly, resistance to plum pox virus and CMV was conferred in *Prunus domestica* (Ravelonandro et al. 1997), *Capsicum annuum* (Zhu et al. 1996) and *Solanum lycopersicum* (Yang et al. 1997), utilising viral coat protein (cp). Other advances in the realm of virus-resistant technologies include the use of the PRSV replicase (rep) gene and antisense technology to confer resistance in papaya (Guo et al. 2009) and common bean (Faria et al. 2006), respectively.

10.4.4 Biofortification

Biofortification is the process of the development of crops with enhanced nutritional values either by conventional selective breeding or by genetic engineering. As per the estimation, more than 800 million people across the globe are malnourished, of which 98% are in developing countries (Sinha et al. 2019). Besides, more than two billion people are experiencing hidden hunger that is caused by the poor intake of key micronutrients via their daily diet (Gillespie et al. 2016). Henceforth, the most feasible and cost-effective way that could help in providing micronutrients to the population of the developing countries is by the development of biofortified crops. Another major benefit that could be achieved by the production of biofortified crops is that biofortified seeds are found to have an indirect impact on agriculture. This is because higher concentrations of minerals are found to provide better protection against diverse biotic and abiotic stresses that have been affecting the crops and their productivity (Welch and Graham 2004). According to the World Health Organisation (WHO), malnutrition includes different forms of undernutrition, such as wasting, stunting, being underweight, inadequate minerals or vitamins, and diet-related non-communicable diseases. As per WHO, 1.9 billion people are either obsessed or overweight, 462 million are underweight, 149 million children below the age of 5 are stunted, 45 million are wasted, and 38.9 million were obese in the year 2020. In addition to this, nearly 45% of the deaths among the children who were under the age of 5 years were related to undernutrition. Various methods have been used for biofortifying the crops, such as via agronomic practices, plant breeding, and genetic engineering. Amid all these methods, genetic engineering is highly popular as it aids in the development of new cultivars with traits of interest. In addition to this, this technique also uses an enlarged gene pool for facilitating the transfer and expression of desirable traits from one species to another. Apart from this, if a particular micronutrient cannot be produced naturally in the crops, then the only method that could aid in fortifying those crops is genetic engineering (Pérez-Massot et al. 2013). Moreover, the transgenic approach provides the scope to insert novel genes, overexpress, or down-regulate the genes that are present in the plants. This transgenic approach has been used for the development of iron bio-fortified crops by inserting the iron-binding protein gene lactoferrin in crops like rice under the control of endospermic promoters, such that iron could be accumulated in the seeds that are consumed by humans. This helped in the development of the rice variety that had 120% more iron content in it and was suitable for meeting the iron needs of children but not adults (Suzuki et al. 2001). Parallel to this, to increase the iron content in the rice by two- to threefold, transgenic rice containing soybean ferritin was developed under the control of GluB-1, the rice seed-storage protein glutelin promoter (Lucca et al. 2001), and an increase of 3.7-fold was achieved indica cv. IR68144 seeds (Vasconcelos et al. 2003). Other than this, zinc had been fortified in the plants either by introducing zinc-binding protein-specific coding sequence, by overexpression of the zinc-storage protein and by enhancing the expression of zinc uptaking proteins. In addition to this, by introducing a protein having the capability to reduce anti-nutrient content, it becomes possible to increase the bioavailability of zinc. In this context, overexpression of the exogenous HvNAS1 (barley NAS gene) in *Arabidopsis* and tobacco had helped in increasing the concentration of copper, iron, and zinc in both the plant's seeds (Kim et al. 2005). Similarly, the overexpression of endogenous NAS genes helped in ameliorating the concentration of sodium, iron, and zinc in the endosperm of the transgenic rice (Wirth et al. 2009). Apart from this, plants have been biofortified for iodine (Itoh et al. 2009), vitamin A (Wang et al. 2014), vitamin B (Chen and Xiong 2009), vitamin C (Scholes et al. 2012), and vitamin E (Tanaka et al. 2015). Henceforth, biofortification holds a great promise for improving the nutritional profile of the plants and ultimately in overcoming the issue of malnutrition.

10.5 Technologies Exploited for the Development of GM Crops

The constant efforts made by the plant biotechnologists had led to the development of methods like *Agrobacterium*-mediated transformation or biolistic transformation for introducing the desired gene into the plant cells. After the insertion, if the gene is stable, inherited, and expressed by the succeeding generations, then such plants are referred to as transgenic plants. Another gene-editing method that has been in the spotlight is RNA interference (RNAi), a technique where the structure and function of the hosts' genes are modified by the insertion of guide RNA. Other than this, the success achieved by the scientists in developing programmable nucleases, like (ZFNs), transcription activator-like effector nucleases (TALENs), and clustered regularly interspaced short palindromic repeat (CRISPR)–Cas-associated nucleases, has opened new dimensions in the field of gene editing.

In this section of the chapter, an overview of the diverse gene-editing technology and its impact on the development of GM plants has been provided.

10.5.1 Agrobacterium and Biolistic Methods

The natural ability of *Agrobacterium* species like *A. tumefaciens* and *A. rhizogenes* to transform the plants and induce crown gall tuber and hairy root disease, respectively, had let the researchers mimic the process for achieving the first plant transformation breakthrough in 1983. However, among the two species, it was *A. tumefaciens* that helped the scientists significantly to understand the mechanism of plant transformation by tracking the activities of Ti (tumour-inducing) plasmid. It is a soil-dwelling bacterium that transforms the plant by injecting ssDNA (also referred to as T-DNA) that leads to the induction of the tumour via the synthesis of the phytohormones. Followed by this, the bacterium also compels the plants to synthesise opines that are used as nutrients by the pathogens (Flores-Mireles et al. 2012). Further, the genes required for the pathogenic function, along with the T-DNA, are encoded on the Ti plasmid. Moreover, the virulence or *vir* genes plays a crucial role in the translocation of the T-DNA within the host nucleus, such that the integration of the tumorigenic genes into the host chromosomes could be achieved. However, during plant transformation, disarmed Ti plasmids are constructed by deleting the oncogenes while keeping opine biosynthetic genes (Pratiwi and Surya 2020). Henceforth, in the *Agrobacterium*-mediated transformation (AMT), requisite foreign DNA substitutes the T-DNA, and the bacterium itself is used as the vehicle for inserting the gene of interest into the host (Chilton et al. 1977). Additionally, while constructing Ti plasmid with genes of interest, selectable marker genes are also inserted for distinguishing transformed cells from normal cells (Matveeva and Lutova 2014). The application of the AMT for the development of transgenic plants has helped in the development of insect-resistant crops, like cotton, and herbicide-tolerant plants, like soybeans and corn. Some of the economically important plants that have been developed with the application of the AMT method include papaya lines resistant to ringspot virus. GM papaya played a vital role in saving the US papaya industry. In addition, AMT has also been used for enhancing the nutritional value of the plants, like beta-carotene in canola (Fujisawa et al. 2009), oil composition in corn (Mohammed and Abalaka 2011), and vitamin A in rice (Dubock 2019). The major advantage associated with the application of AMT is that it ameliorates the probability to achieve single-copy insertions successfully. However, certain limitations associated with the process include the demand for a long tissue culture period for recovering the transgenic plants, the low frequency of stably transformed plants, and a narrow range of genotypes that could be transformed within a species (Rahangdale et al. 2020).

The biolistic method also referred to as particle bombardment aids in delivering the desired DNA directly into the plant cell. In this method, small metal particles (either tungsten or gold) coated with DNA are fired into the plant cells by accelerating them to high speed and releasing high-pressure helium in the gene gun (Bhatia et al. 2015). The target cells for gene gun bombardment are usually the totipotent cells that consist of either embryogenic suspension culture or embryogenic callus that is derived from the recipient plant (George et al. 2009). The metals used in the process are not lethal and safely deliver the DNA on the construct within the nucleus of the cell, aiding in its integration by recombination with the chromosomes. Thereafter, the transformed cells are induced to form plants under selection. This helps only those plants to survive that are expressing the selectable marker and the gene of interest (George et al. 2009). The exogenous DNA used for transformation consists of a plant expression cassette that is inserted in a vector based on a high-copy number bacterial cloning plasmid. With the aid of this process, Datta et al. (2003) were able to develop Golden Indian Rice lines that consisted of genes required for extending the existing carotenoid metabolic pathway (psy, crtI, and lcy), along with selectable marker gene (phospho-mannose isomerase or hygromycin phosphotransferase). Apart from this, the gene gun method was also used for the development of transgenic soybean, cotton plants, and beans by promoting multiple shoot induction from the embryonic axes of the mature seeds (Homrich et al. 2012). Further, the biolistic method is also being readily used in breeding tropical and subtropical fruit trees. One such breeding had been done for transforming banana cultivars, where several genes were transferred. One of the major advantages of using the biolistic method is that it aided in overcoming the host-range limitation of AMT. This is the reason that has led to the genetic transformation of a large number of plants, like corn, cotton, soybean, and wheat (Nicholas et al. 2017). The other advantages of the technique include the ability to transform a large number of cells and tissues, delivery of multiple plasmids for achieving high frequencies of co-transformation, delivery of large DNA fragments, and delivery of mRNA or protein (Nicholas et al. 2017). Additionally, in the biolistic technique, delivery of the desired genes could be done without the availability of a vector. Despite such advantages, the technique also suffers from certain limitations, like messy integration patterns, high input cost, low throughput, and inefficiency in controlling the cellular target.

10.5.2 RNA Interference

Another scientific breakthrough that helped in improving the crops genetically is RNA interference (RNAi). This technique was developed in 1998 and since then has become an approach of choice for plant scientists, as both desirable and undesirable genes could be manipulated for improving the novel traits of the plants. One of the significant features of RNAi is that it could be used for predicting the effects of off-target silencing making this technique more specific as well as effective. The degradation of RNA is triggered by introducing double-stranded RNA through transgenes. This dsRNA is thereafter cleaved by an enzyme called dicer that leads to the formation of duplexes of 21 nucleotides (nt) with symmetric 2 nt 3′ overhangs. These duplexes are termed small interfering RNAs (siRNAs) and are responsible for the degradation process as well as suppression or alteration of the gene expression. This technique had been first utilized for developing plants resistant to viruses as the engineered antiviral strategies are found to mimic the natural RNA silencing process. This was first revealed in the case of potato virus Y-resistant plants that expressed RNA transcripts of a viral proteinase gene (Mansoor et al. 2006). Thereafter, immunity in plants against viruses, such as *Tomato spotted wilt virus, Rice tungro bacilliform virus*, *Cucumber mosaic virus, Tobacco mosaic virus*, *Bean golden mosaic virus*, and others, were observed (ISAAA, Pocket K No. 34). The significant usage of this improved technology has been demonstrated by enhancing the nutritional value, quality, and resistance towards pests and diseases in different crop plants. For example, *Lycopersicon esculentum* (tomato) with enhanced carotenoid and flavonoid was developed using the RNAi technique. In this regard, Davuluri et al. (2005) combined the RNAi technique with a fruit-specific promoter for suppressing *DET1*, an endogenous photomorphogenesis regulatory gene in tomatoes. Further, the RNAi approach was used for enhancing the amylose content by >70% in wheat by suppressing expression of SBEIIa and SBEIIb simultaneously (Regina et al. 2006). Similarly, the amylose content of sweet potato (*Ipomoea batatas*) was also increased using this approach. Furthermore, Sunilkumar et al. 2006, made a successful attempt to develop gossypol-free cottonseed by disrupting gossypol biosynthesis in cottonseed. This disruption was made possible by the interference in the expression of the δ-cadinene synthase gene at the time of seed development. In this process, a tissue-specific promoter was used. The transgenic cottonseed with 99% less gossypol made the crop suitable for human consumption This success led to the application of the RNAi approach in food sources, like *Lathyrus sativus*, cassava, and fava beans (Tang et al. 2007). Additionally, plants can also be triggered to silence essential genes by altering them to produce dsRNA in response to insect and parasitic nematodes. Using this approach, resistant varieties, like root-knot nematode (Huang et al. 2006), cotton bollworm (Baum et al. 2007), and corn rootworm (Mao et al. 2007), had been developed.

10.5.3 Genome-Editing Technologies

Extensive research is being conducted for improving the genetic makeup of the plants such that better outputs could be achieved for meeting the growing demand for agricultural output. In this regard, various gene-editing techniques have been developed and put into practice for remodelling the future of crops. Some of these include zinc-finger nucleases (ZFNs), engineered endonucleases/meganucleases, TAL effector nucleases (TALENs), and clustered regularly interspaced to short palindromic repeats (CRISPR). Among these techniques, ZFNs stirred up the genome manipulation research area as they aided in targeting the protein reagents. ZFNs are the DNA binding domains that can identify three base pairs at the target site (Ahmar et al. 2020). The next site-driven mutagenesis genome-editing system is TALENs, which resembles the technique of ZFNs. However, the major difference between the two techniques is that TALENs can target one site (Table 10.2). Further, research in the field of genome editing led to the development of CRISPR technology, such as CRISPR/Cas9 and CRISPR/Cpf1 (Nadakuduti and Enciso-Rodríguez 2021). Although the technique was first used in the prokaryotes, its application in the eukaryotes in the latter stages had revolutionized crop genome editing by facilitating specific changes in the crops. The simplicity of the tool and the minimal requirement of an RNA guide concerning the target DNA make it very efficient in terms of usage and genome modification (Ahmar et al. 2020).

10.5.3.1 Zinc-Finger Nucleases (ZFNs)

Zinc-finger nucleases (ZFNs) were first developed by using chemically engineered nucleases and since then it has been one of the most potential genome-editing tools that target the double-strand breaks (Durai et al. 2005). The application of this technique led to the discovery of the functional Cys2-His2 zinc-finger domain (Gaj et al. 2013). Structurally, ZFNs are composed of two domains. Firstly, the DNA-binding domain consists of 300–600 zinc-finger repeats. These repeats can monitor and read between 9 and 18 bp (Carlson et al. 2012). Secondly, a DNA cleavage domain (non-specific) of the type II restriction endonuclease enzyme Fok1 (Carroll et al. 2006). Further, ZFNs consist of two monomers that flank reversely between 5 and 6 bp of the target DNA and are attributed to their respective target sequence (Carroll et al. 2006). During the cleavage process, the flanking sequences of Fok1 domains slice the DNA. The zinc-finger domain monitors sequences of 24–30 bp that have either specific or rare targeting sites within the genome (Gaj et al. 2012).

Table 10.2 Comparison between ZFN, TALEN, and CRISPR/Cas9 technology

	ZFN	TALEN	CRISPR/Cas9
Recognition site	Zinc-finger protein	RVD tandem repeat region of TALE protein	Single-strand guide RNA
Modification pattern	Fok1 nuclease	Fok1 nuclease	Cas9 nuclease
Target sequence size	Typically, 9–18 bp per ZFN monomer, 18–36 bp per ZFN pair	Typically, 14–20 bp per TALEN monomer, 28–40 bp per TALEN pair	Typically, 20 bp guide sequence + PAM sequence
Specificity	Tolerating a small number of positional mismatches	Tolerating a small number of positional mismatches	Tolerating positional/multiple consecutive mismatches
Mode of action	Double-strand breaks in the target DNA	Double-strand breaks in the target DNA	Double-strand breaks or single-strand nicks in the target DNA
Target recognition Efficiency	High	High	High
Length of the target Sequence (bp)	24–36	24–59	20–22
Targeting limitations	Difficult to target non-G-rich sites	5′ targeted base must be a T for each TALEN monomer	Targeted site must precede a PAM sequence
Mutation rate	High	Middle	Low
Off-target effects	Low off-target effect	Least off-target Activities	Shows least off-target activities
Difficulties of engineering	Requiring substantial protein engineering	Requiring complex molecular cloning methods	Using standard cloning procedures and oligo synthesis
Difficulties of delivering	Relatively easy as the small size of ZFN expression elements is suitable for a variety of viral vectors	Difficult due to the large size of functional components	Moderate as the commonly used SpCas9 is large and may cause packaging problems for viral vectors such as AAV, but smaller orthologs exist
Cost of Development	High	Higher	Low

[a]The table was modified from Li et al. 2020, and Ahmar et al. 2020

This gene-editing technique has been successfully used to alter plants like maize, *Arabidopsis*, *Glycine max*, *Nicotiana*, petunia, rice, apple, rapeseed, and fig (Martínez-Fortún et al. 2017; Ahmar et al. 2020). In one such application, disruption of maize gene *ZmIPK1* was performed by inserting the PAT gene cascade. This resulted in the development of maize seeds that were tolerant to herbicides and possessed an altered inositol phosphate profile (Shukla et al. 2009). Furthermore,

Cantos et al. (2014) identified safe regions for gene integration in rice such that they could serve as reliable loci for gene integration as well as trait stacking. However, the design of ZFNs is highly complicated and challenging. Moreover, the efficacy associated with ZFNs is low (Zhang et al. 2018a, b).

10.5.3.2 Transcriptional Activator-like Effector Nucleases (TALENs)

The amalgamation of the FokI cleavage domain with the transcription activator-like effectors (TALE) protein's DNA-binding domain led to the development of TALEN. In the TALEN system, the involved proteins have the central domain that is responsible for DNA binding and nuclear localization sequence (Schornack et al. 2006). The significance of the proteins in binding to the DNA was first observed in 2007. This protein consists of a repeated sequence of 34 amino acids, where each repeat perceives three nucleotides in the target DNA (Römer et al. 2007). Like ZFNs, this gene-editing technique also targets DSBs for initiating pathways responsible for DNA damage and modification (Gaj et al. 2013). It has been used in plants like *Arabidopsis*, *Brachypodium*, barley, flax, maize, *Nicotiana*, potato, tomato, sugarcane, rice, rapeseed, soybean, and wheat (Martínez-Fortún et al. 2017; Jansing et al. 2019). Amid all these plants, the first application of TALEN was done for improving rice by disrupting OsSWEET14, a bacterial blight susceptibility gene (Li et al. 2012). Similarly, three TaMLO homeologs were knocked out for creating powdery mildew-resistant wheat (Wang et al. 2014). Other achievements include the alteration of nutritional profiles in crops using the TALEN genome-editing system. Examples include the generation of soybean by disrupting fatty acid desaturase (FAD) genes such that oleic acid content could be alleviated and linoleic acid could be reduced. This, in turn, helped in improving the shelf life as well as heat stability of the soybean oil (Haun et al. 2014; Demorest et al. 2016). Furthermore, the vacuolar invertase (*VInv*) gene was knocked out for developing potato tubers with low or negligible levels of reducing sugar such that their quality does not get influenced during cold storage (Clasen et al. 2016).

10.5.3.3 CRISPR/Cas Technology

The discovery of CRISPR/Cas9 technology is considered to be one of the most prominent breakthroughs of the twentieth century as it is highly efficient as well as a simple tool that could lead to gene modification both in animals and plants (Barrangou and Doudna 2016). The tool relies on the signal of RNA-guided nucleases and has gained stardom due to its versatility, adequacy, potency, and simplicity (Gasiunas et al. 2012). The complex formed by Cas9 protein and guide RNA is responsible for locating and cleaving target DNA. DNA cleavage at the target site is usually 3 bp upstream of the protospacer adjacent motif (PAM) site. This process of breakage of double-stranded DNA leads to the activation of the DNA repair mechanism, non-homologous end joining (NHEJ), and homology-directed repair (HDR) (Symington and Gautier 2011). However, NHEJ mediates the relegation of the broken DNA directly in the absence of a homologous template. This, in turn, leads to insertions and deletions (InDels), or substitutions at the site of breakage. On the contrary, HDR is capable of adding new alleles, inserting new sequences of insert, and correcting the existing ones in the presence of the donor DNA (Zha et al. 2009). As the integration of the DNA within the plant genome occurs at a low frequency, the expression of CRISPR/Cas9 through transgenesis offers a better scope (Hilscher et al. 2016). After the transformation, two methods of selection are considered, viz. antibiotic and herbicide resistance, which aids in the regeneration of the explants expressing the CRISPR/Cas9 system functionally. This technique is being frequently used for gene knockouts and production of null alleles either by the insertion of indels, such that frameshift mutation could be achieved, or by introducing premature stop codons. With the application of this system, genome modification of plants, like cotton, rice, maize, wheat, soybean, grapefruit, potato, tomato, lettuce, oranges, and watermelon, have been done successfully (Zhang et al. 2016; Ricroch et al. 2017). Some of the prominent plant breeding activities conducted by the researchers using CRISPR technology include *LAZY1* gene knockout in rice for generating tiller-spreading phenotype for increasing the crop yield (Miao et al. 2013), mutation of *Gn1a, DEP1*, and *GS3* genes of the rice cultivar Zhonghua11 was achieved for enhancing the grain size and number along with dense erect panicles (Li et al. 2016). Another attempt in rice was made by Sun et al. (2017) for improving the fine structure of the rice grain along with nutritional properties. In this context, the SBEIIb gene was mutated for achieving a long chain of amylopectin. Other than this, with the aid of CRISPR/Cas9, the GW2 gene was disrupted, which led to the increase in the grain weight and protein content of the wheat (Zhang et al. 2018a, b). In addition to this, Jiang et al. 2017, used this gene-editing technology to improve oleic acid content in *Camelina sativa* by targeting FAD2. This also helped in lessening polyunsaturated fatty acids in the oilseed plant. Further, maize waxy gene Wx1 was knocked out for eliminating the expression of the granule-bound starch synthase (GBSS) gene, leading to the development of amylopectin rather than amylose. This led to improved digestibility of the maize along with made the maize species potential for bio-industrial application (Pioneer Dupont 2016). Additionally, wheat resistant to powdery mildew (Zhang et al. 2017), rice resistant to bacterial blight (Wang et al. 2016; Zhou et al. 2015), and tomatoes resistant to powdery mildew (Ortigosa et al. 2019) were developed.

10.5.3.4 New Tools for Genome Editing

Recently, there have been many new additions to the classic CRISPR toolbox, which have offered a multitude of applications in genome editing and beyond. For instance, the adoption of Cas9 variants, like SpCas9-VQR (PAM: NGA), SpCas9-EQR (PAM: NGAG), Cas9 NG (PAM: NG), and xCas9 3.7(PAM: NG/GAA/GAT), helped to overcome the limitation displayed by CRISPR/Cas9 system, i.e. the "NGG" sequence as PAM requirement, which reduces the target recognition sites.

These variants have been successfully used in plant-like *Physcomitrella*, *Arabidopsis*, rice, tomato, and potato (Zhang et al. 2019). The use of nucleases like Cas12/Cpf1 (class 2, type V CRISPR systems) has led to flexibility in base editing and epigenetic modulation. Cpf1 has several advantages over Cas9, including its ability to target T-rich motifs (PAM: TTTV, where V = A, C, or G), the lack of a necessity for trans-activating crRNA, its ability to generate a staggering double-strand break (4–5 nt 5′ overhangs), and the capability for both RNA processing and DNA nuclease activity (Safari et al. 2019). Another recent addition is the CRISPR-CasΦ protein (type V, CRISPR system), which displays the advantage of being smaller in size (~70 kilodalton) and requiring minimal PAM sequence of 5′-TBN-3 (where B = G, T, or C) for execution. However, CRISPR-CasΦ protein showed a low frequency of editing (0.85%) when it was used to edit the phytoene desaturase (*PDS*) gene in *Arabidopsis* protoplasts (Pausch et al. 2020).

Further, the recent development of base editors has allowed all combinations of precise base conversions without requiring DSB of DNA. Base editors use a catalytically hindered dead Cas9, dCas9 (D10A and H840A), or, usually a nickase, nCas9 (D10A), to precisely convert one target DNA nucleotide to another (Nadakuduti and Enciso-Rodríguez 2021). Individual nicks generated by base editors are repaired by a more precise base excision repair pathway, which minimizes the error-prone gene editing mediated by DSBs and NHEJ (Dianov and Hübscher 2013). The classic base editors are the cytosine base editors that catalyse C-to-T using a cytosine deaminase bound to nCas9 (Komor et al. 2016; Nishida et al. 2016) and adenine base editors that catalyse A-to-G conversions using an evolved DNA processing deoxyadenosine deaminase tethered to nCas9 (Gaudelli et al. 2017). Both cytosine base editors and adenine base editors have been optimized and utilized for genome editing in various plant species (Shimatani et al. 2017; Shan and Voytas 2018). The base editing in both cases is limited to transition mutation. However, recently, glycosylase base editors were developed that could mediate mutations, such as C-to-A and C-to-G, making transversion mutation feasible for base editors (Nadakuduti and Enciso-Rodríguez 2021).

Another breakthrough technology that allowed targeted insertions/deletions or a precise transition/transversion mutation at targeted genomic loci is prime editing (PE). The protein component of a prime editor compromises an altered form of the Cas9 enzyme (cuts DNA) and reverse transcriptase (produces complementary DNA from an RNA template). The RNA component here is the prime editing guide RNA (pegRNA) that recognises the targeted DNA site and has information for the desired edit. The section of the pegRNA that encodes the altered DNA sequence is directly copied into the target site by the reverse transcriptase, resulting in a new flap of DNA that carries the edit. The cell replaces the original DNA sequence on both strands of the DNA double helix when it integrates this altered flap (Nadakuduti and Enciso-Rodríguez 2021). The application of the prime editing technique is still in its infant stage. However, it has been greatly exploited in *Oryza sativa* (Zafar et al. 2020) and *Solanum Lycopersicum* (Lu et al. 2021).

10.6 Commercial GM Crops

The increasing awareness about the significance of transgenic crops along with the improvement in the techniques in terms of accuracy and precision has led many transgenic crops to be accepted at the commercial level. Adoption of biotech crop have been increasing in the past 25 years with the USA at 95% (average for soybeans, maize, and canola adoption), Brazil (94%), Argentina (~100%), Canada (90%), and India (94%) (ISAAA 2019a). In the year 2019, 190.4 million hectares of land have been used to plant the GE crops by 1.7 million farmers in 29 countries, compared to 2015 when only 28 countries grew GE crops on nearly 179.7 million hectares (ISAAA 2019b). In the past 22 years, the growth of transgenic plants has increased by several folds. As per the available reports previously in 1996, approximately 1.7 million hectares of land were used for the development of transgenic crops; contrarily in 2018, approximately 191.7 million hectares had been used for growing transgenic crops (ISAAA 2018). Concerning the crops, it was found that 95.9 million hectares were used to grow transgenic soybean which accounts for 50%, 58.9 million hectares were used to grow transgenic maize (31%), transgenic cotton occupied 24.9 million hectares (13%), transgenic canola occupied 5.3% that is 10.1 million hectares, and other transgenic crops were grown in remaining 1.9 million hectares (ISAAA 2018). And in the context of transgenic events, a total of 525 in 32 different crops have been commercialized (ISAAA 2019b). Out of this these, maize is the most exploited crop accounting for the maximum number of events (238 events), followed by cotton (61 events), potato (49 events), Argentine canola (42 events), soybean (41 events), carnation (19 events), and others (Kumar et al. 2020). The details of the most commercially exploited crop (corn) along with their applicants, traits, and date of effectiveness are listed in Table 10.3. The process of commercialization initiated with transgenic tomato called Flavr Savr was launched in the United States (Calgene company) in 1994. The advantage of this transgenic product was that it aided in slowing down the post-harvest ripening of tomatoes (Bruening and Lyons 2000). However, papaya plants resistant to papaya ringspot virus (PRSV) have been declared to be the first successful application of GE technology in the fruit crop at the commercial scale (Gonsalves 1998). The production of papaya was affected adversely in Hawaii due to the actions of the papaya ringspot virus (PRSV), which was detected in the Puna district of Hawaii in 1992 (Gonsalves 1998; Fuchs and Gonsalves 2007). This, in turn, emphasised the development of GE papaya varieties that are resistant to the virus. As a result of which, the first PRSV-resistant papaya plants were obtained by bombarding the PRSV code protein gene (Fitch et al. 1992). These transgenic papaya plants have been commercialised so that they could reach the end users in Hawaii (Tripathi et al. 2008). In addition to this, "Sunset" papaya was transformed with a gene that had been derived from a Hawaiian strain for producing the transgenic papaya "SunUp". This transgenic variety was completely resistant to PRSV. Followed by this, "SunUp" was crossed with "Kapoho," which was a non-engineered cultivar for obtaining yellow fresh papaya called "rainbow". This transgenic line was also

Table 10.3 List of the GM varieties of corn

Applicant	Phenotype
Dow	2,4-D and ACCase-inhibitor tolerant
Bayer/Genective	Herbicide tolerant
Monsanto	Male sterile
Pioneer	Insect resistant and glufosinate tolerant
Stine seed	Herbicide tolerant
Syngenta	Rootworm resistant
Monsanto	Drought tolerant
Pioneer	Male sterile, fertility restored, visual marker
Syngenta	Thermostable alpha-amylase
Syngenta	Moth and butterfly resistant
Pioneer	Herbicide and imidazolinone tolerant
Monsanto	European corn borer resistant
Syngenta	Corn rootworm protected
Monsanto	High lysine
Monsanto	Corn rootworm resistant
Dow	Corn rootworm resistant
Dow	Moth and butterfly resistant and phosphinothricin tolerant
Monsanto	Corn rootworm resistant
Mycogen c/o Dow and Pioneer	Moth and butterfly resistant Phosphinothricin tolerant
Monsanto	Herbicide tolerant
AgrEvo	Phosphinothricin tolerant and male sterile
Pioneer	Male sterile and phosphinothricin tolerant
AgrEvo	Phosphinothricin tolerant and moth and butterfly resistant
Monsanto	Herbicide tolerant
Monsanto	Herbicide tolerant and European corn borer resistant
DeKalb	European corn borer resistant
Northrup king	European corn borer resistant
Monsanto	European corn borer resistant
Plant genetic systems	Male sterileMS3
DeKalb	Glufosinate tolerant
Monsanto	Moth and butterfly resistant
AgrEvo	Glufosinate tolerant
Ciba seeds	Moth and butterfly resistant

[a] Table is adopted from Johnson and O'Connor 2015

resistant to PRSV (Gonsalves et al. 2004). Other transgenic varieties of papaya include Huanong No. 1 papaya, which is resistant to the four predominant PRSV strains, namely, Ys, Vb, Sm, and Lc24, which were found in South China. Moreover, this GM variety of papaya produced bigger and thicker flesh in the fruit (Lobato-Gómez et al. 2021). Besides papaya, in 1996, the first corn variety that was resistant to glyphosate herbicide was commercialised by Monsanto with the name "Roundup Ready Corn" as the corn variety could tolerate the use of roundup (Gutterson 2020). Followed by this, Liberty Link corn resistance to glufosinate was also developed by Bayer CropScience (CASE M.8084). The popularity of GM corn increased gradually; as a result of which, in 2011, 14 countries were involved in the production of herbicide-resistant GM crops, and by the next year, the European Union had authorised 26 varieties of herbicide-resistant GM maize. In 2013, Monsanto also launched the first transgenic drought-tolerance corn hybrids under the trade name of DroughtGard (DiLeo 2012). In the United States, GM corn is being readily used for the production of different ingredients that are used in processing strings in fruits, such as high fructose corn syrup and cornstarch. Apart from this, the major parts of GM corn are used to feed livestock, and some of them are also converted into biofuel. In addition to this, specific maize strains have been GM or engineered for expressing agriculturally desirable traits such as resistance to pests and herbicides. Additionally, GM sweet corn varieties have also been developed by Syngenta and Monsanto and have been commercialised under the trade name "Attribute" (Shelton et al. 2013) and Performance Series™ (Dively et al. 2021), respectively, which are both insect resistant. Other than this, one of the most popular maize varieties resistant to insects is Bt corn. This had been GM for expressing more than one protein-like delta-endotoxin from the bacterium *Bacillus thuringiensis*. The development of Bt corn relied significantly on the success rate of Bt cotton, which was first commercialized in 1996. At that time, Bt cotton cultivation was approved in countries, like the United States, Mexico, and others (Rocha-Munive et al. 2018), and the Bt cotton that produced *Cry1Ac* toxin had high activity on tobacco budworm and pink bollworm (Layton et al. 1997). However, in India, Bt cotton was approved only in 2002 after a thorough study conducted by ICAR. Thereafter, Bt brinjal resistant to brinjal shoot fly received the approval for commercialisation in 2009. Apart from the above-mentioned plants, GM legumes are also being commercialised. Monsanto had developed a GM soybean called Roundup Ready®, which was made commercially available, as it provided tolerance to the herbicide glyphosate (Roundup™). Further, it also could reduce the population of weeds. Gradually, other transgenic soybeans like Liberty Link® soybean (Meyer and Norsworthy 2020), soybeans containing the *Arabpdiosis*csr1–2 gene (Gabard et al. 1989), soybeans resistant to dicamba (Soltani et al. 2020), and many more were developed. Additionally, Bt soybean was also developed for protecting it against lepidopteron species. In 1994, soybean containing synthetic Bt (*Cry1Ac*) was developed (Martins-Salles et al. 2017). The other leguminous plant that has been commercialised is *Phaseolus vulgaris*, which showed resistance against the bean golden mosaic virus (Kumar et al. 2020). Moreover, the GM cowpea, Pod Borer-Resistant Cowpea (PBR Cowpea—event AAT709A) is resistant to *Maruca vitrata* has also been commercialized in Nigeria (ISAAA, GM approval database). The resistance was conferred by the presence of Cry1Ab protein. Another commercially grown fruit GM plant is the apple. Okanagan Specialty Fruits developed Arctic® Apple events GD743 that had limited quinone biosynthesis. This was achieved by targeting four PPO genes of apple using RNAi technology (ISAAA, GM approval database). An intensive amount of research is being carried out for developing transgenic plants with enhanced traits as it holds a great promise for the future of a nation by letting the country overcome the issue of food insecurities. However, it is

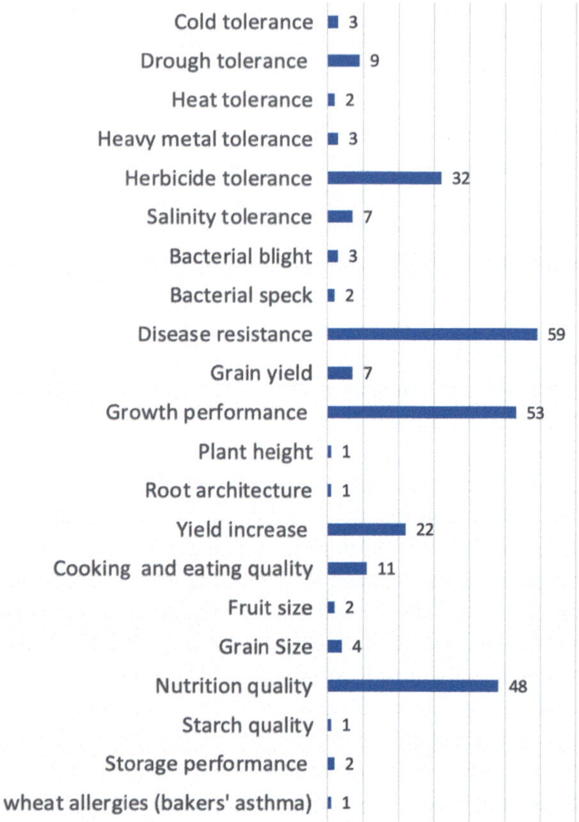

Fig. 10.1 Summary of 21traits (biotic, abiotic stresses, yield and nutritional quality). About 70% of the 273 were focused on enhancing disease resistance, growth performance, and nutrition quality and herbicide tolerance

unfortunate that some of those studies are not being translated into commercialization. Therefore, it is of utmost importance that based on the safety and efficiencies of the improved traits, genetically engineered crops must be commercialised at a larger scale (Fig. 10.1).

10.7 Benefits of GM Crop Cultivation

The gaining popularity of the GM crops itself signifies its potentiality to revolutionise the agricultural sector globally. The development of sophisticated technology like CRISPR and TALEN has helped in improving the agronomic traits of the economically important crops such that they provide benefits beyond being just edible products.

At the farm level, GM crops, like Bt cotton, Bt maize, GM tomatoes, apples, and others, have played a crucial role in reducing the negative impact of herbicides, pesticides, and fertilizers. This, in turn, had not only minimised the million-dollar losses that the farmers had to face but also helped in improving the water quality near the agricultural lands. In addition to this, the development of pest-resistant varieties has also improved the crop yield exponentially (Zilberman et al. 2018). In terms of figures, it has been found that the developing countries had received an extra income of $4.42 in 2018 due to their investment in GM crop seeds. On the other hand, in developed countries, the rise in income was found to be $3.24. The financial upsurge was not only observed at the industrial scale but also at the farm level. Expenses faced by the farmers owing to the utilisation of pesticides and herbicides have declined significantly due to the sowing of GM seeds for eggplant. In 2016, the farmers cultivating Bt eggplant in the 35 districts of Bangladesh were found to achieve direct income gains, as they had to spend 61% less on pesticides in comparison to those who used the conventional varieties (Shelton et al. 2018). Similarly, the sowing of Arctic® apples had improved the benefits for the retailers, because this variety of apples was found to be more suitable for mechanical harvesting. Moreover, they suffered less impact due to bin rubs, finger bruising, and other damages superficially. Thus, fewer amounts of fruit were wasted, and a significant decline was observed in pack-outs. Besides, Arctic® Golden variety also reduced the cost of production as it did not require warm packing. Hence, an ample amount of financial stress faced by the stakeholders of agriculture has been curbed by the cultivation of GM crops (Lobato-Gómez et al. 2021).

Further, the role of transgenic crops in achieving sustainable development has also been recognised. The usage of transgenic crops for biofuel production has helped in switching to the generation of greener energy and decreased the release of greenhouse gases that are responsible for the ongoing climate change. Moreover, as the transgenic crops enable wider utilisation of the conservation tillage systems; henceforth, it is further likely to reduce GHG emissions (Raymond Park et al. 2011). As per the available data, if GM crops were not developed and grown by 2018, then an additional 23 kilograms of carbon dioxide would have been emitted to the atmosphere. This number is equal to the addition of 15.3 million cars to the roads (Brookes 2020). Last but not the least, GE crops are playing a crucial role in helping countries to overcome the issue of food insecurities by minimising the yield loss due to biotic and abiotic factors, thus letting the nations meet the food needs of the people at the local and global level. Over the last 23 years, the utilisation of crop biotechnology had helped in producing 278 million soybeans, 498 million tonnes of corn, 14 million tonnes of canola, and 32.6 million tonnes of cotton lit additionally at the global scale (PG economics, 2014). In addition to this, biofortification has opened the doors to tackle the issue of malnutrition by incorporating essential nutrients within the plants. Thus, it has provided the scope of the farmers to improve the safety and quality of the food being produced in the farms. This, in turn, had further helped in improving both the economic as well as social situations of the farmers (Azadi et al. 2016). Tables 10.4 and 10.5 adopted from ISAAA, Pocket K No. 5, further illustrate the global impact of GM crops. In addition to this, it is also expected that the issue of food security that has been brought in due to the outbreak of the covid-19 pandemic could be improved with the adoption of GM crops (Petrova and

Table 10.4 Global farm income benefits from growing GE crops, 1996–2016 (US$ million)

GM trait	2016 increase in farm income	1996–2016 increase in farm income
HT soybean	4373.3	54,524.4
HT + IR soybean	2490.9	5211.5
HT maize	2104.9	13,108.1
HT cotton	130.1	1916.9
HT canola	509.9	5970.9
IR maize	4809.1	50,565.5
IR cotton	3695.2	53,986.9
Others	81.5	817.9
Totals	**18,194.9**	**186,102.1**

Note: *HT* herbicide tolerant, *IR* insect resistant, *others* virus-resistant papaya and squash and herbicide-tolerant sugar beet

Table 10.5 Impact of changes in the use of herbicides and insecticides in GE crops globally, 1996–2016

GM trait	Change in volume of AI used (million kg)	Change in field EIQ impact (million field EIQ/ha units)	% change in AI use on GE crops	% change in environmental impact associated with herbicide and insecticide use on GE crops
HT soybean	+13.0	-8526	+0.4	−13.4
HT + IR soybean	−7.4	−678	−6.1	−6.3
HT maize	−239.3	−7859	−8.1	−12.5
HT canola	−27.3	−931	−18.2	−29.7
HT cotton	−29.1	−706	−8.2	−10.7
IR maize	−92.1	−4142	−56.1	−58.6
IR cotton	−288.0	−12,762	−29.9	−32.3
HT sugar beet	+1.0	−43	+9.9	−19.4
Totals	**−671.2**	**−35,647**	**−8.2**	**−18.4**

Note: *HT* herbicide tolerant, *IR* insect resistant, *Ai* active ingredient, *EIQ* environmental impact quotient. (Environmental impact quotient (EIQ), a universal indicator where the various environmental impacts of individual pesticides are integrated into a single field value per hectare. This EIQ value is multiplied by the amount of pesticide active ingredient (ai) used per hectare to produce a field EIQ value)

[a]Tables 10.4 and 10.5 *are adopted from isaaa.org, Pocket K No. 5, 2020*

AbouRaya 2020). Moreover, it would also help in overcoming the issue of malnutrition that is being faced by women and children at the global level, especially in underdeveloped and developing countries (Qaim and Kouser 2013).

10.8 Conclusion

We discussed the multiple aspects of GM technologies and their advancements for the development of improved crop varieties. The utilisation of GM tools for quality improvement and yield enhancement in crops is essential to attain food security soon. The modern genetic/genome engineering techniques have revolutionised the trait enhancements in crops as they possess the ability to make precise and quick targeted modifications in genes of interest compared to conventional breeding techniques. Recent technologies like CRISPR/Cas have proved to be a fundamental breakthrough in the field of genome editing and are successfully explored for yield enhancement, quality improvement, and disease resistance. Furthermore, estimations imply that the adoption of genetic engineering technology has aided in the decline of the use of agrochemicals (pesticide and insecticide) as well as a reduction in environmental footprint and an increase in farmer revenue. Moreover, these new technologies have the potential to be adopted as a viable approach to achieve zero hunger and nutritional imbalance for the growing human population.

References

Abbas MA, Haroun SA, Mowafy AM (2021) Dual inoculation of Bradyrhizobium and Enterobacter alleviates the adverse effect of salinity on Glycine max seedling. Not Bot Horti Agrobot Cluj-Napoca 49(3):12461–12461

Ahmar S, Saeed S, Khan MHU, Ullah Khan S, Mora-Poblete F, Kamran M, Jung KH (2020) A revolution toward gene-editing technology and its application to crop improvement. Int J Mol Sci 21(16):5665

Al Atalah B, Smagghe G, Van Damme EJ (2014) Orysata, a jacalin-related lectin from rice, could protect plants against biting-chewing and piercing-sucking insects. Plant Sci 221:21–28

Azadi H, Samiee A, Mahmoudi H, Jouzi Z, Rafiaani Khachak P, De Maeyer P, Witlox F (2016) Genetically modified crops and small-scale farmers: Main opportunities and challenges. Crit Rev Biotechnol 36:434–446

Barrangou R, Doudna JA (2016) Applications of CRISPR technologies in research and beyond. Nat Biotechnol 34(9):933–941

Barry GF, Kishore GM, Padgette SR, Stallings WC (1997) Glyphosate-Tolerant 5-Enolpyruvylshikimate-3-Phosphate Synthases. U.S. Patent No 5,633,435. U.S. Patent and Trademark Office, Washington, DC

Baum JA, Bogaert T, Clinton W, Heck GR, Feldmann P, Ilagan O, Roberts J (2007) Control of coleopteran insect pests through RNA interference. Nat Biotechnol 25(11):1322–1326

Bhatia S, Bera T, Dahiya R, Bera T, Bhatia S, Bera T (2015) Classical and nonclassical techniques for secondary metabolite production in plant cell culture. In: Modern applications of plant biotechnology in pharmaceutical sciences, pp 231–291

Boston RS, Viitanen PV, Vierling E (1996) Molecular chaperones and protein folding in plants. Plant Mol Biol 32(1–2):191–222

Botterman J, Leemans J (1988) Engineering of herbicide resistance in plants. Biotechnol Genet Eng Rev 6(1):321–340

Broadway RM, Duffey SS (1986) The effect of dietary protein on the growth and digestive physiology of larval Heliothis zea and Spodoptera exigua. J Insect Physiol 32(8):673–680

Bruening G, Lyons J (2000) The case of the FLAVR SAVR tomato. Calif Agr 54(4):6–7

Cantos C, Francisco P, Trijatmiko KR, Slamet-Loedin I, Chadha-Mohanty PK (2014) Identification of "safe harbor" loci in indica rice genome by harnessing the property of zinc-finger nucleases to induce DNA damage and repair. Front Plant Sci 5:1–8

Carlson DF, Fahrenkrug SC, Hackett PB (2012) Targeting DNA with fingers and TALENs. Mol Ther–Nucleic Acids 1:e3. https://doi.org/10.1038/mtna.2011.5

Carroll D, Morton JJ, Beumer KJ, Segal DJ (2006) Design, construction and in vitro testing of zinc finger nucleases. Nat Protoc 1(3):1329–1341

Castiglioni P, Warner D, Bensen RJ, Anstrom DC, Harrison J, Stoecker M, Heard JE (2008) Bacterial RNA chaperones confer abiotic stress tolerance in plants and improved grain yield in maize under water-limited conditions. Plant Physiol 147(2):446–455

Chaitanya KV, Sundar D, Masilamani S, Ramachandra Reddy A (2002) Variation in heat stress-induced antioxidant enzyme activities among three mulberry cultivars. Plant Growth Regul 36(2):175–180

Chakroun M, Banyuls N, Bel Y, Escriche B, Ferré J (2016) Bacterial vegetative insecticidal proteins (Vip) from entomopathogenic bacteria. Microbiol Mol Biol Rev 80(2):329–350

Chen H, Xiong L (2009) Enhancement of vitamin B6 levels in seeds through metabolic engineering. Plant Biotechnol J 7(7):673–681

Chen TH, Murata N (2002) Enhancement of tolerance of abiotic stress by metabolic engineering of betaines and other compatible solutes. Curr Opin Plant Biol 5(3):250–257

Cheung AY, Bogorad L, Van Montagu M, Schell J (1988) Relocating a gene for herbicide tolerance: a chloroplast gene is converted into a nuclear gene. Proc Natl Acad Sci U S A 85(2):391–395

Chilton MD, Drummond MH, Merio DJ, Sciaky D, Montoya AL, Gordon MP, Nester EW (1977) Stable incorporation of plasmid DNA into higher plant cells: the molecular bases of crown gall tumorigenesis. Cell 11:263–271

Chrispeels MJ, Sadava DE (2003) Plants, genes, and crop biotechnology. Jones and Bartlett Publisher. SN–9780763715861

Clasen BM, Stoddard TJ, Luo S, Demorest ZL, Li J, Cedrone F, Tibebu R, Davison S, Ray EE, Daulhac A, Coffman A, Yabandith A, Retterath A, Haun W, Baltes NJ, Mathis L, Voytas DF, Zhang F (2016) Improving cold storage and processing traits in potato through targeted gene knockout. Plant Biotechnol J 14(1):169–176

Datta K, Baisakh N, Oliva N, Torrizo L, Abrigo E, Tan J, Rai M, Rehana S, Al-Babili S, Beyer P, Potrykus I, Datta SK (2003) Bioengineered 'golden' indica rice cultivars with β-carotene metabolism in the endosperm with hygromycin and mannose selection systems. Plant Biotechnol J 1:81–90

Davuluri GR, Van Tuinen A, Fraser PD, Manfredonia A, Newman R, Burgess D, Bowler C (2005) Fruit-specific RNAi-mediated suppression of DET1 enhances carotenoid and flavonoid content in tomatoes. Nat Biotechnol 23(7):890–895

Demorest ZL, Coffman A, Baltes NJ, Stoddard TJ, Clasen BM, Luo S et al (2016) Direct stacking of sequence-specific nuclease-induced mutations to produce high oleic and low linolenic soybean oil. BMC Plant Biol 16:1–8

Deng LH, Weng LS, Xiao GY (2014) Optimization of Epsps gene and Development of double herbicide tolerant transgenic PGMS Rice. J Agric Sci Technol 16:217–228

Deutsch CA, Tewksbury JJ, Huey RB, Sheldon KS, Ghalambor CK, Haak DC, Martin PR (2008) Impacts of climate warming on terrestrial ectotherms across latitude. Proc Natl Acad Sci 105(18):6668–6672

Dianov GL, Hübscher U (2013) Mammalian base excision repair: the forgotten archangel. Nucleic Acids Res 41(6):3483–3490

Dively GP, Kuhar TP, Taylor S, Doughty HB, Holmstrom K, Gilrein D et al (2021) Sweet corn sentinel monitoring for lepidopteran field-evolved resistance to Bt toxins. J Econ Entomol 114: 307–319

Duan X, Li X, Xue Q, Abo-El-Saad M, Xu D, Wu R (1996) Transgenic rice plants harboring an introduced potato proteinase inhibitor II gene are insect resistant. Nat Biotechnol 14(4):494–498

Dubock A (2019) Golden rice: to combat vitamin a deficiency for public health. Vitamin A:1–21

Dunse KM, Stevens JA, Lay FT, Gaspar YM, Heath RL, Anderson MA (2010) Coexpression of potato type I and II proteinase inhibitors gives cotton plants protection against insect damage in the field. Proc Natl Acad Sci 107(34):15011–15015

Durai S, Mani M, Kandavelou K, Wu J, Porteus MH, Chandrasegaran S (2005) Zinc finger nucleases: custom-designed molecular scissors for genome engineering of plant and mammalian cells. Nucleic Acids Res 33:5978–5990

FAO (2020) The state of world fisheries and aquaculture. Sustainability in action. Rome

Faria JC, Albino MMC, Dias BBA, Cançado LJ, da Cunha NB, de Silva ML, Aragão FJL (2006) Partial resistance to bean golden mosaic virus in a transgenic common bean (*Phaseolus vulgaris* L.) line expressing a mutated rep gene. Plant Sci 171(5):565–571

Ferreira SA, Pitz KY, Manshardt R, Zee F, Fitch M, Gonsalves D (2002) Virus coat protein transgenic papaya provides practical control of papaya ring-spot virus in Hawaii. Plant Dis 86:101–105

Fitch MMM, Manshardt RM, Gonsalves D, Slightom JL, Sanford JC (1992) Virus resistant papaya derived from tissues bombarded with the coat protein gene of papaya ringspot virus. Biotechnol 10:1466–1472

Flores-Mireles AL, Eberhard A, Winans SC (2012) *Agrobacterium tumefaciens* can obtain Sulphur from an opine that is synthesized by octopine synthase using S-methylmethionine as a substrate. Mol Microbiol 84(5):845–856

Fraley RT, Rogers SG, Horsch RB, Sanders PR, Flick JS, Adams SP, Bittner ML et al (1983) Expression of bacterial genes in plant cells. Proc Natl Acad Sci U S A 80(15):4803–4807

Fuchs M, Gonsalves D (2007) Safety of virus resistant transgenic plants two decade after their introduction: lessons from realistic field risk assessment studies. Annu Rev Phytopathol 45:173–202

Fujisawa M, Takita E, Harada H, Sakurai N, Suzuki H, Ohyama K, Misawa N (2009) Pathway engineering of Brassica napus seeds using multiple key enzyme genes involved in ketocarotenoid formation. J Exp Bot 60(4):1319–1332

Gabard JM, Charest PJ, Iyer VN, Miki BL (1989) Cross-resistance to short residual sulfonylurea herbicides in transgenic tobacco plants. Plant Physiol 91(2):574–580

Gaj T, Guo J, Kato Y et al (2012) Targeted gene knockout by direct delivery of zinc-finger nuclease proteins. Nat Methods 9:805–807

Gaj T, Liu J, Anderson KE, Sirk SJ, Barbas CF (2013) Protein delivery using Cys2-His2 zinc-finger domains. ACS Chem Biol 9:1662–1667

Gasiunas G, Barrangou R, Horvath P, Siksnys V (2012) Cas9–crRNA ribonucleoprotein complex mediates specific DNA cleavage for adaptive immunity in bacteria. Proc Natl Acad Sci 109(39): E2579–E2586

Gastélum-Estrada A, Serna-Saldívar SO, Jacobo-Velázquez DA (2021) Fighting the COVID-19 pandemic through biofortification: innovative approaches to improve the Immunomodulating capacity of foods. ACS Food Sci Technol 1(4):480–486

Gaudelli NM, Komor AC, Rees HA, Packer MS, Badran AH, Bryson DI, Liu DR (2017) Programmable base editing of A•T to G•C in genomic DNA without DNA cleavage. Nature 551(7681):464–471

Gbashi S, Oluwafemi A, Janet AA, Sarem T, Shandry T, Oluwaseun MA, Bunmi O, Julianah OO, Patrick N (2021) Food safety, food security and genetically modified organisms in Africa: a current perspective. Biotechnol Genet Eng Rev 37(1):30–63

George T, Graham PH, Roger H (2009) Genetically modified plants

Gillespie S, Hodge J, Yosef S, Pandya-Lorch R (2016) Nourishing millions: stories of change in nutrition. Intl Food Policy Res Inst

Gonsalves D (1998) Control of papaya ringspot virus in papaya: a case study. Annu Rev Phytopathol 36(1):415–437

Gonsalves D, Gonsalves C, Ferreira S, Pitz K, Fitch M, Manshardt R, Slightom J (2004) Transgenic virus resistant papaya: from hope to reality for controlling of papaya ringspot virus in Hawaii. APS net:1–12. feature story for July, 2004

Green JM, Castle LA (2010) Transitioning from single to multiple herbicide resistant crops. In: Nandula VK (ed) Glyphosate resistance in crops and weeds: history, development, and management. Wiley, Hoboken, NJ, pp 67–91

Griffiths AJF, Wessler SR, Lewontin RC, Gelbart WM, Suzuki DT, Miller JH (2005) Introduction to genetic analysis, 8th edn. W.H. Freeman, New York

Gullner G, Kömives T, Rennenberg H (2001) Enhanced tolerance of transgenic poplar plants overexpressing gamma-glutamylcysteine synthetase towards chloroacetanilide herbicides. J Exp Bot 52(358):971–979

Guo J, Yang L, Liu X, Guan X, Jiang L, Zhang D (2009) Characterization of the exogenous insert and development of event-specific PCR detection methods for genetically modified Huanong no. 1 papaya. J Agric Food Chem 57:7205–7212

Gutterson N (2020) Commercialization and applications of agricultural biotechnology. In: Biotechnology entrepreneurship. Academic Press, pp 385–398

Haun W, Coffman A, Clasen BM, Demorest ZL, Lowy A, Ray E, Retterath A, Stoddard T, Juillerat A, Cedrone F, Mathis L, Voytas DF, Zhang F (2014) Improved soybean oil quality by targeted mutagenesis of the fatty acid desaturase 2 gene family. Plant Biotechnol J 12(7): 934–940

Herrera-Estrella L, Ann Depicker MVM, Schell J (1983) Expression of Chimaeric genes transferred into plant cells using a Ti-plasmid-derived vector. Nature 303(5914):209–213

Hilder VA, Gatehouse AM, Sheerman SE, Barker RF, Boulter D (1987) A novel mechanism of insect resistance engineered into tobacco. Nature 330(6144):160–163

Hilscher J, Bürstmayr H, Stoger E (2016) Targeted modification of plant genomes for precision crop breeding. Biotechnol J 12(1):1600173

Homrich MS, Wiebke-Strohm B, Weber RLM, Bodanese-Zanettini MH (2012) Soybean genetic transformation: a valuable tool for the functional study of genes and the production of agronomically improved plants. Genet Mol Biol 35(4):998–1010

Huang G, Allen R, Davis EL, Baum TJ, Hussey RS (2006) Engineering broad root-knot resistance in transgenic plants by RNAi silencing of a conserved and essential root-knot nematode parasitism gene. Proc Natl Acad Sci 103(39):14302–14306

Islam SMF, Karim Z (2019) World's demand for food and water: the consequences of climate change. In: Farahani MHDA, Vatanpour V, Taheri AH (eds) Desalination–challenges and opportunities. IntechOpen

Itoh N, Toda H, Matsuda M, Negishi T, Taniguchi T, Ohsawa N (2009) Involvement of S-adenosylmethionine-dependent halide/thiol methyltransferase (HTMT) in methyl halide emissions from agricultural plants: isolation and characterization of an HTMT-coding gene from Raphanus sativus (daikon radish). BMC Plant Biol 9(1):1–10

Jagdish T, Morris JJ, Wade BD, Blount ZD (2020) Probing the deep genetic basis of a novel trait in Escherichia coli. In: Evolution in action: past, present and future. Springer, Cham, pp 107–122

James C (1998) Global review of commercialized transgenic crops: the International Service for the Acquisition of Agri-biotech applications. ISAAA 8

Jansing J, Schiermeyer A, Schillberg S, Fischer R, Bortesi L (2019) Genome editing in agriculture: technical and practical considerations. Int J Mol Sci 20(12):2888

Jiang WZ, Henry IM, Lynagh PG, Comai L, Cahoon EB, Weeks DP (2017) Significant enhancement of fatty acid composition in seeds of the allohexaploid, Camelina sativa, using CRISPR/Cas9 gene editing. Plant Biotechnol J 15(5):648–657

Jones RA, Naidu RA (2019) Global dimensions of plant virus diseases: current status and future perspectives. Ann Rev Virol 6:387–409

Jones RA (2009) Plant virus emergence and evolution: origins, new encounter scenarios, factors driving emergence, effects of changing world conditions, and prospects for control. Virus Res 141(2):113–130

Jones RAC (2021) Global plant virus disease pandemics and epidemics. Plan Theory 10:233

Juma C (2011) Preventing hunger: biotechnology is key. Nature 479:471–472

Khan MS, Yu X, Kikuchi A, Asahina M, Watanabe KN (2009) Genetic engineering of glycine betaine biosynthesis to enhance abiotic stress tolerance in plants. Plant Biotechnol 26(1): 125–134

Kim S, Takahashi M, Higuchi K, Tsunoda K, Nakanishi H, Yoshimura E, Nishizawa NK (2005) Increased nicotianamine biosynthesis confers enhanced tolerance of high levels of metals, in particular nickel, to plants. Plant Cell Physiol 46(11):1809–1818

Klas E, Fuchs M, Gonsalves D (2011) Fruit yield of virus-resistant transgenic summer squash in simulated commercial plantings under conditions of high disease pressure. J Hortic For 3(2): 46–52

Komor AC, Kim YB, Packer MS, Zuris JA, Liu DR (2016) Programmable editing of a target base in genomic DNA without double-stranded DNA cleavage. Nature 533(7603):420–424

Kumar K, Gambhir G, Dass A, Tripathi AK, Singh A, Jha AK et al (2020) Genetically modified crops: current status and future prospects. Planta 251:1–27

Kundu PALLOB, Mandal RK (2001) Transgenic approaches for producing Viru resistant plants. Proc-Ind Nat Sci Acad Part A 67(1/2):53–79

Labbé P, Jean-Philippe David H, Alout PM, Djogbenou L, Pasteur N, Weill M (2017) Evolution of resistance to insecticide in disease vectors. In: Genetics and evolution of infectious diseases: Second Edition, pp 313–339

Lamichhane JR, Reay-Jones FP (2021) Editorial: impacts of COVID-19 on global plant health and crop protection and the resulting effect on global food security and safety. Crop Protection (Guildford, Surrey) 139:105383

Layton MB, Williams MR, Stewart S (1997) BT-cotton in Mississippi: the first year. In: 1997 Proceedings Beltwide Cotton Conferences, vol 2. National Cotton Council, New Orleans, LA, USA, pp 861–863

Lermontova I, Grimm B (2000) Overexpression of Plastidic Protoporphyrinogen IX oxidase leads to resistance to the diphenyl-ether herbicide Acifluorfen. Plant Physiol 122(1):75–84

Li T, Liu B, Spalding MH, Weeks DP, Yang B (2012) High-efficiency TALEN-based gene editing produces disease-resistant rice. Nat Biotechnol 30(5):390–392

Li M, Li X, Zhou Z, Wu P, Fang M, Pan X et al (2016) Reassessment of the four yield-related genes Gn1a, DEP1, GS3, and IPA1 in rice using a CRISPR/Cas9 system. Front Plant Sci 7:1–13

Li B, Clohisey SM, Chia BS et al (2020) Genome-wide CRISPR screen identifies host dependency factors for influenza A virus infection. Nat Commun 11:164. https://doi.org/10.1038/s41467-019-13965-x

Littlejohn P, Finlay BB (2021) When a pandemic and an epidemic collide: COVID-19, gut microbiota, and the double burden of malnutrition. BMC Med 19:31

Lobato-Gómez M, Hewitt S, Capell T, Christou P, Dhingra A, Girón-Calva PS (2021) Transgenic and genome-edited fruits: background, constraints, benefits, and commercial opportunities. Hortic Res 8

Lu Y, Tian Y, Shen R, Yao Q, Zhong D, Zhang X, Zhu JK (2021) Precise genome modification in tomato using an improved prime editing system. Plant Biotechnol J 19(3):415

Lucca P, Hurrell R, Potrykus I (2001) Genetic engineering approaches to improve the bioavailability and the level of iron in rice grains. Theor Appl Genet 102(2):392–397

Lutz KA, Knapp JE, Maliga P (2001) Expression of bar in the plastid genome confers herbicide resistance. Plant Physiol 125(4):1585–1590

Macedo MLR, Oliveira CF, Oliveira CT (2015) Insecticidal activity of plant lectins and potential application in crop protection. Molecules 20(2):2014–2033

Majeed H, Gillor O, Kerr B, Riley MA (2011) Competitive interactions in Escherichia coli populations: the role of bacteriocins. ISME J 5(1):71–81

Manosathiyadevan M, Bhuvaneshwari V, Latha R (2017) Impact of insects and pests in loss of crop production: a review. In: Sustainable agriculture towards food security, pp 57–67

Mansoor S, Amin I, Hussain M, Zafar Y, Briddon RW (2006) Engineering novel traits in plants through RNA interference. Trends Plant Sci 11(11):559–565

Mao YB, Cai WJ, Wang JW, Hong GJ, Tao XY, Wang LJ, Chen XY (2007) Silencing a cotton bollworm P450 monooxygenase gene by plant-mediated RNAi impairs larval tolerance of gossypol. Nat Biotechnol 25(11):1307–1313

Martínez-Fortún J, Phillips DW, Jones HD (2017) Potential impact of genome editing in world agriculture. Emerg Top Life Sci 1(2):117–133

Martins-Salles S, Machado V, Massochin-Pinto L, Fiuza LM (2017) Genetically modified soybean expressing insecticidal protein (Cry1Ac): management risk and perspectives. Facets 2(1): 496–512

Matveeva TV, Lutova LA (2014) Horizontal gene transfer from agrobacterium to plants. Front Plant Sci 5:326

Meyer CJ, Norsworthy JK (2020) Timing and application rate for sequential applications of glufosinate are critical for maximizing control of annual weeds in LibertyLink® soybean. Int J Agron 2020:*1*

Miao J, Guo D, Zhang J, Huang Q, Qin G, Zhang X et al (2013) Targeted mutagenesis in rice using CRISPR-Cas system. Cell Res 23:1233–1236

Mohammed A, Abalaka ME (2011) Agrobacterium transformation: a boost to agricultural biotechnology. J Med Genet Genomics 3(8):126–130

Nadakuduti SS, Enciso-Rodríguez F (2021) Advances in genome editing with CRISPR systems and transformation technologies for plant DNA manipulation. Front Plant Sci 11:637159

Nahar K, Hasanuzzaman M, Fujita M (2016) Roles of osmolytes in plant adaptation to drought and salinity. In: Osmolytes and plants acclimation to changing environment: emerging omics technologies. Springer, New Delhi, pp 37–68

Nicholas D, Rodríguez-Bravo B, Watkinson A, Boukacem-Zeghmouri C, Herman E, Xu J, Świgoń M (2017) Early career researchers and their publishing and authorship practices. Learned Publishing 30(3):205–217

Nishida K, Arazoe T, Yachie N, Banno S, Kakimoto M, Tabata M, Mochizuki M, Miyabe A, Araki M, Hara KY, Shimatani Z, Kondo A (2016) Targeted nucleotide editing using hybrid prokaryotic and vertebrate adaptive immune systems. Science *353*(6305):aaf8729

Odipio J, Ogwok E, Taylor NJ, Halsey M, Bua A, Fauquet CM, Alicai T (2014) RNAi-derived field resistance to cassava brown streak disease persists across the vegetative cropping cycle. GM Crops Food 5(1):16–19

OECD, Woodhouse D (1999) Quality and quality assurance. Quality and internationalisation in higher education

Ortigosa A, Gimenez-Ibanez S, Leonhardt N, Solano R (2019) Design of a bacterial speck resistant tomato by CRISPR/Cas9-mediated editing of SlJAZ2. Plant Biotechnol J 17:665–673

Osendarp S, Kweku Akuoku J, Black RE, Headey D, Ruel M, Scott N, Shekar M, Walker N, Flory A, Haddad L, Laborde D, Stegmuller A, Thomas M, Heidkamp R (2021) The COVID-19 crisis will exacerbate maternal and child undernutrition and child mortality in low- and middle-income countries. Nature Food 2:476–484

Padgette SR, Kolacz KH, Delannay X, Re DB, LaVallee BJ, Tinius CN, Rhodes KW, Otero YI, Barry GF, Eichholtz DA, Peschke VM, Nida DL, Taylor NB, Kishore GM (1995) Development, identification, and characterization of a glyphosate tolerant soybean line. Crop Sci 35(5): 1451–1461

Parmar N, Singh KH, Sharma D, Singh L, Kumar P, Nanjundan J, Khan YJ, Chauhan DK, Thakur AK (2017) Genetic engineering strategies for biotic and abiotic stress tolerance and quality enhancement in horticultural crops: a comprehensive review. Biotech 7(4):239

Pausch P, Al-Shayeb B, Bisom-Rapp E, Tsuchida CA, Li Z, Cress BF et al (2020) CRISPR-Cas8 from huge phages is a hypercompact genome editor. Science 369:333–337

Pérez-Massot E, Banakar R, Gómez-Galera S, Zorrilla-López U, Sanahuja G, Arjó G, Zhu C (2013) The contribution of transgenic plants to better health through improved nutrition: opportunities and constraints. Genes Nutr 8(1):29–41

Petersen C, Woods HA, Kingsolver JG (2000) Stage-specific effects of temperature and dietary protein on growth and survival of *Manduca sexta* caterpillars. Physiol Entomol 25:35–40
Petrova P, AbouRaya MM (2020) Global economic consequences and possible solutions to COVID-19 Великотърновски университет „Св. св. Кирил и Методий" 191–198
Pratiwi RA, Surya MI (2020) Agrobacterium-mediated transformation. In: Genetic transformation in crops. IntechOpen
Qaim M, Kouser S (2013) Genetically modified crops and food security. PLoS One 8(6):e64879
Quilis J, López-García B, Meynard D, Guiderdoni E, San Segundo B (2014) Inducible expression of a fusion gene encoding two proteinase inhibitors leads to insect and pathogen resistance in transgenic rice. Plant Biotechnol J 12(3):367–377
Rahangdale S, Singh Y, Katkani D, Surjaye N (2020) Chapter 2: Gene Transfer Methods in Plants. In: Gene transfer methods in plants. Integrated Publications, New Delhi
Ravelonandro M, Scorza R, Bachelier JC, Labonne G, Levy L, Damsteegt V, Dunez J (1997) Resistance of transgenic Prunus domestica to plum pox virus infection. Plant Dis 81(11): 1231–1235
Raymond Park J, McFarlane I, Hartley Phipps R, Ceddia G (2011) The role of transgenic crops in sustainable development. Plant Biotechnol J 9(1):2–21
Regina A, Bird A, Topping D, Bowden S, Freeman J, Barsby T, Morell M (2006) High-amylose wheat generated by RNA interference improves indices of large-bowel health in rats. Proc Natl Acad Sci 103(10):3546–3551
Rengasamy P (2005) World salinisation with emphasis on Australia. In: Comparative biochemistry and physiology a-molecular & integrative physiology. (*141*(3), p. S337-S337. Elsevier science Inc, 360 Park Ave South, New York, USA
Ricroch A, Clairand P, Harwood W (2017) Use of CRISPR systems in plant genome editing: toward new opportunities in agriculture. Emerg Top Life Sci 1(2):169–182
Rocha-Munive MG, Soberón M, Castañeda S, Niaves E, Scheinvar E, Eguiarte LE, Mota-Sánchez D, Rosales-Robles E, Nava-Camberos U, Martínez-Carrillo JL, Blanco CA, Bravo A, Souza V (2018) Evaluation of the impact of genetically modified cotton after 20 years of cultivation in Mexico. Front Bioeng Biotechnol 6:82
Romeis J, Driesche RGV, Barratt BI, Bigler F (2008) Insect-resistant transgenic crops and biological control. In: Integration of insect-resistant genetically modified crops within IPM programs. Springer, Dordrecht, pp 87–117
Römer P, Hahn S, Jordan T, Strauß T, Bonas U, Lahaye T (2007) Plant pathogen recognition mediated by promoter activation of the pepper Bs3 resistance gene. Science 318(5850):645–648
Safari F, Zare K, Negahdaripour M, Barekati-Mowahed M, Ghasemi Y (2019) CRISPR Cpf1 proteins: structure, function and implications for genome editing. Cell Biosci 9:36. https://doi.org/10.1186/s13578-019-0298-7. PMID: 31086658; PMCID: PMC6507119
Scholes J, Endacott R, Biro M, Bulle B, Cooper S, Miles M (2012) Clinical decision-making: midwifery students' recognition of, and response to, post-partum haemorrhage in the simulation environment. BMC Pregnancy Childbirth 12:19
Schornack S, Meyer A, Römer P, Jordan T, Lahaye T (2006) Gene-for-gene-mediated recognition of nuclear-targeted AvrBs3-like bacterial effector proteins. J Plant Physiol 163(3):256–272
Schütte G, Eckerstorfer M, Rastelli V, Reichenbecher W, Restrepo-Vassalli S, Ruohonen-Lehto M, Wuest Saucy A, Mertens M (2017) Herbicide resistance and biodiversity: agronomic and environmental aspects of genetically modified herbicide-resistant plants. Environ Sci Eur 29:5
Shan Q, Voytas DF (2018) Editing plant genes one base at a time. Nat Plants 4:412–413
Shelton AM, Olmstead DL, Burkness EC, Hutchison WD, Dively G, Welty C, Sparks AN (2013) Multi-state trials of Bt sweet corn varieties for control of the corn earworm (Lepidoptera: Noctuidae). J Econ Entomol 106(5):2151–2159
Shelton AM, Hossain MJ, Paranjape V, Azad AK, Rahman ML, Khan ASMMR, Prodhan MZH, Rashid MA, Majumder R, Hossain MA, Hussain SS, Huesing JE, McCandless L (2018) Bt eggplant project in Bangladesh: history, present status, and future direction. Front Bioeng Biotechnol 3(6):106

Shimatani Z, Kashojiya S, Takayama M, Terada R, Arazoe T, Ishii H, Teramura H, Yamamoto T, Komatsu H, Miura K, Ezura H, Nishida K, Ariizumi T, Kondo A (2017) Targeted base editing in rice and tomato using a CRISPR-Cas9 cytidine deaminase fusion. Nat Biotechnol 35(5): 441–443

Shukla VK, Doyon Y, Miller JC, DeKelver RC, Moehle EA, Worden SE, Mitchell JC, Arnold NL, Gopalan S, Meng X, Choi VM, Rock JM, Wu YY, Katibah GE, Zhifang G, McCaskill D, Simpson MA, Blakeslee B, Greenwalt SA, Butler HJ, Urnov FD (2009) Precise genome modification in the crop species Zea mays using zinc-finger nucleases. Nature 459(7245): 437–441

Siminszky B, Corbin FT, Ward ER, Fleischmann TJ, Dewey RE (1999) Expression of a soybean cytochrome P450 monooxygenase cDNA in yeast and tobacco enhances the metabolism of phenylurea herbicides. Proc Natl Acad Sci U S A 96(4):1750–1755

Singh M, Kumar J, Singh S, Singh VP, Prasad SM (2015) Roles of osmoprotectants in improving salinity and drought tolerance in plants: a review. Rev Environ Sci Biotechnol 14(3):407–426

Singla J, Krattinger SG (2016) Biotic stress resistance genes in wheat. In: Wrigley C, Corke H, Seetharaman K, Faubion J (eds) Encyclopedia of food grains (Second Edition). Academic Press, pp 388–392

Sinha S, Sandhu K, Bisht N, Naliwal T, Saini I, Kaushik P (2019) Ascertaining the paradigm of secondary metabolism enhancement through gene level modification in therapeutic plants. J Young Pharm 11(4):337

Soltani N, Shropshire C, Sikkema PH (2020) Weed control in Dicamba-resistant soybean with glyphosate/Dicamba applied at various doses and timings. Int J Agron

Song A, Zhu X, Chen F, Gao H, Jiang J, Chen S (2014) A chrysanthemum heat shock protein confers tolerance to abiotic stress. Int J Mol Sci 15(3):5063–5078

Subramanyam K, Sailaja KV, Subramanyam K, Muralidhara Rao D, Lakshmidevi K (2011) Ectopic expression of an osmotin gene leads to enhanced salt tolerance in transgenic chilli pepper (*Capsicum annum* L.). Plant Cell Tissue Organ Cult (PCTOC) 105(2):181–192

Sun YW, Jiao GA, Liu ZP, Zhang X, Li JY, Guo XP, Du WM, Du JL, Francis F, Zhao YD, Xia LQ (2017) Generation of high-amylose rice through CRISPR/Cas9-mediated targeted mutagenesis of starch branching enzymes. Front Plant Sci 8:298. https://doi.org/10.3389/fpls.2017.00298

Sunilkumar G, Campbell LM, Puckhaber L, Stipanovic RD, Rathore KS (2006) Engineering cottonseed for use in human nutrition by tissue-specific reduction of toxic gossypol. Proc Natl Acad Sci 103(48):18054–18059

Suzuki Y, Makino A, Mae T (2001) Changes in the turnover of rubisco and levels of mRNAs of rbcL and rbcS in rice leaves from emergence to senescence. Plant Cell Environ 24(12): 1353–1360

Symington LS, Gautier J (2011) Double-strand break end resection and repair pathway choice. Annu Rev Genet 45:247–271

Taheri F, Azadi H, D'Haese M (2017) A world without hunger: organic or GM crops? Sustainability 9(4):580

Takabe T, Nakamura T, Nomura M, Hayashi Y, Ishitani M, Muramoto Y, Tanaka A (1998) Glycinebetaine and the genetic engineering of salinity tolerance in plants. In: Stress responses of photosynthetic organisms: molecular mechanisms and molecular regulations. Elsevier Science, Amsterdam, pp 115–131

Tanaka H, Yabuta Y, Tamoi M, Tanabe N, Shigeoka S (2015) Generation of transgenic tobacco plants with enhanced tocotrienol levels through the ectopic expression of rice homogentisate geranylgeranyl transferase. Plant Biotechnol 32:233–238

Tang GL, Galili G, Zhuang X (2007) RNAi and microRNA: breakthrough technologies for the improvement of plant nutritional value and metabolic engineering. Metabolomics 3:357–369

Tien P, Wu G (1991) Satellite RNA for the biocontrol of plant disease. Adv Virus Res 39:321–339

Tohidfar M, Khosravi S (2015) Transgenic crops with an improved resistance to biotic stresses. A review. Biotechnol Agron Soc Environ 19:62–70

Tripathi S, Suzuki JY, Ferreira SA, Gonsalves D (2008) Papaya ringspot virus-P: characteristics, pathogenicity, sequence variability and control. Mol Plant Pathol 9(3):269–280

UNCTAD (2017) The role of science, technology and innovation in ensuring food security by 2030. United Nations, New York and Geneva

Vasconcelos M, Datta K, Oliva N, Khalekuzzaman M, Torrizo L, Krishnan S, Datta SK (2003) Enhanced iron and zinc accumulation in transgenic rice with the ferritin gene. Plant Sci 164(3): 371–378

Vincelli P (2016) Genetic engineering and sustainable crop disease management: opportunities for case-by-case decision-making. Sustainability 8(5):495

Waltz E (2014) Beating the heat. Nat Biotechnol 32:610–613

Wang F, Wang C, Liu P, Lei C, Hao W, Gao Y, Liu YG, Zhao K (2016) Enhanced Rice blast resistance by CRISPR/Cas9-targeted mutagenesis of the ERF transcription factor gene OsERF922. PLoS One 11(4):e0154027

Wang L, Samac DA, Shapir N, Wackett LP, Vance CP, Olszewski NE, Sadowsky MJ (2005) Biodegradation of atrazine in transgenic plants expressing a modified bacterial atrazine chlorohydrolase (atzA) gene. Plant Biotechnol J 3(5):475–486

Wang Y, Cheng X, Shan Q, Zhang Y, Liu J, Gao C, Qiu JL (2014) Simultaneous editing of three homoeoalleles in hexaploid bread wheat confers heritable resistance to powdery mildew. Nat Biotechnol 32(9):947–951

Welch RM, Graham RD (2004) Breeding for micronutrients in staple food crops from a human nutrition perspective. J Exp Bot 55(396):353–364

Wesseler J, Purnhagen K (2020) Is the Covid-19 pandemic a game changer in GMO regulation? EuroChoices 19(3):49–52

Wirth J, Poletti S, Aeschlimann B, Yakandawala N, Drosse B, Osorio S, Sautter C (2009) Rice endosperm iron biofortification by targeted and synergistic action of nicotianamine synthase and ferritin. Plant Biotechnol J 7(7):631–644

Wisniewski M, Fuchigami L, Wang Y, Srinivasan C, Norilli J (2002) Overexpression of a cytosolic ascorbate peroxidase gene in apple improves resistance to heat stress. In: XXXVI International horticultural congress and exhibition, p 147

Yang Y, Al-Khayri JM, Anderson EJ (1997) Transgenic spinach plants expressing the coat protein of cucumber mosaic virus. In Vitro Cellular & Developmental Biology-Plant 33(3):200–204

Zafar K, Sedeek KEM, Rao GS, Khan MZ, Amin I, Kamel R et al (2020) Genome editing Technologies for Rice Improvement: Progress, prospects, and safety concerns. Front Genome Ed 2:1–16

Zha S, Boboila C, Alt FW (2009) Mre11: roles in DNA repair beyond homologous recombination. Nat Struct Mol Biol 16(8):798–800

Zhang HX, Zhang Y, Yin H (2019) Genome editing with mRNA encoding ZFN, TALEN, and Cas9. Mol Ther 27:735–746

Zhang Y, Bai Y, Wu G, Zou S, Chen Y, Gao C, Tang D (2017) Simultaneous modification of three homoeologs of TaEDR1 by genome editing enhances powdery mildew resistance in wheat. Plant J 91(4):714–724

Zhang Y, Li D, Zhang D, Zhao X, Cao X, Dong L et al (2018a) Analysis of the functions of TaGW2 homoeologs in wheat grain weight and protein content traits. Plant J 94:857–866

Zhang Y, Liang Z, Zong Y, Wang Y, Liu J, Chen K et al (2016) Efficient and transgene-free genome editing in wheat through transient expression of CRISPR/Cas9 DNA or RNA. Nat Commun 7:1–8

Zhang Y, Massel K, Godwin ID et al (2018b) Applications and potential of genome editing in crop improvement. Genome Biol 19:210

Zhou J, Peng Z, Long J, Sosso D, Liu B, Eom JS et al (2015) Gene targeting by the TAL effector PthXo2 reveals cryptic resistance gene for bacterial blight of rice. Plant J 82:632–643

Zhu YX, Ou-Yang WJ, Zhang YF, Chen ZL (1996) Transgenic sweet pepper plants from agrobacterium mediated transformation. Plant Cell Rep 16(1):71–77

Zilberman D, Holland TG, Trilnick I (2018) Agricultural GMOs—what we know and where scientists disagree. Sustainability 10(5):1514

URLs

Brookes G (2020) Crop biotechnology continues to provide higher farmer income and significant environmental benefits. Available at https://pgeconomics.co.uk/press+releases/25/Crop+biotechnology+continues+to+provide+higher+farmer+income+and+significant+environmental+benefits

DiLeo M (2012) Monsanto's GM drought tolerant corn. Available at https://biofortified.org/2012/08/monsantos-gm-drought-tolerant-corn/

DuPont Pioneer. Announces Intentions to Commercialize First CRISPR-Cas Product. Press Release (2016). Available at https://www.pioneer.com/home/site/about/news-media/news-releases/template.CONTENT/guid.1DB8FB71-1117-9A56-E0B6-3EA6F85AAE92

FAO (2019) New standards to curb the global spread of plant pests and diseases. Available at https://www.fao.org/news/story/en/item/1187738/icode/#:~:text=FAO%20estimates%20that%20annually%20between,insects%20around%20US%2470%20billion

ISAAA (2018) Global status of commercialized biotech/GM crops in 2018: biotech crops continue to help meet the challenges of increased population and climate change. ISAAA Brief No.54. ISAAA, Ithaca, NY. Available at https://www.isaaa.org/resources/publications/briefs/54/executivesummary/default.asp

ISAAA (2019a) Executive summary, ISAAA Brief 55-2019: global status of commercialized biotech/GM crops: 2019

ISAAA (2019b) Biotech crop highlights in 2019. ISAAA Publication Pocket K No. 16. ISAAA: Ithaca, NY. Available at https://www.isaaa.org/resources/publications/pocketk/16/

Johnson YD, O'Connor S (2015) These charts show every genetically modified food people already eat in the U.S. TIMES newsletter. https://time.com/3840073/gmo-food-charts/

The World Bank Report (2021) Learning losses from COVID-19 could cost this generation of students close to $17 trillion un lifetime earnings. https://www.worldbank.org/en/news/press-release/2021/12/06/learning-losses-from-covid-19-could-cost-this-generation-ofstudents-close-to-17-trillion-in-lifetime-earnings

SPRINGER NATURE

GPSR Compliance

The European Union's (EU) General Product Safety Regulation (GPSR) is a set of rules that requires consumer products to be safe and our obligations to ensure this.

If you have any concerns about our products, you can contact us on ProductSafety@springernature.com

In case Publisher is established outside the EU, the EU authorized representative is:

Springer Nature Customer Service Center GmbH
Europaplatz 3
69115 Heidelberg, Germany

The manufacturer's authorised representative in the EU is Springer Nature Customer Service Centre GmbH, Europaplatz 3, 69115 Heidelberg, Germany. If you have any concerns regarding our products, please contact ProductSafety@springernature.com

Printed and bound by CPI Group (UK) Ltd, Croydon, CR0 4YY
01/10/2026
02225899-0006